Multimedia Communications

ACADEMIC PRESS SERIES IN COMMUNICATIONS, NETWORKING, AND MULTIMEDIA

Editor-in-Chief
Jerry D. Gibson
Southern Methodist University

This series has been established to bring together a variety of publications that cover the latest in applications and cutting-edge research in the fields of communications and networking. The series will include professional handbooks, technical books on communications systems and networks, books on communications and network standards, research books for engineers, and tutorial treatments of technical topics for non-engineers and managers in the worldwide communications industry. The series is intended to bring the latest in communications, networking, and multimedia to the widest possible audience.

Books in the Series:

Handbook of Image and Video Processing, Al Bovik, editor
Nonlinear Image Processing, Sanjit Mitra, Giovanni Sicuranza, editors
The E-Commerce Book, Second Edition, Steffano Korper and Juanita Ellis

Multimedia Communications
Directions and Innovations

JERRY D. GIBSON, EDITOR
Department of Electrical Engineering
Southern Methodist University
Dallas, Texas

ACADEMIC PRESS

A Harcourt Science and Technology Company

SAN DIEGO / SAN FRANCISCO / NEW YORK / BOSTON / LONDON / SYDNEY / TOKYO

ACADEMIC PRESS
A Harcourt Science and Technology Company
525 B Street, Suite 1900, San Diego, CA 92101-4495, USA
http://www.academicpress.com

Academic Press
Harcourt Place, 32 Jamestown Road, London NW1 7BY, UK

Library of Congress Catalog Card Number: 00-104377
International Standard Book Number: 0-12-282160-2

Printed in the United States of America
00 01 02 03 04 QW 9 8 7 6 5 4 3 2 1

To Ruth Elaine

Contents

Chapter 4: Audio Coding Standards 45
Chi-Min Liu and Wen-Whei Chang

Chapter 5: Still Image Compression Standards 61
Michael W. Hoffman and Khalid Sayood

Chapter 9: ATM Network Technology 129
Yoichi Maeda and Koichi Asatani

Chapter 10: ISDN 143
Koichi Asatani and Toshinori Tsuboi

Preface

This book is a collection of invited chapters on multimedia communications contributed by experts in the field. We use the term *multimedia communications* to encompass the delivery of multiple media content such as text, graphics, voice, video, still images, and audio over communications networks to users. Note that several of these media types may be part of a particular interaction between (or among) users, and thus we are not simply considering networks that support different traffic types. We are specifically interested in applications that incorporate multiple media types to deliver the desired information. Example applications of interest include two-way, multipoint videoconferencing and one-way streaming of video and audio in conjunction with text or graphical data.

The topics covered in the book were carefully selected to provide critical background material on multimedia communications and to expose the reader to key aspects of the hottest areas in the field. Chapter 1, *Multimedia Communications: Source Representations, Networks, and Applications,* provides a context for the rest of the book, but each chapter is intended to stand alone and the chapters can be read in any order so that readers may get the necessary information as efficiently as possible. Among the topics discussed are wireline network technologies and services, compression standards, video-on-demand, IP telephony, wideband wireless data, IP over wireless, transcoding of multimedia content, and multicasting. It would be difficult to find a more timely collection of topics in a single volume anywhere.

The book is intended for beginners and experts alike, and the chapters are descriptive in nature, focused primarily on the presentation of results, insights, and key concepts, with a minimum of mathematical analyses and abstraction. The beginner will be able to get a good overview of the field and an introduction to fundamental ideas, while the expert will be able to discern very quickly what technologies are critical to current applications and what technologies will form the basis for future services and products.

The authors are chosen from both industry and academia in order to give the reader as clear a view of current practices and future directions as possible. In reading these chapters myself, I am amazed at how much content the authors have been able to include in so few pages. I am most appreciative of these authors and their efforts, and I want to thank Joel Claypool at Academic Press for his guidance and patience. I hope that each reader finds this book of great value.

List of Contributors

Koichi Asatani Dr. Eng, Kogakain University, Nishi-Shinjuku, Shinjuku-ku, Tokyo 163-8677 JAPAN. Tel: +81 3 3340 2845 (direct) +81 3 3342-1211 ex 2638, Fax: +81 3 3348 3486

Steven W. Carter Computer Science Department, University of California, Santa Cruz

Wen-Whei Chang Associate Professor, Department of Communications Engineering, National Chiao Tung University, Hsinchu, Taiwan, ROC. Tel: 0021 886 3 5731826, Fax: 0021 886 3 5710116, e-mail: wwchang@cc.nctu.edu.tw

Justin Chuang AT&T Labs–Research, 100 Schulz Drive, Room 4-140, Red Bank, NJ 07701, U.S.A. Tel: (732) 345-3125, Fax: (732) 345-3038

Leonard J. Cimini Jr. AT&T Labs–Research, 100 Schulz Drive, Room 4-140, Red Bank, NJ 07701, U.S.A. Tel: (732) 345-3125, Fax: (732) 345-3038

Jerry D. Gibson Chair, Department of Electrical Engineering, Caruth Hall, Room 331, 3145 Dyer Street, School of Engineering and Applied Science, Southern Methodist University, Dallas, TX 75275-0338, U.S.A. Tel: (214) 768-3133, Fax: (214) 768-3573

Richard Han IBM Thomas J. Watson Research Center, IBM Research, 30 Saw Mill River Road, Hawthorne, NY 10532, U.S.A. Tel: (914) 784-7608, Fax: (941) 784-6079, e-mail: rhan@us.ibm.com

Michael Hoffman Department of Electrical Engineering, University of Nebraska-Lincoln, 209N WSEC, Lincoln, NE 68588-0511, U.S.A. Tel: (402) 472-1979, Fax: (402) 472-4732, e-mail: mhoffman1@unl.edu

Leonid Kazovsky Professor, Stanford University, 81 Riverside Drive, Los Altos, CA 94024, U.S.A. Tel: (650) 725-3813, Fax: (650) 723-9251, e-mail: kazovsky@stanford.edu

Giok-Djan Khoe Professor, Technical University Eindhoven

Dr. Bernard S. Ku WorldCom, 2400 North Glenville Drive, 1225/107, Richardson, TX 75082, U.S.A. Tel: (972) 729-5770, Fax: (972) 729-6038, e-mail: bernard.ku@wcom.com

Chin-Min Liu Professor, Department of Computer Science and Information Engineering, National Chiao Tung University, Hsinchu, Taiwan, ROC. Tel: 0021 886 3 5731826, Fax: 0021 886 3 5710116

David Lindbergh   Picture Tel Corp, 70 Dividence Road, Reading, MA 01867, U.S.A. Tel: (781) 942-8808, Fax: (781) 944-1267, e-mail: dave_lindbergh@yahoo.com

Darrell D.E. Long   Professor, Computer Science Department, University of California, 1156 High Street, Santa Cruz, CA 95064, U.S.A. Tel: (831) 459-2616, Fax: (831) 459-4829, e-mail: darrell@cs.ucsc.edu

Tom Lookabaugh   Hermonic Inc., 1545 Country Club Drive, Los Altos, CA 94024, U.S.A. Tel: (650) 917-1704, Fax: (650) 917-8663, e-mail: christic_tom@email.msn.com

Yoichi Maeda   NTT Service Intergration Laboratories, 3-9-11 Midori-cho, Musashino-shi, Tokyo 180-8585 JAPAN. Tel: +81 422 60 7429, Fax: +81 422 60 7429, e-mail: maeda.yoichi@lab.ntt.co.jp

Jehan-François Pâris   Professor, Computer Science Department, University of Houston, Houston, TX, 77204, U.S.A.

George Polyzos   Center for Wireless Communications and Computer Systems Laboratory, Department of Computer Science and Engineering, University of California, San Diego, La Jolla, CA 92093-0114, U.S.A. e-mail: xgeorge@cs.ucsd.edu

Khalid Sayood   Department of Electrical Engineering, University of Nebraska-Lincoln, 209N WSEC, Lincoln, NE 68588-0511, U.S.A. Tel: (402) 472-1979, Fax: (402) 472-4732

John R. Smith   IBM Thomas J. Watson Research Center, IBM Research, 30 Saw Mill River Road, Hawthorne, NY 10532, U.S.A. Tel: (914) 784-7608, Fax: (941) 784-6079

Nelson Sollenberger   AT&T Labs–Research, 100 Schulz Drive, Room 4-140, Red Bank, NJ 07701, U.S.A. Tel: (732) 345-3125, Fax: (732) 345-3038, e-mail: nelson@research.att.com

Andreas Spanias   Arizona State University, U.S.A. Tel: (480) 965-1837, Fax: (480) 965-8325, e-mail: spanias@asu.edu

Toshinori Tsuboi   Tokyo University of Technology

M. Oskar van Deventer   KPN Research Leidschendam

Dr. Upkar Varshney   Department of Computer Information Systems, Georgia State University, 35 Broad Street, Room #936, Atlanta, GA 30303-4015, U.S.A. Tel: (404) 463-9139, Fax: (404) 651-3842, e-mail: uvarshney@aus.edu

George Xylomenos   Center for Wireless Communications and Computer Systems Laboratory, Department of Computer Science and Engineering, University of California, San Diego

CHAPTER 1

Multimedia Communications: Source Representations, Networks, and Applications

JERRY D. GIBSON

1.1 INTRODUCTION

Universal access to multimedia information is now the principal motivation behind the design of next-generation computer and communications networks. Furthermore, products are being developed to extend the capabilities in all existing network connections to support multimedia traffic. This is a profound paradigm shift from the original analog-voice telephony network developed by the Bell System and from the packet-switched, data-only origins of the Internet. The rapid evolution of these networks has come about because of new technological advances, heightened public expectations, and lucrative entrepreneurial opportunities.

In this chapter and in this book as a whole, we are interested in multimedia communications; that is, we are interested in the transmission of multimedia information over networks. By *multimedia,* we mean data, voice, graphics, still images, audio, and video, and we require that the networks support the transmission of multiple media, often at the same time. Two observations can be made at the outset. The media to be transmitted, often called sources, are represented in digital form, and the networks used to transmit the digital source representations may be classified as digital communications networks, even though analog modulation is often used for free-space propagation or for multiplexing advantages. In addition to the media sources and the networks, we will find that the user terminals, such as computers, telephones, and personal digital assistants (PDAs), also have a large impact on multimedia communications and what is actually achievable.

The development here breaks the multimedia communications problem down into the components shown in Figure 1.1. Components shown there are the Source, the Source Terminal, the Access Network, the Backbone Network, the Delivery Network, and the Destination Terminal. This categorization allows us to consider two-way, peer-to-peer communications connections, such as videoconferencing or telephony, as well as asymmetric communications situations,

1

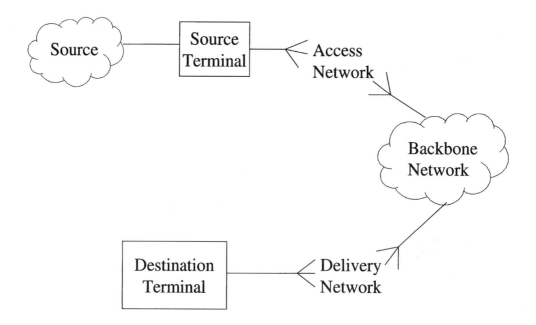

FIGURE 1.1

Components of a multimedia communications network.

such as broadcasting or video streaming. In Figure 1.1, the Source consists of any one or more of the multimedia sources, and the job of the Source Terminal is to compress the Source such that the bit rate delivered to the network connection between the Source Terminal and the Destination Terminal is at least approximately appropriate. Other factors may be considered by the Source Terminal as well. For example, the Source Terminal may be a battery-power-limited device or may be aware that the Destination Terminal is limited in signal processing power or display capability. Further, the Source Terminal may packetize the data in a special way to guard against packet loss and aid error concealment at the Destination Terminal. All such factors impinge on the design of the Source Terminal. The Access Network may be reasonably modeled by a single line connection, such as a 28.8 Kbit/s modem, a 56 Kbit/s modem, a 1.5 Mbit/s Asymmetric Digital Subscriber Line (ADSL) line, and so on, or it may actually be a network that has shared capacity, and hence have packet loss and delay characteristics in addition to certain rate constraints. The Backbone Network may consist of a physical circuit-switched connection, a dedicated virtual path through a packet-switched network, or a standard best-effort Transmission Control Protocol/Internet Protocol (TCP/IP) connection, among other possibilities. Thus, this network has characteristics such as bandwidth, latency, jitter, and packet loss, and may or may not have the possibility of Quality of Service (QoS) guarantees. The Delivery Network may have the same general set of characteristics as the Access Network, or one may envision that in a one-to-many transmission that the Delivery Network might be a corporate intranet. Finally, the Destination Terminal may have varying power, mobility, display, or audio capabilities.

The source compression methods and the network protocols of interest are greatly determined by international standards, and how these standards can be adapted to produce the needed connectivity is a challenge. The terminals are specified less by standards and more by what users have available now and are likely to have available in the near future. The goal is clear, however—ubiquitous delivery of multimedia content via seamless network connectivity.

We will first present discussions of the various components in Figure 1.1, and then we elaborate by developing common examples of multimedia communications and highlight the challenges and state-of-the-art. We begin our discussions with the Networks and Network Services.

1.2   NETWORKS AND NETWORK SERVICES

We focus in this section on everything between the Source Terminal and the Destination Terminal in Figure 1.1. Two critical characteristics of networks are transmission rate and transmission reliability. The desire to communicate using multimedia information affects both of these parameters profoundly. Transmission rate must be pushed as high as possible, and in the process, transmission reliability may suffer. This becomes even more true as we move toward the full integration of high-speed wireless networks and user mobility. A characterization of networks and network services according to rate is shown in Table 1.1. These networks and services not only show a wide variation in available transmission rates, but also the underlying physical transmission media vary dramatically, as do the network protocols. The additional considerations of wireless local area networks (LANs), cellular data, and mobility add a new dimension to network reliability, through the physical layer channel reliability, that makes the problem even more challenging.

The Access and Delivery Networks are often characterized as the "last-mile" network connections and are often one of the first five entries in Table 1.1. Certainly, most people today connect to the Internet through the plain old telephone system (POTS) using a modem that operates

Table 1.1 Networks and Network Services

Service/Network	Rate
POTS	28.8–56 Kbit/s
ISDN	64–128 Kbit/s
ADSL	1.544–8.448 Mbit/s (downstream)
	16–640 Kbit/s (upstream)
VDSL	12.96–55.2 Mbit/s
CATV	20–40 Mbit/s
OC-N/STS-N	$N \times 51.84$ Mbit/s
Ethernet	10 Mbit/s
Fast Ethernet	100 Mbit/s
Gigabit Ethernet	1000 Mbit/s
FDDI	100 Mbit/s
802.11 (wireless)	1, 2, 5.5, and 11 Mbit/s in 2.4 GHz band
802.11a (wireless)	6–54 Mbit/s in 5 GHz band

Abbreviations: CATV, cable television; FDDI, Fiber Distributed Data Interface; OC-N/STS-N, optical cable-number of times the single link bandwidth/synchronous transport protocol-number of times the single link bandwidth; VDSL, very high rate digital subscriber line.

at 28.8 Kbit/s up to 56 Kbit/s. While relatively low speed by today's standards and for the needs of multimedia, these connections are reliable for data transmission. For transporting compressed multimedia, however, these lower speeds can be extremely limiting and performance limitations are exhibited through slow download times for images, lower frame rates for video, and perhaps noticeable errors in packet voice and packet video. Of course, as we move to the higher network speeds shown in the table, users experience some of the same difficulties if the rates of the compressed multimedia are increased proportionately or the number of users sharing a transport connection is increased. For example, when users move from POTS to Integrated Services Digital Network (ISDN) to ADSL, they often increase the rate of their multimedia transmissions and thus continue to experience some packet losses even though they have moved to a higher rate connection. Further, the higher speed connectivity in the Access Networks increases the pressure on the Backbone Networks or servers being accessed to keep up. Thus, even though a user has increased Access Network bandwidth, packet losses and delay may now come from the Backbone Network performance. Additionally, although POTS, ISDN, and ADSL connections are usually not shared, CATV services are targeted to support multiple users. Therefore, even though there is 20 Mbits/s or more available, a few relatively high rate users may cause some congestion.

Many people are seeking to upgrade their individual Access Network transmission rate, and higher speed modems, ISDN, Digital Subscriber Line (xDSL), and cable modems are all being made available in many areas. As users obtain higher Access Network rates, possible bottlenecks move to the Backbone Networks. This potential bottleneck can be viewed on a couple of levels. First, it is not unusual today for users to experience delays and congestion due to a lower rate Delivery Network or server. If delays are experienced when accessing a remote web site, the user does not know whether the difficulty is with the Access Network speed, the Backbone Network, the Delivery Network speed (in either direction), or the server being accessed (the server would be the Destination Terminal in Figure 1.1). Of course, commercial servers for web sites have a financial motivation for maintaining adequate network rates and server speeds.

Notice that there are substantial differences in the protocols used for several of the network services mentioned in Table 1.1; generally, these protocols were developed around the concept of transmitting non-time-critical data as opposed to time-sensitive multimedia traffic. Fortunately, the protocols have been designed to interoperate, so that network interfaces do not pose a problem for data-only traffic. The concept of internetworking is not too overwhelming if one considers only a set of isolated networks interconnected for the use of (say) one company. But when one contemplates the global internetwork that we call the Internet (capital I), with all of its diverse networks and subnetworks, having relatively seamless internetworking is pretty amazing. The Internet Protocol (IP) achieves this by providing connectionless, best-effort delivery of datagrams across the networks. Additional capabilities and functionality can be provided by transport layer protocols on top of IP. Transport layer protocols may provide guaranteed message delivery and/or correctly ordered message delivery, among other things.

In the Internet, the most commonly used transport layer protocol is the Transmission Control Protocol (TCP). TCP provides reliable, connection-oriented service and is thus well suited for data traffic as originally envisioned for the Internet [via Advanced Research Projects Agency Network (ARPANET)]. Unfortunately, one of the ways reliable delivery is achieved is by retransmission of lost packets. Since this incurs delay, TCP can be problematical for the timely delivery of delay-sensitive multimedia traffic. Therefore, for multimedia applications, many users often employ another transport layer protocol called User Datagram Protocol (UDP). Unlike TCP, UDP simply offers connectionless, best-effort service over the Internet, thus avoiding the delays associated with retransmission, but not guaranteeing anything about whether data will be reliably delivered.

When considering packet-switched networks like the Internet, we usually have in mind the situation where the Source Terminal wants to send packets to a single Destination Terminal. This is called *unicast*. There are situations, however, where the Source Terminal needs to send a message to all terminals on a network, and this is called *broadcast*. Since broadcasting sends one copy of the message for each end node or Destination Terminal, this type of transmission may flood a network and cause congestion. An alternative to broadcasting when it is desired to send a message to a subset of network nodes is *multicast*. Multicast allows the Source Terminal to send one copy of the message to a multicast address called a *multicast group*. The message finds the appropriate terminals by the destination terminals knowing this address and joining the multicast group. Since multicast currently is not supported by many of the routers in the Internet, multicast is currently achieved by using the Multicast Backbone or Mbone. The MBone is implemented on top of the current Internet by a technique called *tunneling*, where a standard unicast IP address is used to encapsulate the MBone multicast transmissions. Routers that do not support multicast see only a unicast packet, but routers that have multicast capability can implement multicast. The Mbone is extremely popular and is used to transmit Internet Engineering Task Force (IETF) meetings, among other multimedia applications.

Although multimedia applications may have their own protocols, recently the IETF has developed a protocol for multimedia communications called the Real-time Transport Protocol (RTP) and its associated control protocol, Real-time Transport Control Protocol (RTCP). The fundamental goal of RTP is to allow multimedia applications to work together. There are several aspects to achieving such interoperability. For one thing, there should be a choice of audio and video compression techniques negotiated at the beginning of the session. Further, since packets can arrive out of order, there needs to be a timing relationship. Of course, if audio and video are both present, there must be a method for synchronization of the multiple media. Realtime multimedia applications cannot absorb the time delays involved with retransmission under TCP, but many applications can respond to known packet losses in various ways. Thus, it is desirable to provide the sender with some indication of packet loss. In achieving these traits, as well as others, it is critical that bandwidth be effectively utilized, which implies short headers in the protocols.

All of these characteristics are incorporated in RTP and RTCP. Since RTCP is a data flow control protocol, we would like for it not to take away too much bandwidth from the actual multimedia applications. The control messages may cause difficulties as users are added, and so there are adjustments made, according to rules that we will not elaborate here, that attempt to limit the RTCP traffic to 5% of the RTP traffic.

For multimedia traffic, we need reliable delivery, but not necessarily perfect delivery, and we usually need low delay. Furthermore, we may need to put bounds on the variations in data arrival delay, often called *jitter*. One approach to achieving this level of service in today's packet-switched networks is to develop connection-oriented protocols that have reservable bandwidth guarantees. In recent years, network requirements of this type have all been lumped under the term Quality of Service (QoS). Thus, users may often request QoS guarantees from network service providers. One protocol that has been developed for the Internet that allows a receiver to request certain performance requirements is the Resource Reservation Protocol (RSVP). RSVP supports both unicast and multicast, and Figure 1.2 shows RSVP in a multicast application. RSVP takes requests for performance reservations and passes them up the tree (shown as RESV in Figure 1.2). For multicast, RSVP allows reservations to be merged; thus, if Receiver A's request is inclusive of Receiver B's, the reservation at the router where the two paths first come together (when going upstream) will be only that of Receiver A.

In the dial-up videoconferencing applications that use H.320 or H.324, a number of users agree on a common videoconference time and conference bridge dial-in numbers are distributed

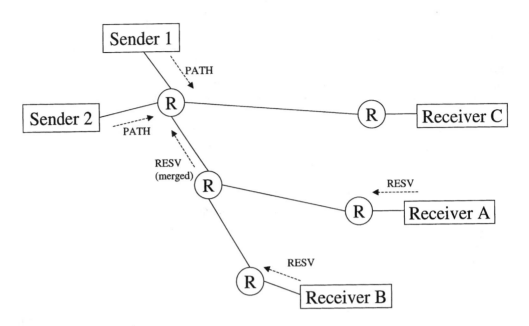

FIGURE 1.2
RSVP and multicast. (Adapted from Peterson and Davie, 2000.)

to the conference participants. Notice that this application differs from many of the Internet-oriented multimedia applications we have noted in that there are a limited number of very specific conference participants, and we need to facilitate their joining the conference. Protocols for this functionality perform what is called *session control,* and the IETF has been working on developing such protocols. At the same time, the International Telecommunications Union (ITU), which is responsible for the H.320 and H.324 standards, has also turned its attention to developing similar protocols. There has been good collaboration with the IETF and the result is the H.323 standard that includes many important details, especially the H.245 call control protocol. Details are elaborated in Chapter 6, but we simply note here that H.323 has become very popular for Internet multimedia applications, especially Internet telephony.

1.3 MULTIMEDIA SOURCES

Now that we have some idea of network bandwidths (or rates), services, and protocols, we examine how these network capabilities match the multimedia sources that we wish to transmit through the networks. Table 1.2 lists several common multimedia sources, their bandwidths, the common sampling rates, and typical uncompressed bit rates. When we compare the uncompressed bit rate requirements in the rightmost column of Table 1.2 with the available network rates shown in Table 1.1, we observe immediately that source compression is a necessity. This particular realization is one of the principal driving forces behind all of the standards activity in compression over the last 15 years. Fortunately, numerous excellent source compression standards have been developed.

Table 1.2 Multimedia Sources and Typical Uncompressed Bit Rates

Source	Bandwidth (Hz)	Sampling Rate	Bits per Sample	Bit Rate
Telephone voice	200–3400	8000 samples/s	12	96 Kbit/s
Wideband speech	50–7000	16,000	14	224 Kbit/s
Wideband audio (two channels)	20–20,000	44.1 Ks/s	16 per channel	1.412 Mbit/s (two channels)
Color image		512×512	24	6.3 Mbit/s
CCIR TV		$720 \times 576 \times 30$	24	300 Mbit/s
HDTV		$1280 \times 720 \times 60$	24	1327 Mbit/s

Abbreviations: CCIR, Comité International des Radiocommuncations; HDTV, high-definition television.

In Table 1.3 are listed several of the more prominent telephone bandwidth speech coder standards, including their designation, bit rate, quality, and complexity. The speech coding standards listed are not exhaustive and the reader is referred to Chapter 3, which gives a more extensive table and supporting discussion. Table 1.3 requires a few explanatory notes. In the Quality column, "Toll" refers to toll quality, which is taken to be equivalent to log-PCM (logarithmic pulse code modulation) at 64 Kbit/s. The acronym MOS means mean opinion score, which is a subjective listening test score from 1 to 5, with the value for log-PCM taken as being equivalent to toll quality. Any MOS testing should always include log-PCM as the anchor point for toll

Table 1.3 Telephone Bandwidth Speech Coding Standards

Coder	Bit Rate (Kbit/s)	Quality	Complexity (MIPS)
Log PCM (G.711)	64	Toll 4–4.3 MOS	0.01
ADPCM (G.726)	16–40	Toll at 32 Kbits/s 4.1 MOS	2
LD-CELP (G.728)	16	4.0 MOS	30
RPE-LTP (GSM)	13	3.5 MOS	6
QCELP (IS-96)	0.8–8.5 (variable)	3.3 MOS	15
VSELP (IS-54)	7.95	3.5 MOS	14
EFR (IS-641)	8	3.8 MOS	14
EVRC (IS-127)	1.2–9.6 (variable)	3.8 MOS	20
CS-ACELP (G.729)	8	4.0 MOS	20
CS-ACELP (G.729A)	8	3.75 MOS	11
MPC-MLQ (G.723.1)	5.3–6.4	3.5 MOS	16
CELP (FS 1016)	4.8	3.2 MOS	16

Abbreviations: ADPCM, Adaptive Differential Pulse Code Modulation; CELP, Code Excited Linear Prediction; CS-CELP, Conjugate Structure Code Excited Linear Prediction; CS-ACELP, Conjugate Structure Algebraic Code Excited Linear Prediction; EFR, Enhanced Full Rate; EVRC, Enhanced Variable Rate Coder; IS, Interim Standard; LD-CELP, Low-Delay Code Excited Linear Prediction; MPC-MLQ, Multipulse Coder, Maximum Likelihood Quantization; QCELP, Qualcomm Code Excited Linear Prediction; RPE-LTP, Regular Pulse Excitation, Long-Term Prediction.

Table 1.4 Selected Videoconferencing Standards (Basic Modes)

Standard	Network	Video	Audio
H.320 (1990)	ISDN	H.261	G.711
H.323 (1996)	LANs/Internet	H.261	G.711
H.324 (1995)	PSTN	H.263	G.723.1
H.310 (1996)	ATM/B-ISDN	H.262	MPEG-1

Abbreviations: ATM/B-ISDN, Asynchronous Transfer Mode/Broadband ISDN; MPEG-1, standard for videotape quality video and high quality audio on a CD; PSTN, Public Switched Telephone Network.

quality since MOS for any coder varies according to test conditions. This is why a range of MOS values is given for G.711. The complexity measure is millions of instructions per second (MIPs), and is also approximate. In the bit rate column, a range of bit rates is shown for G.726, which indicates that this coder has selectable rates at 16, 24, 32, and 40 Kbit/s. In contrast, IS-96 and IS-127 show a range of rates, but these rates are adaptively varied by the coder based upon the input speech. These coders are therefore designated variable rate coders. Some of the coders shown are for wireline telephony, some are for digital cellular, and several are used in video-conferencing and multimedia applications. Key information presented in Table 1.3 is typical bit rates that are possible and the substantial level of complexity that has become common for these telephone bandwidth speech coders.

There are also wideband speech compression methods, such as G.722, and newly evolving standards in this area (G.722.1), plus wideband audio, such as MP3. We leave further discussion of these topics to later chapters. For completeness, Table 1.4 lists important videoconferencing standards. The video and audio codecs listed in Table 1.4 are only for the basic modes, and several alternatives are part of the more complete standard and are often implemented—see Chapter 6. Note that the standards listed in Table 1.4 are systems standards, and as such, have multiplexing and control as part of the specification. These are not shown in the table, but are left for discussion in Chapter 6. Considerable effort is also being expended in developing the Motion Picture Experts Group standards MPEG-4 (for object-based audio-visual representation) and MPEG-7 (for multimedia content description interface), which will be important to multimedia communications over networks, and these are discussed in Chapter 8.

1.4   SOURCE AND DESTINATION TERMINALS

In this chapter, we use the word *terminal* to refer to any device that connects a user to the network. For voice communications over the PSTN, the terminal may simply be a telephone handset, or for the Internet, it may be a desktop computer. However, today and certainly for the future, terminals are going to take a host of shapes and sizes and be asked to accommodate a full range of tasks. Terminals will thus be classified according to characteristics such as: sources handled (messaging, voice, data, images, video), size and weight, battery power and battery life, input devices (keyboards, microphone, or handset), output devices (handset audio, low-resolution black-and-white display, high-resolution color display), input/output (I/O) and processor speeds, special signal processing capabilities, mobility, and portability. One or more of these characteristics can dictate what can and cannot be done in a particular multimedia application.

At the time of this writing, there are several trends in evidence. First, central processing unit (CPU) speeds for desktop machines are at 850 MHz, and 1.2 GHz speeds are on the near horizon. In fact, Intel projects that by 2011, chips will have 1 billion transistors, 10 GHz clock speeds, and an additional 10 times increase in performance. The increases in performance will be due to innovations such as the use of increasing parallelism. Therefore, the 10-GHz chips will be 100 times more powerful than the 1-GHz chips soon to be available. Second, mobile, highly portable terminals such as laptop personal computers (PCs) and palm computing devices are evolving rapidly in terms of weight, power dissipation, battery life, and display capabilities, with palm computing devices certain to be important terminals of the future. Third, it is evident that special-purpose signal processing will be a significant part of tomorrow's terminals, and considering the compression needs as outlined in Tables 1.2–1.4, Digital Signal Processors (DSPs) are going to be a major component in these terminals. This trend will be accelerated for wireless connectivity such as in third-generation digital cellular and evolving wireless LAN standards, since considerable signal processing will be needed to mitigate the effects of the time-varying wireless channels and to achieve reliable data throughput. DSP processor speeds are increasing at a rate that tracks rates of increases in CPU speeds, and DSP designs are also exploiting techniques to extract more MIPS per MHz of processor speeds. There is also substantial effort being expended to develop low-power DSP designs in response to the need for DSPs in wireless devices.

Many different types of terminals will be connected to a network at any one time; furthermore, several kinds of terminals will often be involved in a single multimedia communications session. It is easy to imagine such a scenario for streaming applications. For example, a presentation by a chief executive officer (CEO), a seminar by a researcher, or a training session for a piece of equipment may all be available via streamed audio and video from a server. Users interested in any of these may be spread throughout their organization, out of the office on a business trip, or simply working at home. The streamed session would then be accessed by desktop PCs, laptop PCs, possibly PDAs, or perhaps, through a wireless phone with audio-only capability. Because of the great popularity of the Internet, the issues involved with transcoding multimedia content for access via a variety of different terminals is under way.

The widely varying types of terminals and the diverse types of physical layer channels will create an even more heterogeneous environment than we have today in the Internet. This will keep the pressure on the development of new protocols and network interfaces to maintain interoperability at the high level expected.

1.5   APPLICATIONS OF MULTIMEDIA COMMUNICATIONS NETWORKS

The heterogeneity mentioned in the previous section becomes explicitly visible when one considers the Multimedia Communications Network shown diagrammatically in Figure 1.3. Network connection speeds range from a few tens of Kilobits per second to more than 100 Mbit/s, and the media that comprise these channels range from optical fiber and coaxial cable through copper wire pairs and free space. The terminals may be high-end workstations with large displays, desktop personal computers, battery-powered laptops, and personal digital assistants (shown as personal communicators in the figure) that have small black-and-white displays.

The two most common multimedia communications applications today and that are foreseen in the near future are video streaming to multiple users and party-to-party or multiparty videoconferencing. In this section, we present typical scenarios for these two applications and highlight some of the issues and challenges.

1.5.1   Video Streaming to Multiple Users

For this application, we assume that the Multimedia Server in the upper left-hand corner of Figure 1.3 wishes to stream a video lecture to any user on the network. The video may be stored in MPEG-2 format (standard for movie quality video and audio), which is very high quality, but direct streaming of MPEG-2 video requires a variable rate of from 4 to 10 Mbit/s. Such rates would be incompatible with many of the network connections shown in Figure 1.3, so the first inclination might be to transcode this information down to a rate that is commensurate with all, or a majority of, the network links. The video compression method of choice to do this might be H.263, which offers a wide range of rates and frame sizes, and is widely supported. The result of the transcoding, however, would be that we would have a least common denominator encoding, so that even those users with higher rate network connections would be forced to accept the quality produced for the low-rate users.

   One approach to working around the lowest common denominator limitation would be to use a layered coder with multicasting. That is, you would choose a video coder that allows multiple compression rates that can be obtained by incrementally improving the base layer. Coders that have this capability are sometimes said to be *scalable*. MPEG-2 has several scalability options, including signal-to-noise ratio (SNR), spatial, and frame rate scalability. One or more of these options could be used and combined with multicasting to create (say) three multicast groups. The first group would be the baseline coded layer, and the other two would use scalability to create incremental improvements in output quality as users join the remaining two multicast groups. There might be another good reason to use multicast in this application. Specifically, if the desired number of viewers is large, unicast transmissions to each of them could flood various links in the network. If multicast is employed, users on congested links could reduce the rate by reducing the number of multicast groups that they join.

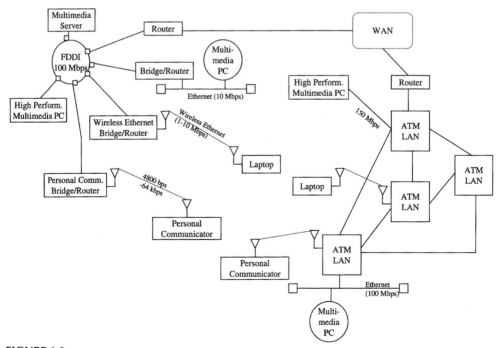

FIGURE 1.3

Multimedia communications network. (ATM = Asynchronous Transfer Mode.)

Another approach to establishing interoperability between networks would be to transcode at network gateways. There are three disadvantages usually cited for transcoding: (1) complexity, (2) delay, and (3) added distortion. For video streaming, because it is one way, delay is not a serious concern. Complexity would be an issue at some network interfaces, and in those cases, the streamed video might not be transcoded, thus yielding degraded network performance and poor delivered video quality. If complexity does not preclude transcoding, then the remaining issue is the distortion added during the transcoding process. Of course, transcoding to a lower rate will yield lower quality, smaller frame size, and/or slower frame rate, so it is key to add as little additional distortion as possible. This implies that we would prefer to not completely decode back to video and then re-encode the video at a lower rate (notice that this has implications in terms of complexity and delay, too). We would add less distortion if we could directly map the encoded stream, or at least the decoded parameters, directly into a lower-rate-coded version. As you can imagine, there are numerous possible options corresponding to any given network interface and the particular compression method, and so we do not go into further detail here.

1.5.2 Videoconferencing

One-to-one videoconferences (single party to single party) are relatively easy to deal with. Compared to video streaming, the serious new issue that arises is *latency* or round trip delay. During call setup, the users can negotiate the particular compression methods to be used and the desired transmitted data rates. Usually the two participants agree on these issues and the conference can be initiated. Each user thus sees and hears the other participant, subject to distortions or breakup in the video or audio due to packet losses. In the situation where the total video and audio transmitted data rate cannot be sustained without packet losses, it is often desirable to give priority to the audio signal. This is because participants are more forgiving of breakups in video than in audio.

Notice in this case that it does not make sense for a participant with a high data rate connection to request or send high-rate video if the other participant has a much lower rate channel. The lower-rate channel cannot send or receive data at the higher rate. Thus, one-to-one videoconferences are negotiated to preferences that reflect the lowest common denominator in transmitted data rate. In Internet videoconferencing, the principal question to be answered before attempting a conference is whether each user supports a common videoconferencing tool. Compared to the familiar PSTN videoconferencing applications, Internet videoconferencing tools offer more diversity and less standardization.

For multiparty videoconferences, a number of new issues appear. It is desirable that all participants receive all of the audio. This can be accomplished by mixing the audio from all of the participants at a central location, called a bridge or multipoint control unit (MCU), and then retransmitting the combined audio to each participant. The drawbacks of such an approach are that all of the audio has to be decoded, combined, and re-encoded, resulting in high complexity and possible performance loss. Another alternative is for each participant to transmit audio to all of the other participants. This approach requires that all participants either use the same audio coding method, so that an appropriate decoder is available at every location, or that all participants be able to decode all compression schemes in the videoconference. Note that the bit rate over the links does not increase linearly with the number of conference participants since there is usually only a single speaker at any time instant.

The question of what video is displayed at each location is also important. Preferably each participant should be able to see all other participants at the same time. As long as bandwidth is not a problem, multiple received video streams can be presented on the computer display in what is often called a "Hollywood Squares" arrangement. With reference to Figure 1.3, notice that this may not be possible for all participants because of their individual bandwidth limitations or because of the resolution of their terminal's display. This problem can be simplified by adopting the approach that participants are provided with only the video of the current speaker. This approach makes sense because there should be only a single speaker at any one time instant; however, this speaker can be at any of the locations, and the speaker may change fairly often. There are three standard approaches to accommodating the need for switching the video supplied to participants. The first is what is called *director control,* where one participant in the videoconference is designated the director and manually switches between locations as a participant begins to speak. For this to work well, it is desirable that the director have access to all of the audio. A second scenario is where the video is switched automatically by a volume cue that selects the location with the loudest audio. To prevent rapid, but inadvertent, switching due to coughing, dropped objects, doors slamming, etc., there is usually a requirement that the loudest audio be present for some minimum time interval. A third alternative is for each participant to choose which video stream is displayed on his or her individual terminal. All three of these approaches are in use today.

As in video streaming, network heterogeneity and the variety of devices serving as user terminals present challenges to videoconferencing. Since there is already considerable delay due to video compression and possible audio combining, additional delay due to transcoding at network gateways becomes difficult to tolerate. At the same time, it is much preferred for each participant to be able to choose the bit rate over its local network link so that the quality of the videoconference can be kept as high as possible.

Multicast transmission can be particularly useful in this environment. Of course, the situation where all participants are multicasting could also aggravate traffic congestion. However, it would be unusual for all participants to be able to multicast their audio and video, either because they lack multicast capability or because their local link bandwidth precludes sending more than a single data stream. Multicasting combined with layered video compression methods presents the best videoconferencing quality to participants, since each participant can choose the quality received in proportion to either available link bandwidth or the processing and display capabilities of the participant's terminal.

1.6   CONCLUSIONS

The stated goal of "ubiquitous delivery of multimedia content via seamless network connectivity" is becoming a reality. Certainly there are many technical challenges, but new solutions are being developed every day, and the commercial demand for the resulting services is large and growing steadily. One can expect to see continued technological innovation and a host of new products and services that facilitate multimedia communications. The remaining chapters of this book develop key topics in greater detail and describe many of the particular applications of multimedia communications over networks that are available today and will be available in the near future.

1.7   FOR FURTHER READING

The chapters in the remainder of this book elaborate on many of the issues raised here. Two additional references of interest are:

J. D. Gibson, T. Berger, T. Lookabaugh, D. Lindbergh, and R. L. Baker, *Digital Compression for Multimedia: Principles and Standards,* Morgan Kaufmann Publishers, San Francisco, CA, 1998.

L. L. Peterson and B. S. Davie, *Computer Networks: A Systems Approach,* 2nd ed., Morgan Kaufmann Publishers, San Francisco, CA, 2000.

CHAPTER 2

Future Telecommunication Networks: Traffic and Technologies

LEONID G. KAZOVSKY
GIOK-DJAN KHOE
M. OSKAR VAN DEVENTER

The efficient transport of information is becoming a key element in today's society. This transport is supported by a complex communications infrastructure that, if properly implemented and operated, is invisible to end users. These end users seem to be primarily interested in services and costs only. As new services evolve and the needs of users change, the industry must adapt by modifying existing infrastructures or by implementing new ones. Telecommunication experts are therefore challenged to produce roadmaps for the development of future infrastructures. This is a difficult task because of incomplete knowledge of trends in users' demands and of how technology will evolve.

The purpose of this chapter is to provide one view of the future based on technologies that are likely to be implemented in the future, four hypotheses of traffic types, and competition. To be somewhat concrete, the study is based on the situation in a compact country with a high population density, such as the Netherlands. The following characteristics apply for the Netherlands:

- Size: 42,000 km^2
- Population: 15.2 million
- Population density:
  —Average: 450 persons/km^2
  —Peak: 10 times average
- Backbone node spacing: 70 km

The boundary conditions of our study can be applied to many other regions as well. For example, the population of the Netherlands is similar to the population of California. The strategy used to support the arguments in this chapter can also be applied for regions or countries that are not similar to the above example, when details are properly adapted.

Another boundary condition is the projection's time scale, which is set for 10–15 years. Developments that are happening within 2 or 3 years are not of interest here, and issues that will emerge in 50 years from now will not be speculated upon.

We begin by reviewing key electronic and optical technologies. We map the key features of the technologies and show that success or failure of a technology depends on the particular type of traffic. Further, we discuss the impact of competition on the future of telecommunications. We show that the combined impact of open competition and rapid technological progress will force telecom companies to adopt service packages, in addition to the conventional means of better service and lower cost. We review main strategies that telecommunication companies may adopt to cope with this trend.

We next outline four possible hypotheses of future traffic growth and discuss attributes and requirements of these four types of traffic. We also indicate the effects of each hypothesis on the traffic mix in telecommunications networks in 2012.

Finally, we integrate synergistically the issues considered to develop a view of the future that appears to be probable, and based on these projections, review the technologies that have a good chance to be implemented. Finally, we discuss possible factors which may influence the course of the trends as predicted.

2.1　KEY TECHNOLOGIES

In this section we review technologies and their four attributes. The key technologies are electronic and optical. Examples of electronic technologies are Synchronous Optical Network (SONET), Asynchronous Transfer Mode (ATM), Internet, Switched Multimega-bit Data Service (SMDS), frame relay, Integrated Services Digital Network (ISDN), Broadband Integrated Services Digital Network (B-ISDN), analog [television (TV)], and wireless. Optical technologies include Wavelength Division Multiplexing (WDM) point-to-point, Optical Time-Division Multiplexing (OTDM), solitons, WDM static networking (Add-Drop and Cross-Connect), and WDM dynamic networking.

WDM point-to-point basically uses different wavelengths to carry information from one switch to another. All processing within these switches is performed electronically. WDM static networking means that the information is carried from one point to another using different wavelengths. In addition it is also possible to drop one (or more) of those wavelengths at the node, so that a certain conductivity is provided between the nodes on that particular wavelength. However, the network is not reconfigurable. In dynamic network configurations, it is possible to rearrange the Add-Drop wavelengths.

In Table 2.1 some of the technologies discussed earlier are listed along with their traffic attributes, such as bit rate, latency, and burstiness. The list of attributes will allow us to match these technologies to some of the hypotheses on traffic types presented later in the chapter. The match between telephony and SONET is not really surprising: SONET was designed explicitly to carry telephone traffic, so it has a properly controlled latency and it can accommodate high bit rates when required. It is probably not suitable for bursty traffic because it was not designed to do so.

In the case of the Internet, it is possible to have a huge latency, and it is ideally suitable for bursty traffic. Hence, if future networks will be dominated by Internet traffic, it will be desirable to construct them entirely from Internet Protocol (IP) routers, interconnected by (possibly SONET) links. On the other hand, if the future traffic will be dominated by voice, it is not desir-

Table 2.1 Technologies and Traffic Attributes: Feature Mapping

	Latency	Bit rate	Holding time	Burstiness	Directionality
Telephony	Sensitive	64 kb/s	Minutes–hours	Low	Bidirectional
Internet	Not sensitive	56 kb/s and up	One hour	High	Highly directional
Digital video distribution	Not sensitive	Several Mb/s	Hours	Medium/high	Highly directional
Digital video communications	Sensitive	110 kb/s-1 Mb/s	Minutes–hours	Medium/high	Bidirectional

able to have IP routers switch the traffic because of the latency requirements. Table 2.1 indicates that each technology works well with a specific kind of traffic. An attempt to map traffic to technology produces the following list:

- Telephone: SONET, ATM, ISDN
- Internet: IP, ATM, ISDN
- Digital video distribution: SONET, ATM, IP
- Digital video communication: SONET, ATM
- Backbones: optics

There is an almost unique match between *technologies* and the *traffic* they are capable of carrying adequately. If the traffic will be dominated by voice, the key components in the network will be SONET, ATM, and ISDN. The other technologies will probably be less important.

If the future network will be dominated by Internet traffic, it may become essential to have IP switches instead of SONET Add-Drop Multiplexers. ATM may or may not be needed, but ISDN is not really necessary, although it may be used as well.

If the future view is pointing toward digital video networks, narrowband ISDN is probably not becoming part of the picture because it is not satisfying broadband needs. In that case it should be anticipated that the upgrade from ISDN to broadband ISDN will take place very soon. If, however, it is assumed that digital telephony will dominate the future traffic, ISDN will probably be implemented for a long time, along with SONET and ATM. Backbones in all these cases will be optical.

Let's consider now the interrelationship between ATM, IP, SONET, and WDM. ATM and IP compete directly with SONET in switching, but not in transmission. It is possible to implement Add-Drop Multiplexing (ADM) functions in ATM, IP, or WDM. In fact, if a completely new network had to be constructed today there would be no real need to use SONET ADM networking. However, it can be assumed that SONET transmission will be used for a long time. For switching, however, it is possible to use ATM switches or IP routers. On the other side of the technology spectrum WDM static networking also squeezes SONET networking because it is also possible to implement ADM functions in the WDM domain. SONET networking is thus being challenged from both sides by WDM, IP, and ATM.

Hence, the future network might develop as follows: ATM switches and IP routers are implemented, and all users are connected to ATM switches or IP routers. The ATM switches and IP routers are further interconnected to each other by SONET links without any SONET networking. WDM can be used to add more links as needed and also to provide ADM. This scenario will provide good flexibility and cost-effectiveness.

Because our studies focus on a compact country or region, we can state that technologies driven by very long distances such as soliton or optical Time Division Multiplexing (TDM) are not likely to become important. For that reason, it would not be desirable for a company focusing on compact countries or regions to invest much in those technologies.

2.2   IMPACT OF COMPETITION

Competition is traditionally perceived as a purely economic issue that has little relevance to technology, but this view must be challenged. Clearly there are traditional competition tools like better services. An example is the presence of somebody at the telephone company who answers the phone and knows what happened to the bills. Another factor considered important by users is initial *costs*. The company can certainly get more appreciation for a high bit rate, but it is much more important to offer what we call *service bundles*. A typical example of a service bundle occurs when a cable company providing cable TV also offers an additional box for telephone service. Similar situations are beginning to happen with the Internet. A user connected to Internet can also obtain telephone service through the same connection. Hence in the future it will be impossible to be a pure telephone company because the competitor who is going to provide cable or the Internet will offer customers telephone service for just a little more money. To be able to compete with others, each company has to provide more services. In the next section we will argue that service bundles are strongly influenced by technology. Examples of possible service bundles are:

- Internet over voice network
- Voice over Internet network
- Video over Internet network
- Video over wireless network

Developments in technology have made it possible not only to provide the Internet over a voice network, which has been done for a long while, but also to provide the other bundles listed above. Basically, the important issue is no longer that of different services but rather of different bits. Telephone service over the Internet used to be an unusual combination, but an announcement in early 1997 from Lucent Technology seems to contradict that view. Users of the Lucent Internet Telephone Server do not need any fancy equipment. We expect that more combinations like this will be offered in the future.

At this point we may ask how companies can adequately adapt to the expected trend. Basically, two options are possible. A company can build a heterogeneous network. One network is used for voice, another network is used for the Internet, and yet another network is used for video. In that case, users will perhaps have three or four different wall sockets. However, it may be possible to physically combine the different networks by using WDM. Alternatively, it may be possible to build one integrated network, using IP or ATM, that carries all traffic over the same network. The network is user transparent but not physically transparent. The two options are quite different from each other and both approaches are being used today.

An example of the first option is the Sprint/Stanford WDM Ring Research Testbed. The idea behind the Sprint/Stanford WDM ring testbed is to provide a backbone network around the San Francisco Bay Area; telephone traffic can be carried over one wavelength and video over the other wavelength, and so on. WDM is used here not only as a multiplier of bit rates but also as a service integrator. Another approach is Pacific Bell's superhighway configuration. The approach essentially starts with an existing telephone server, which is then combined electronically with a video server in a digital host. The combined signals are transported to a remote node and subsequently distributed by coaxial cable to the homes of individual subscribers.

Both options are possible and companies are pursuing different configurations. It is difficult to predict which options will finally prevail. However, we will next attempt to develop a vision of the future based on the above considerations.

2.3   FOUR TRAFFIC HYPOTHESES

We consider four hypotheses of future traffic growth and show how they impact the picture of telecommunications network traffic in the year 2012.

2.3.1   Hypothesis 1: Conventional Growth

The first hypothesis is one of conventional growth. According to this hypothesis, telephone traffic will continue to dominate the telecom network. Figure 2.1 illustrates the expected growth rates under this hypothesis. Growth rates of about 10% for telephone traffic and about 30% for Internet traffic are assumed. In this situation, even though the absolute volume of telephone traffic will continue to grow at a modest rate, the absolute volume of telephone traffic is still much larger than Internet and digital video traffic. Basically, it would be possible to design networks for telephone traffic conditions and subsequently to accommodate all the other traffic in the network.

The precise numbers taken for the growth are rather irrelevant. The important fact is the ratio between telephone traffic and other traffic. As long as the vast majority (e.g., 95%) is telephone traffic, the basic philosophy and technological setup of the network will be oriented to telephone traffic and other traffic will be accorded lower priority. The network will not be designed around a tiny fraction of nonvoice traffic.

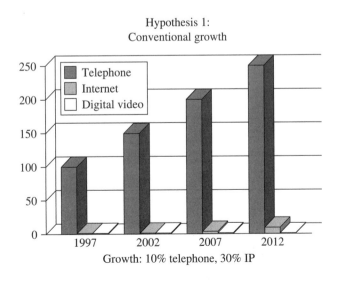

FIGURE 2.1
Expected traffic growth rates assuming conventional growth.

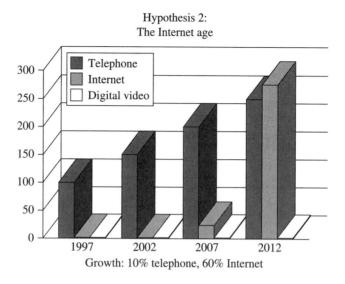

FIGURE 2.2
Expected traffic growth rates in the Internet age.

2.3.2 Hypothesis 2: The Internet Age

The second hypothesis takes into account the impact of the Internet age. We assume in this case that the IP traffic between computers and servers will grow so dramatically that it will become the dominant and perhaps the main force in telecommunications. Hence, most of the traffic in the future will consist of Internet traffic. Figure 2.2 shows that a huge growth is assumed in the Internet network, about 60% per year. Thus, by 2012 the volume of Internet traffic will actually become comparable to that of telephone traffic. Since it is certainly possible to transmit telephone traffic over the Internet, and vice versa, the two are not really mutually exclusive and Internet dominance may be amplified even further by assuming that most of the resources will be devoted to Internet traffic.

Internet domination can be caused by a number of issues. A new software release may cause a sudden increase in the number of providers and users. When Microsoft released the latest version of Windows, including Internet Explorer, the number of Internet providers increased by a factor of 3. Repeating such events drives the traffic volume up, as does new applications such as multimedia. When users play multimedia games and participate in and play multimedia movies, this creates additional traffic volume. There are signs that this is actually happening. Driven by the need to have more bandwidth, ISDN lines are becoming very popular in the United States. It took about 10 years for such a trend to happen, but it is now taking off at an incredible rate. The rest of the world will probably follow soon. Thus, many observers believe that the hypothesis is materializing right now.

2.3.3 Hypotheses 3 and 4: The Digital Video Age

The third and fourth hypotheses make the assumption that a digital video age is coming and that digital video will dominate all other traffic. At the moment digital video is not even in the same category as telephone traffic, but merely rates as noise in terms of traffic flow. However, in the

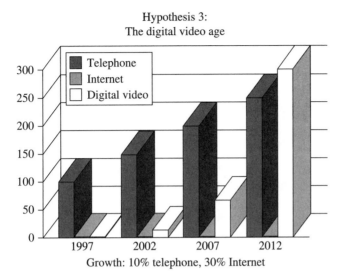

FIGURE 2.3
Expected traffic growth rates assuming that digital video will dominate all other traffic.

digital video age hypothesis illustrated in Figure 2.3, digital video grows at an incredible rate, doubling every year and dominating other traffic by 2012.

Video and the Internet may be interdependent because it is technically possible to send video traffic through the Internet. Yet, if the dominant kind of traffic is the kind currently on the Internet, it is very likely that the entire network will be optimized for this traffic mix and that all other kinds of traffic will adapt to the existing network. If video is dominant, however, the network will be optimized for video, and all other traffic will adapt accordingly.

2.3.3.1   *Hypothesis 3: Digital Video Distribution*
Two different kinds of digital video traffic may develop. One is digital video distribution, basically entertainment video and possibly pay TV that is completely digitized. To provide high quality digital video distribution, a high-bandwidth downstream is required. The precise bandwidth required may vary between a few Mbit/s to maybe 100 Mbit/s, depending on the compression used or the particular type of TV or high-definition television (HDTV) format used.

Some upstream and control transport must be provided, but most of the traffic will be transported from servers downstream to the subscribers. Latency, an important topic in video communication (see following discussion), is not an issue.

2.3.3.2   *Hypothesis 4: Digital Video Communication*
Another possible occurrence in the digital video age is the dominance of digital video communication, which has completely different implications. The traffic type is still video, but it is typically communications by means of video, similar to videophones where one can place a videocall to another user. Even though both fall within the category of digital video, the attributes of video communications are completely different. The bit-rate requirements and the quality required will in general be much lower than for entertainment video. Usually, it is agreed that digital video sessions at 100 or 200 Kbit/s are acceptable. Latency requirements, on the other hand, are extremely stringent. Typically, users want to have pictures synchronized with voice.

Latency requirements for video communication are thus rather different from those in video distribution, where it is acceptable for users to order a movie and have it delivered a few seconds later. Video distribution and video communication are thus very different. In the former, large bandwidths are needed but very long delays can be tolerated. In the latter, the bandwidths needed are moderate but the latency requirements are stringent. Latency may become a serious problem if ATM is used. By the time the transported data hops through several ATM switches, timing disturbances may become intolerable.

2.3.4   HDTV in the United States

The development of digital video in the United States is of particular interest now. After many years of efforts to develop analog-enhanced television in the United States, it was subsequently decided to adopt a digital standard for the future. One reason for that decision may be that Japan developed analog technologies. This may have been a good reason for the United States to adopt digital technology instead. For a long while, however, that decision was not taken seriously, and it was consistently said that digital technology would develop within a few years. That attitude has changed dramatically.

One factor that has influenced recent interest in digital video in the United States is the policy adopted by the Federal Communications Commission (FCC). The FCC recently approved rules giving broadcasters free licenses to provide high-definition digital television. The FCC rules call for 30% of the households to receive the broadcast of at least three digital TV stations by May 1, 1999. However, many stations in the top-10 market have committed to start service sooner. It may thus be expected that very soon some markets in the United States will develop HDTV broadcast. It is not clear what the situation in the rest of the world will be. Technologies developed in the United States may emerge within a few years in Europe.

2.3.5   Traffic Attributes

We next consider traffic attributes. Table 2.2 lists the traffic attributes for the four hypotheses just discussed and formulated: telephony, the Internet, digital video distribution, and digital video communications. The attributes listed are latency, bit rate, holding time, burstiness, and directionality. It is clear that these attributes are vastly different for each type of traffic. For example, telephony is associated with sensitive latency, a very modest bit rate, and a reasonable holding time, depending on whether the user has teenagers in the house. For telephone, burstiness is fairly modest and the traffic is bidirectional. On the other hand, in the case of digital video distribution, latency is not an issue at all. In this case the characteristic attributes are bit rates of many Mbit/s and very long holding times. A movie is generally watched for 2 hours, and the time is extended with commercials. Other characteristics are medium to high burstiness depending on the encoding format and almost unidirectional traffic.

Table 2.2 Traffic Attributes for the Hypotheses Considered

Technology	Latency	Bit rate	Suitable for bursty traffic
SONET	Controlled	High	No
ATM	Variable, small	N/A	Yes
IP	Variable, large	N/A	Yes
ISDN	Low	Low	Yes
Optics	Lowest	Highest	No

Table 2.2 clearly illustrates that the four kinds of traffic are extremely different from each other. A network optimized for one particular type of traffic may be a poor network for another type.

2.4   SYNERGY: FUTURE PROJECTIONS

We first consider general projections concerning topology and technology. As far as topology is concerned, rings are probably suited for backbones. For distribution purposes, stars and double stars are considered to be a better option, not only for telephony but also for video. Technologically, backbones are clearly the domain for optical technologies; interest in wireless technologies for distribution is currently increasing. Copper-based technologies are therefore challenged both in telephone and optical systems, and have clearly become less important. Obviously there are examples, such as the Pacific Bell configuration discussed in the preceding part of the chapter, where the use of copper has been maintained. A more precise technological projection depends on the particular assumptions made for the traffic development, summarized in the preceding discussion. In the following, we review the projections for each of the four hypotheses above in terms of technology.

If conventional growth takes place, the technology will be dominated by SONET/ Synchronous Digital Hierarchy (SDH). Other technologies like IP will remain marginal. ISDN will also grow, but if most of the traffic is in the voice domain, ISDN is not going to be comparable to voice technologies. Switched Multimegabit Data Service (SMDS) is not likely to develop, since there is no reason to use SMDS if the only issue is carrying voice as adequately as possible. In addition, companies need to offer a mixture of services because of competitive pressure. That need is likely to lead to service mixers, which can be either electronic, like ATM, or optical, such as WDM. Service mixing in this case is not simply bit multiplexing but multiplexing of different services.

If, however, the Internet age develops, explosive growth in IP routers can be expected and subsequently all other types of traffic will start gravitating toward the IP domain. Transmission of voice conversations through the Internet will grow. As a consequence, some drop in conventional technologies can be expected, in favor of IP switches and IP technologies in general. In this case IP technology will compete directly with SONET ADMs. SONET networking will decline rapidly but SONET transmission will remain. WDM will become important, mainly for transmission. This scenario, with IP and WDM squeezing out SONET and ATM, is currently very popular.

If the digital video age is expected, traffic will develop along the video distribution route or video communication route, as outlined in the preceding discussion. Technology will then develop either along voice or IP lines of development. ATM will become important because it is a very powerful service mixer. ISDN will become marginal and B-ISDN will develop much sooner because the volume of traffic will be too large for conventional ISDN to handle.

Optical technologies will develop along with the growth of traffic. Link rates and related multiplex and demultiplex technologies will develop at 20 Gbit/s and become commercially important. Subsequently rates of 40 Gbit/s will follow, but it is difficult to foresee the technology going beyond 40 Gbit/s per wavelength, because it is easier at the moment to handle an aggregate bit rate above 40 Gbit/s in the WDM domain than in the TDM domain. WDM is already being implemented in many point-to-point links in the United States. On the other hand, substantial efforts are being spent in Japan to reach 40 Gbit/s in the time domain. The Femtosecond Association is an example of how Japanese efforts in that domain are being supported by the industry.

WDM point-to-point links are being rapidly implemented, and we can expect WDM static networking, with Add-Drop capabilities, to follow. The future of other optical technologies such as solitons and dynamic networking is more difficult to predict. In a compact geographic region there is less need to use such optical technologies. Certainly technologies such as optical memory and optical switches are not an issue for countries like the Netherlands, at least in the time frame are we looking at.

2.5   SUMMARY AND CONCLUSIONS

In the authors' opinion, none of the hypotheses discussed in this chapter is likely to materialize in its pure form. It is unlikely that the world's sole telecommunications network will simply consist of IP switches and that all traffic will go through them, at least not in the time frame considered. The picture will thus become much more complex. Many different kinds of traffic will still need to be carried and therefore a rather heterogeneous mix of technologies will remain. We expect that fiber technologies and wireless will grow tremendously. It is also clear that successful companies that wish to survive will have to offer all four service categories mentioned above, and probably many more.

Wireless technologies are likely to oust copper-based systems from the distribution side, especially in new installations. In many countries without a good infrastructure, last mile has not even been installed; thus the use of wireless will be considered. Another likely development is the tremendous impact of ATM and IP on telephone companies. SONET will remain important for point-to-point transmission, but SONET networking is likely to be squeezed out since many of its functions will be taken over by ATM and WDM.

We can thus envision the following developments. There is likely to be a rapid growth in the use of IP routers and ATM switches interconnected by SONET links. Subsequently, when capacity becomes a bottleneck in SONET links, WDM will be implemented. SONET ADMs will probably not be used much. Service mixers will be used more than anticipated. Further research is required, because it is not clear how to squeeze heterogeneous traffic together into the same network.

Advanced methods of high-speed distribution, perhaps xDSL or SMDS, will likely emerge and become very important in some market segments, such as high-technology areas (e.g., Silicon Valley or similar areas in other countries) where professional people urgently need fast connections. Links at 20 and 40 Gbit/s are likely to emerge. The use of other optical technologies is not so predictable. It is clear that point-to-point WDM links will continue to grow. Past experience has shown that there will be a continuous dispute between higher bit rates and WDM. It has been predicted many times that further upgrades in the time domain will be very difficult to implement, but bit rates have been steadily increasing and will probably reach 40 Gbit/s.

WDM static networking will be implemented within the next few years. WDM dynamic networking is somewhat different. To install dynamic networking, it must be integrated with software and an appropriate control mechanism must be found. It is not clear at the moment whether this is possible, and this constitutes a challenge.

2.6   BIBLIOGRAPHY

1. Announcement of PacBell on a roadmap toward a superhighway configuration in the *San Jose Mercury News*, July 1995.

2. H. Armbuster, "The Flexibility of ATM: Supporting Future Multimedia and Mobile Communications," *IEEE Commun. Mag.,* Vol. 33, Feb. 1995, pp. 76–84.

3. K. P. Davies, "HDTV Evolves for the Digital Era," *IEEE Commun. Mag.,* Vol. 34, June 1996, pp. 110–113.

4. H. D'Hooge, "The Communicating PC," *IEEE Commun. Mag.,* Vol. 34, April 1996, pp. 36–42.

5. M. Figueroa and S. Hansen, "Technology Interworking for SMDS: From the DXISNI to the ATM UNI," *IEEE Commun. Mag.,* Vol. 34, June 1996, pp. 102–108.

6. R. T. Hofmeister, S. M. Gemelos, C. L. Liu, M. C. Ho, D. Wonglumsom, D. T. Mayweather, S. Agrawal, I. Fishman, and L. G. Kazovsky, "Project LEARN—Light Exchangeable, Add/Drop Ring Network," Proc. OFC'97, Feb. 16–21, 1997, Dallas, TX, Post Deadline Paper PD25, pp. PD25.1–PD25.4.

7. M. Hopkins, G. Louth, H. Bailey, R. Yellon, A. Ajibulu, and M. Niva, "A Multi-Faceted Approach to Forecasting Broadband Demand and Traffic," *IEEE Commun. Mag.,* Vol. 33, Feb. 1995, pp. 36–42.

8. A. Jain, W. Fischer, and P-Y. Sibille, "An Evolvable ATM-based Video Network Design Supporting Multiple Access Network," *IEEE Commun. Mag.,* Vol. 33, Nov. 1995, pp. 58–62.

9. Y.-D. Lin, Y.-T. Lin, P.-N. Chen, and M. M. Choy, "Broadband Service Creation and Operations," *IEEE Commun. Mag.,* Vol. 35, Dec. 1997, pp. 116–124.

10. W. Pugh and G. Boyer, "Broadband Access: Comparing Alternatives," *IEEE Commun. Mag.,* Vol. 33, Aug. 1995, pp. 34–47.

11. C.-J. van Driel, P. A. M. van Grinsven, V. Pronk, and W. A. M. Snijders, "The (R)evolution of Access Networks for the Information Superhighway," *IEEE Commun. Mag.,* Vol. 35, June 1997, pp. 104–112.

12. R. S. Vodhanel, L. D. Garrett, S. H. Patel, W. Kraeft, J.-C. Chiao, J. Gamelin, C. A. Gibbons, H. Shirokmann, M. Rauch, J. Young, G.-K. Chang, R. E. Wagner, A. Luss, M. Maeda, J. Pastor, M. Post, C.-C. Chen, J. Wei, B. Wilson, Y. Tsai, R. M. Derosier, A. H. Gnauck, A. R. McCormick, R. W. Tkach, C. L. Allyn, A. R. Chraplyvy, J. Judkins, A. K. Srivastava, J. W. Sulhoff, Y. Sun, A. M. Vengsarkar, C. Wolf, J. L. Zyskind, A. Chester, B. Comission, G. W. Davis, G. Duverney, N. A. Jackman, A. Jozan, V. Nichols, B. H. Lee, R. Vora, A. F. Yorinks, G. Newsome, P. Bhattacharjya, D. Doherty, J. Ellson, C. Hunt, A. Rodriguez-Moral, and J. Ippolito, "National-Scale WDM Networking Demonstration by the MONET Consortium," Proc. OFC'97, Feb. 16–21, 1997, Dallas, TX, Post Deadline Paper PD27, pp. PD27.1–PD27.4.

CHAPTER 3

Speech Coding Standards

ANDREAS S. SPANIAS

ABSTRACT

In this chapter, we provide a survey of speech coding algorithms with emphasis on those methods that are part of voice communication standards. The organization of the chapter is as follows. The first section presents an introduction to speech coding algorithms and standards. The section Speech Analysis—Synthesis and Linear Prediction discusses short- and long-term linear prediction, and the section Linear Prediction and Speech Coding Standards presents standards based on open- and closed-loop linear prediction. The section Standards Based on Subband and Transform Coders discusses standards based on subband coders and transform coders. The chapter concludes with a summary.

3.1 INTRODUCTION

The worldwide growth in communication networks has spurred a renewed interest in the area of speech coding. In addition, advances in very large-scale integration (VLSI) devices along with the emergence of new multimedia applications have motivated a series of algorithm standardization activities. The standardization section of the International Communications Union (ITU), which is the successor of the International Telephone and Telegraph Consultative Committee (CCITT), has been developing compatibility standards for telecommunications. Other standardization committees, such as the European Telecommunications Standards Institute (ETSI) and the International Standards Organization (ISO), have also drafted requirements for speech and audio coding standards. In North America, the Telecommunication Industry Association (TIA) has been developing digital cellular standards and in Japan the Research and Development

25

Center for Radio Systems (RCR) has also formed cellular standards for use in that region. In addition to these organizations there are also committees forming standards for private or government applications such as secure telephony, satellite communications, and emergency radio applications.

The challenge to meet standard specifications has driven much of the research in this area and several speech and audio coding algorithms have been developed and eventually adopted in international standards. A series of competing signal models based on linear predictive coding (LPC) [1–7] and transform-domain analysis-synthesis [8–10] have been proposed in the last 15 years. As it turned out, in speech coding standardization efforts several of the algorithms selected are based on LPC models. In particular, a class of algorithms called analysis-by-synthesis linear prediction [1–6] are embedded in many network and cellular telephony standards. Multimedia and internet audio applications also motivated speech and audio coding work using subband and transform coding algorithms that rely on psychoacoustic signal models [11, 12]. This work yielded a series of audio coding algorithms that are part of wideband and high-fidelity audio standards [13–19]. This chapter provides a survey of speech coding algorithms with emphasis on those methods that are important in communications and multimedia applications. Since the focus is speech coding standards, only occasional reference is made to wideband audio applications.

Speech coding for low-rate applications involves speech analysis-synthesis. In the analysis stage, speech is represented by a compact parametric set that is encoded with a small number of bits. In the synthesis stage, these parameters are decoded and used in conjunction with a reconstruction mechanism to form speech. Analysis can be open-loop or closed-loop. In closed-loop analysis, also called analysis-by-synthesis, the parameters are extracted and encoded by minimizing the perceptually weighted difference between the original and reconstructed speech. Essentially all speech coding algorithms are lossy, that is, the original bit stream of input speech is not maintained. Speech coding algorithms are evaluated based on speech quality, algorithm complexity, delay, and robustness to channel and background noise. Moreover, in network applications coders must perform reasonably well with nonspeech signals such as Dual Tone Multi-Frequency (DTMF) tones, voiceband data, music, and modem. Standardization of candidate speech coding algorithms involves evaluation of speech quality and intelligibility using subjective measures such as the Mean Opinion Score (MOS), the Diagnostic Rhyme Test (DRT), and the Diagnostic Acceptability Measure (DAM). The MOS is quite common in standardization tests and involves rating speech according to a five-level quality scale, that is, MOS relates to speech quality as follows: a MOS of 4–4.5 implies network or toll quality, scores between 3.5 and 4 imply communications quality (cellular grade), and a MOS between 2.5 and 3.5 implies synthetic quality. The simplest coder that achieves toll quality is the 64 Kbit/s ITU G.711 PCM.

Table 3.1 Mean Opinion Score (MOS) Ratings

MOS	Subjective Quality
5	Excellent
4	Good
3	Fair
2	Poor
1	Bad

G.711 has a MOS of 4.3 and is often used as a reference in comparative studies. Several of the new general-purpose algorithms, such as the 8 Kbit/s ITU G.729 [20], also achieve toll quality. Algorithms for cellular standards, such as the IS-54 [21] and the full-rate GSM 6.10 [22], achieve communications quality, and the Federal Standard 1015 (LPC-10) [23] is associated with synthetic quality. These standards and others will be discussed in greater detail in subsequent sections.

This chapter is intended both as a survey and a tutorial. The organization of the chapter is as follows. The section Speech Analysis—Synthesis and Linear Prediction describes linear prediction methods, while the section Linear Prediction and Speech Coding Standards presents standards-based open- and closed-loop LPC. The section Standard Based on Subband and Transform Coders presents standards based on transform and subband coders, and the Summary presents an overview of the chapter along with concluding remarks.

3.2   SPEECH ANALYSIS—SYNTHESIS AND LINEAR PREDICTION

In the human speech production system, speech is produced by the interaction of the vocal tract with the vocal chords in the glottis. Engineering models (Figure 3.1) for speech production typically model the vocal tract as a time-varying digital filter excited by quasi-periodic waveform when speech is voiced (e.g., as in steady vowels) and random waveforms for unvoiced speech (e.g., as in consonants). The vocal tract filter is estimated using linear prediction (LP) algorithms [24, 25].

Linear prediction algorithms are part of several speech coding standards, including Adaptive Differential Pulse Code Modulation (ADPCM) systems [26, 27], Code Excited Linear Prediction (CELP) algorithms [1], as well as other analysis-by-synthesis linear predictive coders [28]. Linear prediction is a process whereby the most recent sample of speech is predicted by a linear combination of past samples. Linear prediction analysis is done using a finite-length impulse response (FIR) digital filter (Figure 3.2) whose output e(n) is minimized. The output of the linear predictor analysis filter is called LP residual or simply residual. Because only short-term delays are considered, the linear predictor in Figure 3.2 is also known as a *short-term predictor.*

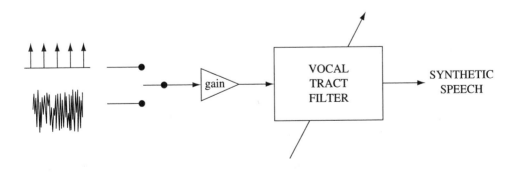

FIGURE 3.1
Engineering model for speech synthesis.

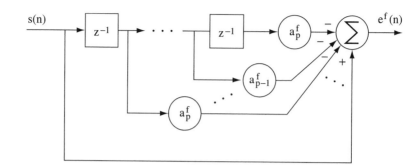

FIGURE 3.2
Linear prediction analysis.

The LP coefficients are chosen to minimize the LP residual. The minimization of the residual error is usually done over a frame of speech, that is,

$$\varepsilon = \sum_{n=1}^{N} e^2(n) \tag{1}$$

where

$$e(n) = x(n) - \sum_{k=1}^{p} a_k x(n-k). \tag{2}$$

The inverse of the LP analysis filter is an all-pole filter called LP synthesis filter and can be considered as a reasonably good model for the human vocal tract. The frequency response associated with the short-term predictor captures the *formant* structure of the short-term speech spectrum. The formants are a set of spectral peaks that are associated with the resonant modes of the human vocal tract. The all-pole filter or vocal tract transfer function is given by

$$H(z) = \frac{g}{1 - \sum_{k=1}^{p} a_k z^{-k}} = \frac{g}{1 - A(z)} \tag{3}$$

The minimization of e or a modified version of e yields a set of equations involving correlation statistics and the LP coefficients are found by inverting a structured matrix. There are many efficient algorithms [24, 25] for inverting this matrix, including algorithms tailored to work well with finite precision arithmetic [29]. Preconditioning of the speech and correlation data using tapered windows improves the numerical behavior of these algorithms. In addition, bandwidth expansion of the LP coefficients is very typical in LPC as it reduces transient effects in the synthesis process.

In LPC the analysis window is typically 20 ms long. To avoid transient effects from large changes in the LP parameters from one frame to the next, the frame is usually divided into subframes (typically 5–7.5 ms long) and subframe parameters are obtained by linear interpolation. The direct form LP coefficients, $\mathbf{a}_k$, are not adequate for quantization and transformed coefficients are typically used in quantization tables. The reflection or lattice prediction coefficients are a by-product of the Levinson recursion and have better quantization properties than direct-form coefficients. Some of the early standards such as the LPC-10 [23] and the IS-54 Vector Sum Exited Linear Prediction (VSELP) [21] encode reflection coefficients for the vocal tract. Transformation of the reflection coefficients can also lead to a set of parameters that are less sensitive to quantization. In particular, the log area ratios (LARs) and the inverse sine transformation have been used in the GSM 6.10 [22] and in the skyphone standard [29]. Most recent LPC-related cellular standards [30–32] quantize a set of parameters called Line Spectrum Pairs (LSPs). The main advantage of the LSPs is that they relate directly to frequency-domain information and hence they can be encoded using perceptual quantization rules. In some of the most recent standards such as the TIA/EIA IS-127 [31] the LSPs are jointly quantized using split vector quantization.

3.2.1 Long-Term Prediction (LTP)

Long-term prediction, as opposed to the short-term prediction presented in the previous section, is a process that captures the long-term correlation in the speech signal. The LTP provides a mechanism for representing the periodicity in speech and as such it represents the fine harmonic structure in the short-term speech spectrum. The LTP requires estimation of two parameters, that is, a delay t and a parameter at. For strongly voiced segments the delay is usually an integer equal to the pitch period. A transfer function of a simple LTP synthesis filter is given below. More complex LTP filters involve multiple parameters and noninteger (fractional) delays [33]. We will provide more information on the estimation of LTP parameters in subsequent sections describing the standards.

$$H_\tau(z) = \frac{1}{1 - a_\tau z^{-\tau}} = \frac{1}{1 - A_L(z)} \tag{4}$$

3.3 LINEAR PREDICTION AND SPEECH CODING STANDARDS

Our discussion of linear prediction is divided in two categories, that is, open-loop LP and closed-loop or analysis-by-synthesis linear prediction. In closed-loop LP the excitation parameters are determined by minimizing the difference between input and reconstructed speech. Unless otherwise stated, the input to all coders discussed in this section is speech sampled at 8 kHz.

3.3.1 Open-Loop Linear Prediction

In our discussion of open-loop LP algorithms, we include ADPCM standards and also source-system algorithms that use open-loop analysis to determine the excitation sequence.

3.3.1.1 The ITU G.726 and G.727 ADPCM Coders

One of the simplest scalar quantizers that uses LP is the Adaptive Differential Pulse Code Modulation (ADPCM) coder [26, 27]. ADPCM algorithms encode the difference between the current and the predicted speech samples. The prediction parameters are obtained by backward estimation, that is, from quantized data, using a gradient algorithm. The ADPCM 32 Kbit/s algorithm in the ITU G.726 standard (formerly known as CCITT G.721) uses a pole-zero adaptive predictor. ITU G.726 also accommodates 16, 24, and 40 Kbit/s with individually optimized

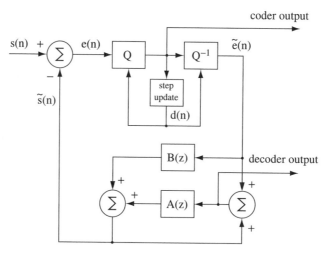

FIGURE 3.3
The ITU G.726 ADPCM encoder.

quantizers. The ITU G.727 has embedded quantizers and was developed for packet network applications. Because of the embedded quantizers, G.727 has the capability to switch easily to lower rates in network congestion situations by dropping bits. The MOS for 32 Kbit/s G.726 is 4.1 and complexity is estimated to be 2 million instructions per second (MIPS) on special-purpose chips.

3.3.1.2 The Inmarsat-B Adaptive Predictive Vocoder

The International Mobile Satellite B (Inmarsat-B) standard [34] uses an ADPCM coder with an open-loop LTP in addition to a sixth-order short-term predictor. The LTP parameters are estimated using the average magnitude difference function (AMDF) (see [23] for details). Both pre-

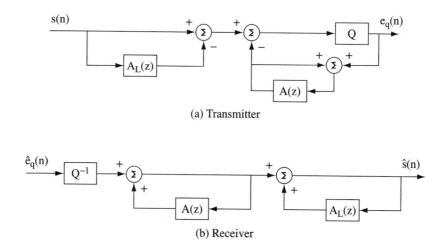

(a) Transmitter

(b) Receiver

FIGURE 3.4
The Inmarsat-B adaptive predictive coder.

dictors are updated every frame of 20 ms. The algorithm operates at 12.8 and 9.6 Kbit/s and its complexity is about 10 MIPS.

3.3.1.3 The LPC-10 Federal Standard 1015

In 1976 the Department of Defense (DoD) adopted an LPC algorithm for secure communications at 2.4 Kbit/s (Figure 3.5). The algorithm, known as the LPC-10 or in an enhanced version LPC10e [35], eventually became the Federal Standard FS-1015 [23, 35, 36]. The LPC-10 uses a 10th-order predictor to estimate the vocal-tract parameters. Segmentation and frame processing in LPC-10 depends on voicing. Pitch information is estimated using the average magnitude difference function (AMDF). Voicing is estimated using energy measurements, zero-crossing measurements, and the maximum to minimum ratio of the AMDF. The excitation signal for voiced speech in the LPC-10 consists of a sequence that resembles a sampled glottal pulse. This sequence is defined in the standard [23] and periodicity is created by a pitch-synchronous pulse-repetition process. The MOS for LPC-10e is 2.3 and complexity is estimated at 7 MIPS.

3.3.1.4 Mixed Excitation Linear Prediction

In 1996 the U.S. government standardized a new 2.4 Kbits/s algorithm called Mixed Excitation LP (MELP) [7, 37]. The development of mixed excitation models in LPC was motivated largely by voicing errors in LPC-10 and also by the inadequacy of the two-state excitation model in cases of voicing transitions (mixed voiced-unvoiced frames). This problem can be solved using a mixed excitation model where the impulse train (buzz) excites the low-frequency region of the LP synthesis filter and the noise excites the high-frequency region of the synthesis filter (Figure 3.6). The excitation shaping is done using first-order FIR filters H1(z) and H2(z) with time-varying parameters. The mixed source model also uses (selectively) pulse position jitter for the synthesis of weakly periodic or aperiodic voiced speech. An adaptive pole-zero spectral enhancer is used to boost the formant frequencies. Finally, a dispersion filter is used after the LPC synthesis filter to improve the matching of natural and synthetic speech away from the formants. The 2.4 Kbit/s MELP is based on a 22.5 ms frame and the algorithmic delay was estimated to be 122.5 ms. An integer pitch estimate is obtained open-loop by searching autocorrelation statistics

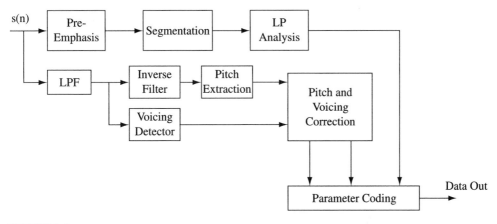

FIGURE 3.5
The LPC-10e encoder.

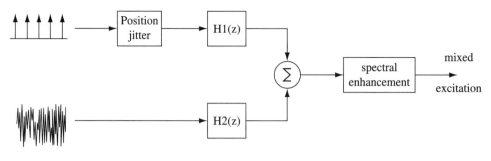

FIGURE 3.6
Mixed Excitation LPC.

followed by a fractional pitch refinement process. The LP parameters are obtained using the Levinson-Durbin algorithm and vector quantized as LSPs. MELP has an estimated MOS of 3.2 and complexity estimated at 40 MIPS.

3.3.2   Standards Based on Analysis-by-Synthesis Linear Prediction

We describe here several standards based on a class of modern source-system coders where system parameters are determined by linear prediction and the excitation sequence is determined by closed-loop or analysis-by-synthesis optimization [1–3]. The optimization process determines an excitation sequence that minimizes the weighted difference between the input speech and synthesis speech. The system consists of a short-term LP synthesis filter, a long-term LP synthesis filter for the pitch (fine) structure of speech, a perceptual weighting filter, $W(z)$ that shapes the error such that quantization noise is masked by the high-energy formants, and the excitation generator. The three most common excitation models for analysis-by-synthesis LPC are: the multipulse model [2, 3], the regular pulse excitation model [5], and the vector or code excitation model [1]. These excitation models are described in the context of standardized algorithms.

3.3.2.1   Multi-Pulse Excited Linear Prediction for the Skyphone Standard
A 9.6 Kbit/s Multi-Pulse Excited Linear Prediction (MPLP) algorithm is used in Skyphone airline applications [29]. The MPLP algorithm (Figure 3.7) forms an excitation sequence that consists of multiple nonuniformly spaced pulses. During analysis both the amplitude and locations of the pulses are determined (sequentially) one pulse at a time such that the weighted mean squared error is minimized. The MPLP algorithm typically uses four to six pulses every 5 ms [2, 3]. The weighting filter is given by

$$W(z) = \frac{1 + \sum_{i=1}^{p} \gamma_1^i a_i z^{-i}}{1 + \sum_{i=1}^{p} \gamma_2^i a_i z^{-i}} \quad 0 < \gamma_2 < \gamma_1 < 1 \tag{5}$$

The role of $W(z)$ is to deemphasize the error energy in the formant regions. This deemphasis strategy is based on the fact that in the formant regions quantization noise is partially masked by speech. Excitation coding in the MPLP algorithm is more expensive than in the classical linear predictive vocoder because MPLP encodes both the amplitudes and the locations of the pulses. The British Telecom International skyphone MPLP algorithm accommodates passenger

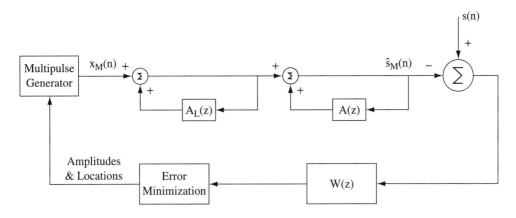

FIGURE 3.7
MPLP for the Skyphone standard.

communications in aircraft. The algorithm incorporates both short- and long-term prediction. The LP analysis window is updated every 20 ms. The LTP parameters are obtained using open-loop analysis. The MOS for the skyphone algorithm is 3.4. A more sophisticated multipulse scheme is used in the dual-rate ITU G.723.1 that will be discussed in the following section.

3.3.2.2 The Regular Pulse Excitation (RPE) Algorithm for the ETSI Full-Rate GSM 6.10 Standard

RPE coders also employ an excitation sequence that consists of multiple pulses. The basic difference of the RPE algorithm from the MPLP algorithm is that the pulses in the RPE coder are uniformly spaced and therefore their positions are determined by specifying the location of the first pulse within the frame and the spacing between nonzero pulses. The analysis-by-synthesis optimization in RPE algorithms represents the LP residual by a regular pulse sequence that is determined by weighted error minimization [5]. A 13 Kbit/s coding scheme that uses RPE with long-term prediction (LTP) was adopted for the full-rate ETSI GSM [22] Pan-European digital

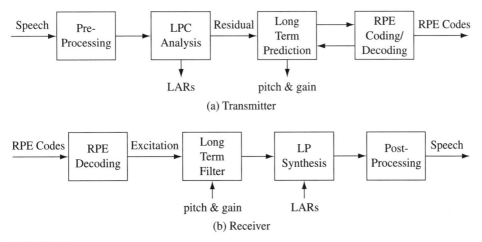

FIGURE 3.8
The ETSI GSM 6.10 RPE-LTP algorithm.

cellular standard (Figure 3.8). The performance of the GSM codec in terms of MOS was reported to be between 3.47 (min) and 3.9 (max) and its complexity is 5–6 MIPS.

3.3.2.3   The FS-1016 Code Excited Linear Prediction (CELP) Algorithm

The Vector or Code Excited Linear Prediction (CELP) algorithm [1] (Figure 3.9) encodes the excitation using vector quantization. The codebook used in a CELP coder contains vector excitation and in each subframe a vector is chosen using an analysis-by-synthesis process. The "optimum" vector is selected such that the perceptually weighted MSE is minimized. A gain factor scales the excitation vector and the excitation samples are filtered by the long- and short-term synthesis filters.

A 4.8 Kbit/s CELP algorithm is used by the DoD for use in the third-generation secure telephone unit (STU-III) [38, 39]. This algorithm is described in the Federal Standard 1016 (FS-1016) and was jointly developed by the DoD and AT&T Bell Laboratories. The synthesis configuration for the FS-1016 CELP is shown in Figure 3.10.

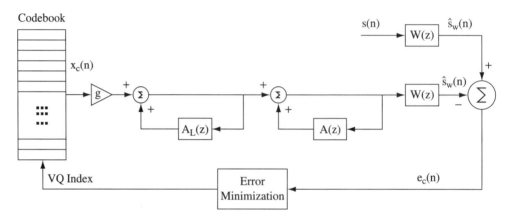

FIGURE 3.9
The analysis-by-synthesis CELP algorithm.

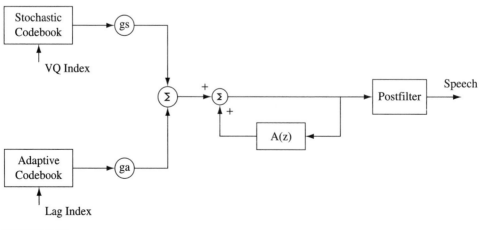

FIGURE 3.10
The FS-1016 CELP algorithm.

Speech in the FS-1016 CELP is sampled at 8 kHz and segmented in frames of 30 ms and each frame is segmented in subframes of 7.5 ms. The excitation in this CELP is formed by combining vectors from an adaptive (LTP) and a stochastic codebook. The excitation vectors are selected in every subframe and the codebooks are searched sequentially starting with the adaptive codebook. The term *adaptive codebook* is used because the LTP lag search can be viewed as an adaptive codebook search where the codebook is defined by previous excitation sequences (LTP state) and the lag, t, determines the vector index. The adaptive codebook contains the history of past excitation signals and the LTP lag search is carried over 128 integer (20–147) and 128 noninteger delays. A subset of lags is searched in even subframes to reduce the computational complexity. The stochastic codebook contains 512 sparse and overlapping codevectors. Each codevector consists of 60 samples and each sample is ternary valued (1, 0, –1) to allow for fast convolution. Ten short-term prediction parameters are encoded as LSPs on a frame-by-frame basis. Subframe LSPs are obtained by linear interpolation. The computational complexity of the FS1016 CELP was estimated at 16 MIPS and a MOS score of 3.2 has been reported.

3.3.2.4   *Vector Sum Excited Linear Prediction (VSELP) and International Standards*

The Vector Sum Excited Linear Prediction (VSELP) algorithm [6] and its variants are embedded in three digital cellular standards, that is, the TIA IS-54 [21], the Japanese standard [40], and the half-rate GSM [41]. An 8 Kbit/s VSELP algorithm was adopted for the IS-54 North American Digital Cellular System. The IS-54 VSELP uses highly structured codebooks that are tailored for reduced computational complexity and increased robustness to channel errors. VSELP excitation is derived by combining excitation vectors from three codebooks, namely, an adaptive codebook and two highly structured stochastic codebooks (Figure 3.11). The frame in the VSELP algorithm is 20 ms long and each frame is divided into four 5 ms subframes. A 10th-order short-term synthesis filter is used and its coefficients are encoded as reflection coefficients once per frame. Subframe LPC parameters are obtained by linear interpolation. The excitation parameters are updated every 5 ms. The 128 40-sample vectors in each stochastic codebook are formed by linearly combining seven basis vectors. The weights used for the basis vectors are allowed to take the values of one or minus one. Hence the effect of changing one bit in the code-

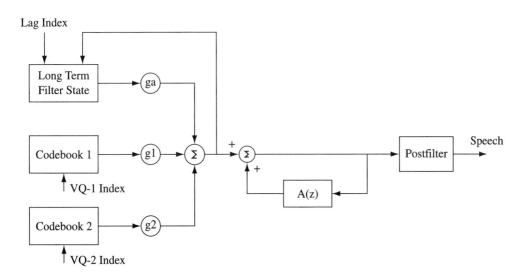

FIGURE 3.11
The IS-54 VSELP algorithm.

word, possibly due to a channel error, is not minimal since the codevectors corresponding to adjacent (gray code-wise) codewords are different only by one basis vector. The search of the codebook is also greatly simplified because the response of the short-term synthesis filter, to codevectors from the stochastic codebook, can be formed by combining filtered basis vectors. The codebook structure lends itself to an efficient recursive search process. The complexity of the 8 Kbit/s VSELP was reported to be around 14 MIPS and the MOS reported were 3.5.

The ETSI 6.20 GSM half-rate VSELP algorithm [41] is a 5.6 Kbit/s algorithm. Additional bits are used for channel error protection. The GSM VSELP also includes a voice activity detector and comfort noise generator.

3.3.2.5   The 16 Kbit/s ITU G.728 Low-Delay CELP

One of the problems in network applications of speech coding is that coding gain is achieved at the expense of coding delay. The one-way delay is basically the time elapsed from the instant a speech sample arrives at the encoder to the instant that this sample appears at the output of the decoder. This definition of one-way delay does not include channel- or modem-related delays. Roughly speaking, the one-way delay is generally between two and four frames. The ITU G.728 low-delay CELP coder [42, 43] achieves low one-way delay by: short frames, backward-adaptive predictor, and short excitation vectors (five samples). In backward-adaptive prediction, the LP parameters are determined by operating on previously quantized speech samples that are also available at the decoder (Figure 3.12). The LD-CELP algorithm does not utilize LTP. Instead, the order of the short-term predictor is increased to 50 to compensate for the lack of a pitch loop.

The frame size in LD-CELP is 2.5 ms and the subframes are 0.625 ms long. The parameters of the 50th-order predictor are updated every 2.5 ms. The perceptual weighting filter is based on 10th-order LP operating directly on unquantized speech and is updated every 2.5 ms. To limit the buffering delay in LD-CELP only 0.625 ms of speech data are buffered at a time. LD-CELP utilizes adaptive short- and long-term postfilters to emphasize the pitch and formant structures of speech. The one-way delay of the LD-CELP is less than 2 ms and MOSs as high as 3.93 and 4.1 were obtained. The speech quality of the LD-CELP was judged to be equivalent or better than the G.726 standard even after three asynchronous tandem encodings. The coder was also shown to be capable of handling voiceband modem signals at rates as high as 2400 baud (provided that perceptual weighting is not used). The coder complexity and memory requirements on a floating-point DSP chip were found to be: 10.6 MIPS and 12.4 kbytes for the encoder and 8.06 MIPS and 13.8 kbytes for the decoder. Fixed-point MIPS are estimated at 30.

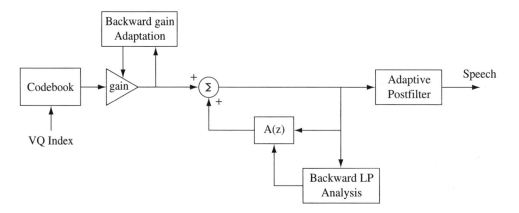

FIGURE 3.12
The G.728 low-delay CELP algorithm.

3.3.2.6 The CDMA IS-96 QCELP

The IS-96 QCELP [44] (Figure 3.13) is a variable bit-rate algorithm and is part of the Code Division Multiple Access (CDMA) standard for cellular communications. The bit rate is variable with four rates supported: 9.6, 4.8, 2.4, and 1.2 Kbit/s. The rate is determined by speech activity. Rate changes can also be initiated upon command from the network. The short-term LP parameters are encoded as LSPs. Lower rates are achieved by allocating fewer bits to LP parameters and by reducing the number of updates of the LTP and fixed codebook parameters. At 1.2 Kbit/s (rate 1/8) the algorithm essentially encodes comfort noise. The MOS for QCELP at 9.6 Kbit/s is 3.33 and the complexity is estimated to be around 15 MIPS.

3.3.2.7 The ITU G.729 and G.729A CS-ACELP

A low-delay 8 Kbit/s conjugate structure algebraic CELP (CS-ACELP) has been adopted as the ITU G.729 [20]. The G.729 is designed for both wireless and multimedia network applications. CS-ACELP is a low-delay algorithm with a frame size of 10 ms, a look-ahead of 5 ms, and a total algorithmic delay of 15 ms. The algorithm is based on an analysis-by-synthesis CELP scheme and uses two codebooks. The short-term prediction parameters are obtained every 10 ms and vector quantized as LSPs. The algorithm uses an algebraically structured fixed codebook that does not require storage. Each 40-sample (5 ms) codevector contains four nonzero, binary valued (−1, 1) pulses. Pulses are interleaved and position-encoded, thereby allowing efficient search. Gains for the fixed and adaptive codebooks are jointly vector quantized in a two-stage, two-dimensional conjugate structured codebook. Search efficiency is enhanced by a preselection process that constrains the exhaustive search to 32 of 128 codevectors. The algorithm comes in two versions, that is, the original G.729 (20 MIPS) and the less complex G.729 Annex A (11 MIPS). The algorithms are interoperable and the lower complexity algorithm has slightly lower quality. The MOS for the G.729 is 4 and for the G.729A, 3.76. G.729 Annex B defines a silence compression algorithm allowing either the G.729 or the G.729A to operate at lower rates, thereby making them particularly useful in digital simultaneous voice and data (DSVD) applications. Extensions to G.729 at 6.4 and 12 Kbit/s are planned by ITU.

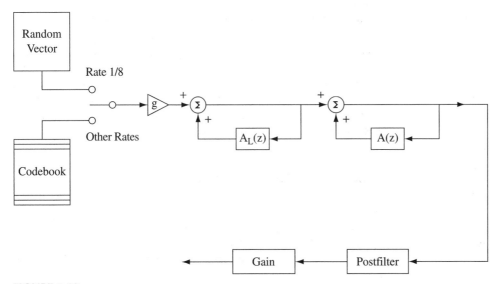

FIGURE 3.13
The IS-96 QCELP decoder.

3.3.2.8   The ITU G.723.1 MP-MLQ/ACELP

The ITU G.723.1 [45] is a dual-rate speech coder intended for audio and videoconferencing/ telephony over public phone [plan over telephone service (POTS)] networks. G.723.1 is part of the ITU H.323 and H.324 audio/video conferencing standards. The standard is dual-rate 6.3 and 5.3 Kbit/s. The excitation is selected using an analysis-by-synthesis process and two excitation schemes are defined, that is, the multipulse maximum likelihood quantization (MP-MLQ) for the 6.3 Kbit/s mode and the ACELP for 5.3 Kbit/s. Ten short-term LP parameters are computed and vector quantized as LSPs. A fifth-order LTP is used in this standard and the LTP lag is determined using a closed-loop process searching around a previously obtained open-loop estimate. The LTP gains are vector quantized. The high-rate MP-MLQ excitation involves matching the LP residual with a set of impulses at restricted positions. The lower-rate ACELP excitation is similar but not identical to the excitation scheme used in G.729. G.723.1 provides a toll-quality MOS of 3.98 at 6.3 Kbit/s and has a frame size of 30 ms with a look-ahead of 7.5 ms. The estimated one-way delay is 37.5 ms. An option for variable rate operation using a voice activity detector (silence compression) is also available. The Voice over IP (VoIP) forum, which is part of The International Multimedia Teleconferencing Consortium (IMTC), recommended G.723.1 to be the default audio codec for voice of the network (decision pending).

3.3.2.9   The ETSI GSM 6.60 Enhanced Full-Rate Standard

The enhanced full-rate (EFR) encoder was developed for use in the full-rate GSM standard [30]. The EFR is a 12.2 Kbit/s algorithm with a frame of 20 ms and a 5 ms look-ahead. A 10th-order short-term predictor is used and its parameters are transformed to LSPs and encoded using split vector quantization. The LTP lag is determined using a two-stage process where open-loop search provides an initial estimate. This is then refined by closed-loop search around the neighborhood of the initial estimate. An algebraic codebook similar to that of the G.729 ACELP is also used. The algorithmic delay for the EFR is 25 ms and the MOS is estimated to be around 4.1. The standard has provisions for a voice activity detector and an elaborate error protection scheme is also part of the standard. The North American PCS 1900 standard uses the GSM infrastructure and hence the EFR.

3.3.2.10   The EFR Algorithm and the IS-641 TDMA Cellular/PCS Standard

The IS-641 [32], a 7.4 Kbit/s EFR algorithm also known as the Nokia/USH codec, is a variant of the GSM EFR. IS-641 is the speech coding standard embedded in the IS-136 Digital-AMPS (D-AMPS) North American digital cellular standard. EFR offers improved quality for this service relative to IS-54. The algorithm is based on a 20 ms frame and a 10th-order LP whose parameters are split vector quantized as LSPs. An algebraic codebook (ACELP) is used and the LTP lag is determined using open-loop search followed by closed-loop refinement with a fractional pitch resolution. The complexity of this coder is estimated at 14 MIPS and the MOS is estimated at 3.8.

3.3.2.11   The IS-127 Enhanced Variable Rate Coder (EVRC) Used in the CDMA Standard

The IS-127 Enhanced Variable Rate Coder (EVRC) [31] is based on Relaxed CELP (RCELP). RCELP uses interpolative coding methods [4] as a means for reducing further the bit rate and complexity in analysis-by-synthesis linear predictive coders. The EVRC encodes the RCELP parameters using a variable-rate approach. There are three possible bit rates for EVRC, that is, 8, 4, and 0.8 Kbit/s or after error protection 9.6, 4.8, and 1.2 Kbit/s, respectively. The rate is determined using a voice activity detection algorithm that is embedded in the standard. Rate changes can also be initiated upon command from the network. At the lowest rate (0.8 Kbit/s) the algorithm does not encode excitation information hence the decoder essentially produces comfort noise. The other rates are achieved by changing the number of bits allotted to LP and excitation parameters. The algorithm is based on a 20 ms frame and each frame is divided into three 6.75 ms subframes. A 10th-order short-term LP is obtained using the Levinson-Durbin algorithm and LP parameters are split vector encoded as LSPs. The fixed codebook structure is ACELP. The LTP parameters are estimated using generalized analysis-by-synthesis where, instead of matching the input speech, a down-sampled version of a modified LP residual that conforms to a pitch contour is matched. The pitch contour is established using interpolative methods. The standard also specifies a Fast Fourier transform (FFT)-based speech enhancement preprocessor that is intended to remove background noise from speech. The MOS for EVRC is 3.8 at 9.6 bit/s and the algorithmic delay is estimated to be 25 ms.

3.3.2.12   The Japanese PDC Full-Rate and Half-Rate Standards

The Japanese Research and Development Center for Radio Systems (RCR) has adopted two algorithms for the Personal Digital Cellular (PDC) full-rate (6.3 Kbit/s) and the PDC half-rate (3.45 Kbit/s) standards. The full-rate algorithm [40] is a variant of the IS-54 VSELP algorithm described in a previous section. The half-rate PDC coder is a high-complexity pitch-synchronous innovation CELP (PSI-CELP) [46]. As the name implies, the codebooks of PSI-CELP depend on the pitch period. PSI-CELP defaults to periodic vectors if the pitch period is less than the frame size. The complexity of the algorithm is about 50 MIPS and the algorithmic delay is about 50 ms.

3.4   STANDARDS BASED ON SUBBAND AND TRANSFORM CODERS

In this section we present two classes of speech coders that rely on subband and sinusoidal representations of the speech waveform.

3.4.1   The ITU G.722 Subband Coder

In subband coders, the signal is divided into frequency subbands using a Quadrature Mirror Filter (QMF) bank (Figure 3.14). The subband coder (SBC) exploits the statistics of the signal and/or perceptual criteria to encode the signal in each band using a different number of bits. For example, the lower-frequency bands are usually allotted more bits than higher-frequency bands

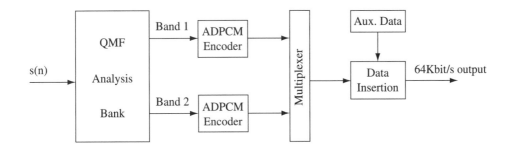

FIGURE 3.14
The ITU G.722 encoder.

to preserve critical pitch and formant information. The design of the filter bank is a very important consideration in SBC. In the absence of quantization noise, perfect reconstruction can be achieved using QMF banks. The ITU G.722 standard [47] for 7 kHz wideband audio at 64 Kbit/s is based on a two-band SBC and was formed primarily for Integrated Services Digital Network (ISDN) teleconferencing. The low-frequency subband is quantized at 48 Kbit/s, while the high-frequency subband is coded at 16 Kbit/s. Provisions for lower rates have been made by quantizing the low-frequency subband at 40 or 32 Kbit/s. The quantizers in G.722 are similar to the ADPCM quantizers specified in G.726. The MOS at 64 Kbit/s is greater than 4 for speech and slightly less than 4 for music, and the G.722 introduces a delay around 1.5 ms. The complexity of the G.722 is estimated to be 5 MIPS. There is currently an ITU standardization effort aimed at a new wideband speech coder at 16, 24, and 32 Kbit/s.

3.4.2   Sinusoidal Transform Coding

The sinusoidal model [8] represents speech by a linear combination of sinusoids. The opportunity to reduce the bit rate using this model stems from the fact that voiced speech is typically periodic and hence it can be represented by a constrained set of sinusoids. In addition, the statistical structure of unvoiced speech can be preserved by a sinusoidal model with appropriately defined random phase. The low-rate sinusoidal transform coder (STC) uses a mixed-voicing model for the frequencies where sinusoids up to a certain frequency are harmonic and above it nonharmonic. The STC amplitude and phase information is efficiently "packed" in an LP spectral envelope whose parameters are vector encoded as LSPs. The STC was considered in several federal and TIA standardization competitions and is being currently considered for wideband and high-fidelity applications. Full-duplex STCs have been implemented on the TI TMS320C30 chip and the complexity reported was 13 MIPS while the MOS scores were 3.52 (4.8 Kbit/s) and 2.9 (2.4 Kbit/s).

3.4.3   The Multiband Excitation Coder and the Inmarsat-M Standard

The Multiband Excitation (MBE) [9], the Improved MBE (IMBE) [10], and the recently introduced low-rate Advanced MBE (AMBE) coders rely on a model that treats the short-time speech spectrum as the product of a mixed harmonic/aharmonic excitation spectrum and a vocal tract envelope. This is done by dividing the speech segments into distinct frequency bands and providing a distinct voiced/unvoiced excitation for each band. The analysis process in MBE consists of determining: a pitch period, voiced and unvoiced envelope parameters, and voicing information for each subband. An integer pitch period is first estimated using an autocorrelation-like method and a pitch tracker is used to smooth the estimate for interframe continuity. The synthesis process involves a bank of sinusoids mixed with appropriately shaped noise. The IMBE is based on MBE analysis-synthesis and employs more efficient methods for quantizing the MBE model parameters. In addition, the IMBE coding scheme is more robust to channel impairments. An IMBE that operates at 6.4 Kbit/s is part of the Australian (AUSSAT) mobile satellite standard and the International Mobile Satellite (Inmarsat-M) standard [48]. The 2250 bit/s of the Inmarsat-M IMBE are used for forward error correction. The remaining 4150 bit/s are used for coding the IMBE parameters. The pitch period in the IMBE is quantized with one-half sample accuracy at 8 bits. A maximum of 12 voiced/unvoiced decisions are encoded, 1 bit per group of three harmonics. The amplitudes are encoded using a combination of block and scalar coding and the DCT. The phases of the harmonic components in IMBE are obtained using a phase prediction algorithm. The 6.4 Kbit/s IMBE has an algorithmic delay of 78.75 ms and a MOS of 3.4.

A variant of the IMBE is also part of the digital standard for public safety radio (APCO Project 25). A new Advanced MBE (AMBE) was developed recently with claims for near-toll quality at 4 Kbit/s. A 2.4 Kbit/s version of the AMBE was licensed for the Motorola IRIDIUM project.

3.5 SUMMARY AND EMERGING STANDARDS

In this chapter, we have provided a review of some of the current methods for speech coding. Speech coding research has come a long way in the last 10 years and several algorithms are rapidly finding their way into consumer products ranging from wireless cellular telephones to computer multimedia systems. Research and development in code-excited linear prediction yielded several algorithms that have been adopted for several cellular standards (see Table 3.2 below). In addition, new algorithms are being considered to improve the capacity and performance of several interface standards. A new algorithm, called the Adaptive Multi-Rate (AMR) coder, is

Table 3.2 Algorithms and Their Properties

Algorithm	Bit Rate (Kbit/s)	MOS	Complexity (FXP MIPS)	Frame size (ms)
PCM G.711	64	4.3	0.01	0
ADPCM G.726	32	4.1	2	0.125
SBC G.722	48/56/64	4.1	5	0.125
LD-CELP G.728	16	4	~30	0.625
CS-ACELP G.729	8	4	~20	10
CS-ACELP-A G.729	8	3.76	11	10
MPC-MLQ G.723.1	6.3/5.3	3.98/3.7	~16	30
GSM FR RPE-LTP	13	3.7 (ave)	5	20
GSM EFR	13	4	14	20
GSM HR VSELP	6.3	~3.4	14	20
IS-54 VSELP	8	3.5	14	20
IS-641 EFR	8	3.8	14	20
IS-96 QCELP	1.2/2.4/4.8/9.6	3.33 (9.6)	15	20
IS-127 EVRC	1.2/4.8/9.6	~3.8 (9.6)	20	20
PDC VSELP	6.3	3.5	14	20
PDC PCI-CELP	3.45	~3.4	~48	40
FS 1015 – LPC 10e	2.4	2.3	7	22.5
FS 1016 – CELP 4.8	4.8	3.2	16	30
MELP	2.4	3.2	~40	22.5
Inmarsat-B APC	9.6/12.8	~3.1/3.4	10	20
Inmarsat-M IMBE	6.3	3.4	~13	20

Note that some of the figures in Table 3.2 are estimates designated with "~". Although MOS and MIPS provide a perspective on quality and complexity, they have been obtained from different tests and correspond to implementations on different platforms. Therefore, these figures may not be adequate for direct comparison between algorithms.

being adopted for the GSM system [49–50]. On the other hand, the evolution of CDMA wireless data services from IS-95 to the so-called CDMA-2000 also includes definitions for new vocoders [51]. Finally, ITU activities are concentrating on the development of toll-quality algorithms that operate at 4 Kbit/s, and several proposals are being evaluated [52–58].

3.6 REFERENCES

[1] M. R. Schroeder and B. Atal, "Code-Excited Linear Prediction (CELP): High Quality Speech at Very Low Bit Rates," *Proc. ICASSP-85,* Tampa, FL, p. 937, April 1985.

[2] S. Singhal and B. Atal, "Improving the Performance of Multi-Pulse Coders at Low Bit Rates," *Proc. ICASSP-84,* p. 1.3.1, 1984.

[3] B. Atal and J. Remde, "A New Model for LPC Excitation for Producing Natural Sounding Speech at Low Bit Rates," *Proc. ICASSP-82,* pp. 614–617, April 1982.

[4] W. B. Kleijn et al., "Generalized Analysis-by-Synthesis Coding and Its Application to Pitch Prediction," *Proc. ICASSP-92,* pp. I-337–I-340, 1992.

[5] P. Kroon, E. Deprettere, and R. J. Sluyeter, "Regular-Pulse Excitation—A Novel Approach to Effective and Efficient Multi-Pulse Coding of Speech," *IEEE Trans. ASSP-34(5),* Oct. 1986.

[6] I. Gerson and M. Jasiuk, "Vector Sum Excited Linear Prediction (VSELP) Speech Coding at 8 kbits/s," *Proc. ICASSP-90,* NM, pp. 461–464, April 1990.

[7] A. McCree and T. Barnwell III, "A New Mixed Excitation LPC Vocoder," *Proc. ICASSP-91,* Toronto, pp. 593–596, May 1991.

[8] R. McAulay and T. Quatieri, "Low-Rate Speech Coding Based on the Sinusoidal Model," in *Advances in Speech Signal Processing,* Ed. S. Furui and M. M. Sondhi, Marcel Dekker, NY, pp. 165–207, 1992.

[9] D. Griffin and J. Lim, "Multiband Excitation Vocoder," *IEEE Trans. ASSP-36,* No. 8, p. 1223, Aug. 1988.

[10] J. Hardwick and J. Lim, "The Application of the IMBE Speech Coder to Mobile Communications," *Proc. ICASSP-91,* pp. 249–252, May 1991.

[11] P. Noll, "Digital Audio Coding for Visual Communications," *Proc. IEEE,* Vol. 83, No. 6, pp. 925–943, June 1995.

[12] T. Painter and A. Spanias, "Perceptual Coding of Digital Audio," *Proc IEEE,* Vol. 88, No. 4, pp. 451–513, April 2000.

[13] G. Davidson, "Digital Audio Coding: Dolby AC-3," in *The Digital Signal Processing Handbook,* CRC Press and IEEE Press, p. 41-1, 1998.

[14] C. Todd et. al., "AC-3: Flexible Perceptual Coding for Audio Transmission and Storage," in Proc. 96th Conv. Aud. Eng. Soc., preprint #3796, Feb. 1994.

[15] D. Sinha, J. Johnston, S. Dorward, and S. Quackenbush, "The Perceptual Audio Coder (PAC)," in *The Digital Signal Processing Handbook,* CRC Press and IEEE Press, p. 42-1, 1998.

[16] K. Akagiri et al., "Sony Systems," in *The Digital Signal Processing Handbook,* CRC Press and IEEE Press, p. 43-1, 1998.

[17] K. Brandenburg and G. Stoll, "ISO-MPEG-1 Audio: A Generic Standard for Coding of High-Quality Digital Audio," *J. Audio Eng. Soc.,* pp. 780–792, Oct. 1994.

[18] A. Hoogendoorn, "Digital Compact Cassette," *Proc. IEEE,* pp. 1479–1489, Oct. 1994.

[19] T. Yoshida, "The Rewritable MiniDisc System," *Proc. IEEE,* pp. 1492–1500, Oct. 1994.

[20] ITU Study Group 15 Draft Recommendation G.729, "Coding of Speech at 8 kbit/s using Conjugate-Structure Algebraic-Code-Excited Linear-Prediction (CS-ACELP)," International Telecommunication Union, 1995.

[21] TIA/EIA-PN 2398 (IS-54), "The 8 kbit/s VSELP Algorithm," 1989.

[22] GSM 06.10, "GSM Full-Rate Transcoding," Technical Rep. Vers. 3.2, ETSI/GSM, July 1989.

[23] Federal Standard 1015, Telecommunications: Analog to Digital Conversion of Radio Voice By 2400 Bit/Second Linear Predictive Coding, National Communication System—Office Technology and Standards, Nov. 1984.

[24] J. Makhoul, "Linear Prediction: A Tutorial Review," *Proc. IEEE,* Vol. 63, No. 4, pp. 561–580, April 1975.

[25] J. Markel and A. Gray, Jr., *Linear Prediction of Speech,* Springer-Verlag, NY, 1976.

[26] N. Benevuto et al., "The 32Kb/s coding standard," *AT&T Technical Journal,* Vol. 65, No. 5, pp. 12–22, Sept.–Oct. 1986.

[27] ITU Recommendation G.726 (formerly G.721),"24, 32, 40 kb/s Adaptive Differential Pulse Code Modulation (ADPCM)," *Blue Book,* Vol. III, Fascicle III.3, Oct. 1988.

[28] A. Spanias, "Speech Coding: A Tutorial Review," *Proc. IEEE,* Vol. 82, No. 10, pp. 1541–1582, Oct. 1994.

[29] I. Boyd and C. Southcott, "A Speech Codec for the Skyphone Service," *Br. Telecom. Techn. J.,* Vol. 6, No. 2, pp. 51–55, April 1988.

[30] GSM 06.60, "GSM Digital Cellular Communication Standards: Enhanced Full-Rate Transcoding," ETSI/GSM, 1996.

[31] TIA/EIA/IS-127, "Enhanced Variable Rate Codec," Speech Service Option 3 for Wideband Spread Spectrum Digital Systems," TIA, 1997.

[32] TIA/EIA/IS-641, "Cellular/PCS Radio Interface—Enhanced Full-Rate Speech Codec," TIA, 1996.

[33] P. Kroon and B. Atal, "Pitch Predictors with High Temporal Resolution," *Proc. ICASSP-90,* NM, pp. 661–664, April 1990.

[34] INMARSAT-B SDM Module 1, Appendix I, Attachment 1, MAC/SDM/BMOD1/ATTACH/ISSUE 3.0.

[35] J. Campbell and T. E. Tremain, "Voiced/Unvoiced Classification of Speech with Applications to the U.S. Government LPC-10e Algorithm," *Proc. ICASSP-86,* pp. 473–476, Tokyo, 1986.

[36] T. E. Tremain, "The Government Standard Linear Predictive Coding Algorithm: LPC-10," *Speech Technology,* pp. 40–49, April 1982.

[37] J. Makhoul et al., "A Mixed-Source Model for Speech Compression and Synthesis," *Acoustical Society of America,* Vol. 64, pp. 1577–1581, Dec. 1978.

[38] J. Campbell, T. E. Tremain, and V. Welch, "The Proposed Federal Standard 1016 4800 bps Voice Coder: CELP," *Speech Technology,* pp. 58–64, April 1990.

[39] Federal Standard 1016, "Telecommunications: Analog to Digital Conversion of Radio Voice By 4800 Bit/Second Code Excited Linear Prediction (CELP)," National Communication System— Office Technology and Standards, Feb. 1991.

[40] I. Gerson, "Vector Sum Excited Linear Prediction (VSELP) Speech Coding for Japan Digital Cellular," Meeting of IEICE, RCS90-26, Nov. 1990.

[41] GSM 06.20, "GSM Digital Cellular Communication Standards: Half Rate Speech; Half Rate Speech Transcoding," ETSI/GSM, 1996.

[42] ITU Draft Recommendation G.728,"Coding of Speech at 16 kbit/s Using Low-Delay Code Excited Linear Prediction (LD-CELP)," 1992.

[43] J. Chen, R. Cox, Y. Lin, N. Jayant, and M. Melchner, "A Low-Delay CELP Coder for the CCITT 16 KB/s Speech Coding Standard," *IEEE Trans. on Selected Areas in Communications, Special Issue on Speech and Image Coding,* Ed. N. Hubing, pp. 830–849, June 1992.

[44] TIA/EIA/IS-96, "QCELP," Speech Service Option 3 for Wideband Spread Spectrum Digital Systems," TIA, 1992.

[45] ITU Recommendation G.723.1,"Dual Rate Speech Coder for Multimedia Communications Transmitting at 5.3 and 6.3 kbit/s," 1995.

[46] T. Ohya, H. Suda, and T. Miki, "The 5.6 kb/s PSI-CELP of the Half-Rate PDC Speech Coding Standard," *Proc. IEEE Vehic. Tech. Conf.,* pp. 1680–1684, 1993.

[47] ITU Recommendation G.722,"7 KHz Audio Coding within 64 kbits/s," *Blue Book,* Vol. III, Fascicle III, Oct. 1988.

[48] Inmarsat Satellite Communications Services, Inmarsat-M System Definition, Issue 3.0—Module 1: System Description, Nov. 1991.

[49] ETSI AMR Qualification Phase Documentation, 1998.

[50] R. Ekudden, R. Hagen, I. Johansson, and J. Svedburg, "The Adaptive Multi-Rate Speech Coder," *Proc. IEEE Workshop on Speech Coding,* pp. 117–119, 1999.

[51] D. N. Knisely, S. Kumar, S. Laha, and S. Navda, "Evolution of Wireless Data Services: IS-95 to cdma2000," *IEEE Communications Magazine,* pp. 140–146, Oct. 1998.

[52] Conexant Systems, *"Conexant's ITU-T 4 kbit/s Deliverables,"* ITU-T Q21/SG16 Rapportuer meeting, AC-99-20, Sept. 1999.

[53] Matsushita Electric Industrial Co. Ltd., *"High Level Description of Matsushita's 4-kbit/s Speech Coder,"* ITU-T Q21/SG16 Rapportuer meeting, AC-99-19, Sept. 1999.

[54] Mitsubishi Electric Corporation, *"High Level Description of Mitsubishi 4 kb/s Speech Coder,"* ITU-T Q21/SG16 Rapportuer meeting, AC-99-016, Sept. 1999.

[55] NTT, *"High Level Description of NTT 4 kb/s Speech Coder,"* ITU-T Q21/SG16 Rapportuer meeting, AC-99-17, Sept. 1999.

[56] Rapportuer (Mr. Paul Barrett, BT/UK), "Q.21/16 Meeting Report," ITU-T Q21/SG16 Rapportuer meeting, Temporary Document –E (3/16), Geneva, 7–18 Feb. 2000.

[57] Texas Instruments, *"High Level Description of TI's 4 kb/s Coder,"* ITU-T Q21/SG16 Rapportuer meeting, AC-99-25, Sept. 1999.

[58] Toshiba Corporation, *"Toshiba Codec Description and Deliverables (4 kbit/s Codec),"* ITU-T Q21/SG16 Rapportuer meeting, AC-99-15, Sept. 1999.

CHAPTER 4

Audio Coding Standards

CHI-MIN LIU
WEN-WHEI CHANG

4.1 INTRODUCTION

With the introduction of the compact disc (CD) in 1982, digital audio media have quickly replaced analog audio media. However, the significant amount of uncompressed data (1.41 million bits per second) required for digital audio has led to a large transmission and storage burden. The advances of audio coding techniques and the resultant standards have greatly eased the burden. Ten years ago, nearly nobody believed that 90% of the audio data could be deleted without affecting audio fidelity. Nowadays, the fantasy has become reality and ongoing coding technologies are inspiring new dreams. This chapter reviews some international and commercial product audio coding standards, including the International Organization for Standardization/Moving Pictures Experts Group (ISO/MPEG) family [ISO, 1992, 1994, 1997, 1999], the Philips Precision Adaptive Subband Coding (PASC) [Lokhoff, 1992], the Sony Adaptive Transform Acoustic Coding (ATRAC) [Tsutsui et al., 1992], and the Dolby AC-3 [Todd et al., 1994] algorithms.

4.2 ISO/MPEG AUDIO CODING STANDARDS

The Moving Pictures Experts Group (MPEG) within the International Organization for Standardization (ISO) has developed a series of audio coding standards for storage and transmission of various digital media. The ISO standard specifies a syntax for only the coded bit streams and the decoding process; sufficient flexibility is allowed for encoder implementation. The MPEG first-phase (MPEG-1) audio coder operates in single- or two-channel stereo mode at sampling rates of 32, 44.1, and 48 kHz. In the second phase of development, particular emphasis is placed on the multichannel audio support and on an extension of MPEG-1 to lower

45

sampling rates and lower bit rates. MPEG-2 audio consists mainly of two coding standards: MPEG-2 BC [ISO, 1994] and MPEG-2 AAC [ISO, 1997]. Unlike MPEG-2 BC, which is constrained by its backward compatibility (BC) with MPEG-1 format, MPEG-2 AAC (Advanced Audio Coding) is unconstrained and can therefore provide better coding efficiency. The most recent development is the adoption of MPEG-4 [ISO, 1999] for very-low-bit-rate channels, such as those found in Internet and mobile applications. Table 4.1 lists the configurations used in MPEG audio coding standards.

4.2.1   MPEG-1

The MPEG-1 standard consists of three layers of audio coding schemes with increasing complexity and subjective performance. These layers were developed mainly in collaboration with AT&T, CCETT, FhG/University of Erlangen, Philips, IRT, and Thomson Consumer Electronics. MPEG-1 operates in one of four possible modes: mono, stereo, dual channel, and joint stereo. With a joint stereo mode, further compression can be realized through some intelligent exploitation of either the correlation between the left and right channels or the irrelevancy of the phase difference between them.

Table 4.1 Comparison of ISO/MPEG Audio Coding Standards

Standards	Audio Sampling Rate (kHz)	Compressed Bit Rate (Kbit/s)	Channels	Standard Approved
MPEG-1 Layer I	32, 44.1, 48	32–448	1–2 channels	1992
MPEG-1 Layer II	32, 44.1, 48	32–384	1–2 channels	1992
MPEG-1 Layer III	32, 44.1, 48	32–320	1–2 channels	1993
MPEG-2 Layer I	32, 44.1, 48	32–448 for two BC channels	1–5.1 channels	1994
	16, 22.05, 24	32–256 for two BC channels		
MPEG-2 Layer II	32, 44.1, 48	32–384 for two BC channels	1–5.1 channels	1994
	16, 22.05, 24	8–160 for two BC channels		
MPEG-2 Layer III	32, 44.1, 48	32–384 for two BC channels	1–5.1 channels	1994
	16, 22.05, 24	8–160 for two BC channels		
MPEG-2 AAC	8, 11.025, 12, 16, 22.05, 24, 32, 44.1, 48, 64, 88.2, 96	Indicated by a 23-bit unsigned integer	1–48 channels	1997
MPEG-4 T/F coding	8, 11.025, 12, 16, 22.05, 24, 32, 44.1, 48, 64, 88.2, 96	Indicated by a 23-bit unsigned integer	1–48 channels	1999

4.2.1.1 MPEG-1 Layers I and II

Block diagrams of Layer I and Layer II encoders are given in Figure 4.1. An analysis filterbank splits the input signal with sampling rate F_S by dividing it into 32 equally spaced subband signals with sampling rate $F_S/32$. In each of the 32 subbands, 12 consecutive samples are assembled into blocks with the equivalent of 384 input samples. All of the samples within one block are normalized by a scale factor so that they all have absolute values less than one. The choice of a scale factor is done by first finding the sample with the maximum absolute value, and then comparing it to a scale factor table of 63 allowable values. After normalization, samples are quantized and coded under the control of a psychoacoustic model. Detailed psychoacoustic analysis is performed through the use of a 512 (Layer I) or 1024 (Layer II) point fast Fourier transform (FFT) in parallel with the subband decomposition. The bit-allocation unit determines the quantizer resolution according to the targeted bit rate and the perceptual information derived from the psychoacoustic model. Layer II introduces further compression with respect to Layer I through three modifications. First, the overall information is reduced by removing redundancy and irrelevance between the scale factors of three adjacent 12-sample blocks. Second, a quantization table with improved precision is provided. Third, the psychoacoustic analysis benefits from better frequency resolution because of the increased FFT size.

4.2.1.2 MPEG-1 Layer III

The MPEG-1 Layer III audio coder introduces many new features, in particular a hybrid filterbank that is a cascade of two filterbanks. For notational convenience, the first filterbank is labeled as the Layer III first hybrid level and the second as the Layer III second hybrid level. A block diagram of the Layer III encoder is given in Figure 4.2. Although its first level is based on

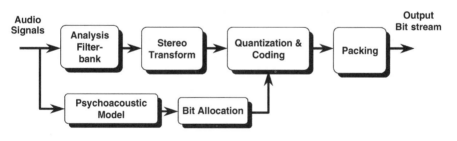

FIGURE 4.1
MPEG-1 Layer I or II audio encoder.

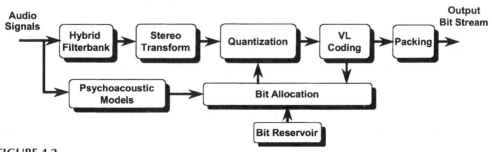

FIGURE 4.2
MPEG-1 Layer III audio encoder.

the same filterbank found in the other Layers, Layer III provides a higher frequency resolution by subdividing each of the 32 subbands with an 18-point modified discrete cosine transform (MDCT). (Key terms are defined in the section "Definitions of Key Terms" at the end of the chapter.) Furthermore, the transform block size adapts to signal characteristics to ensure dynamic tradeoffs between time and frequency resolution. It also employs nonuniform quantization in conjunction with variable-length coding for further savings in bit rate. One special feature of Layer III is the bit reservoir; it provides the vehicle to better fit the encoder's time-varying demand on code bits. The encoder can donate bits to a reservoir when it needs less than the average number of bits to code the samples in a frame. But in case the audio signals are hard to compress, the encoder can borrow bits from the reservoir to improve the fidelity.

4.2.2   MPEG-2

MPEG-2 differs from MPEG-1 in that it supports up to 5.1 channels, including five full-bandwidth channels of the 3/2 stereo, plus an optional low-frequency enhancement channel. This multichannel extension leads to an improved realism of auditory ambience not only for audio-only applications, but also for high-definition television (HDTV) and digital versatile disc (DVD). In addition, initial sampling rates can be extended downward to include 16, 22.05, and 24 kHz. Two coding standards within MPEG-2 are defined: the BC standard preserves the backward compatibility with MPEG-1, and the AAC standard does not.

4.2.2.1   MPEG-2 BC

Regarding syntax and semantics, the differences between MPEG-1 and MPEG-2 BC are minor, except in the latter case for the new definition of a sampling frequency field, a bit rate index field, and a psychoacoustic model used in bit allocation tables. In addition, parameters of MPEG-2 BC have to be changed accordingly. With the extension of lower sampling rates, it is possible to compress two-channel audio signals to bit rates less than 64 Kbit/s with good quality. Backward compatibility implies that existing MPEG-1 audio decoders can deliver two main channels of the MPEG-2 BC coded bit stream. This is achieved by coding the left and right channels as MPEG-1, while the remaining channels are coded as ancillary data in the MPEG-1 bit stream.

4.2.2.2   MPEG-2 AAC

MPEG-2 AAC provides the highest quality for applications where backward compatibility with MPEG-1 is not a constraint. While MPEG-2 BC provides good audio quality at data rates of 640–896 Kbit/s for five full-bandwidth channels, MPEG-2 AAC provides very good quality at less than half of that data rate. A block diagram of an AAC encoder is given in Figure 4.3. The gain control tool splits the input signal into four equally spaced frequency bands, which are then flexibly encoded to fit into a variety of sampling rates. The pre-echo effect can also be alleviated through the use of the gain control tool. The filterbank transforms the signals from the time domain to the frequency domain. The temporal noise shaping (TNS) tool helps to control the temporal shape of the quantization noise. Intensity coding and the coupling reduce perceptually irrelevant information by combining multiple channels in high-frequency regions into a single channel. The prediction tool further removes the redundancies between adjacent frames. Middle/Side (M/S) coding removes stereo redundancy based on coding the sum and difference signal instead of the left and right channels. Other units, including quantization, variable-length coding, psychoacoustic model, and bit allocation are similar to those used in MPEG Layer III.

MPEG-2 AAC offers flexibility for different quality-complexity tradeoffs by defining three profiles: the main profile, the low-complexity profile, and the sampling rate scalable (SRS) pro-

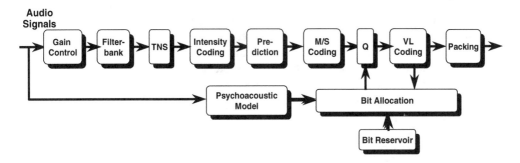

FIGURE 4.3
MPEG-2 AAC audio encoder.

file. Each profile builds on some combinations of different tools as listed in Table 4.2. The main profile yields the highest coding efficiency by incorporating all the tools with the exception of the gain control tool. The low-complexity profile is used for applications where memory and computing power are constrained. The SRS profile offers a scalable complexity by allowing partial decoding of a reduced audio bandwidth.

4.2.3 MPEG-4

The MPEG-4 standard, which was finalized in 1999, integrates the whole range of audio from high-fidelity speech coding and audio coding down to synthetic speech and synthetic audio. The MPEG-2 ACC tool set within the MPEG-4 standard supports the compression of natural audio at bit rates ranging from 2 up to 64 Kbit/s. The MPEG-4 standard defines three types of coders: parametric coding, code-excited linear predictive (CELP) coding, and time/frequency (T/F) coding. For speech signals sampled at 8 kHz, parametric coding is used to achieve targeted bit rates between about 2 and 6 Kbit/s. For audio signals sampled at 8 and 16 kHz, CELP coding offers good quality at medium bit rates between about 6 and 24 Kbit/s.

Table 4.2 Coding Tools Used in MPEG-2 AAC

Tools	Main	Low Complexity	SRS
Variable-length decoding	✓	✓	✓
Inverse quantizer	✓	✓	✓
M/S	✓	✓	✓
Prediction	✓	✗	✗
Intensity/coupling	✓	✗	✗
TNS	✓	Limited	Limited
Filterbank	✓	✓	✓
Gain Control	✗	✗	✓

T/F coding is typically applied to the bit rates starting at about 16 Kbit/s for audio signals with bandwidths above 8 kHz. T/F coding is developed based on the coding tools used in MPEG-2 AAC with some add-ons. One is referred to as the twin-vector quantization (VQ), which makes combined use of an interleaved VQ and Linear Predictive Coding (LPC) spectral estimation. In addition, the introduction of bit-sliced arithmetic coding (BSAC) offers noiseless transcoding of an AAC stream into a fine-granule scalable stream between 16 and 64 Kbit/s per channel. BSAC enables the decoder to stop anywhere between 16 Kbit/s and the bit rate arranged in 1 Kbit/s steps.

4.3   OTHER AUDIO CODING STANDARDS

The audio data on a compact disc is typically sampled at 44.1 kHz that requires an uncompressed data rate of 1.41 Mbit/s for stereo sound with 16-bit pulse code modulation (PCM). Lower bit rates than those given by the 16-bit PCM format are mandatory to support a circuit realization that is compact and has low-power consumption—two key enabling factors of equipment portability for the user. The digital compact cassette (DCC) developed by Philips is one of the first commercially available forms of perceptually coded media. To offer backward compatibility for playback of analog compact cassettes, DCC's combination of tape speed and symbol spacing yields a raw data rate of only 768 Kbit/s, and half of that is used for error-correcting redundancy. Another example is Sony's MiniDisc (MD), which allows one to store a full CD's worth of music on a disc only half the diameter. DCC and MD systems make use of perceptual coding techniques to achieve the necessary compression ratios of 4:1 and 5:1, respectively. Dolby AC-3 is currently the audio coding standard for the U.S. Grand Alliance HDTV system and has been widely adopted for DVD films. Dolby AC-3 can reproduce various playback configurations from one channel up to 5.1 channels: left, right, center, left-surrounding, right-surrounding, and low-frequency enhancement channels.

4.3.1   Philips PASC

Philips' DCC incorporates the Precision Adaptive Subband Coding (PASC) algorithm that is capable of compressing two-channel stereo audio to 384 Kbit/s with near-CD quality [Lokhoff,1992]. PASC can be considered as a simplified version of ISO/MPEG-1 Layer I; it does not require a side-chain FFT analysis for the estimation of the masking threshold. The PASC encoder creates 32 subband representations of the audio signal, which are then quantized and coded according to the bit allocation derived from a psychoacoustic model. The first-generation PASC encoder performs a very simple psychoacoustic analysis based on the outputs of the filterbank. By measuring the average power level of 12 samples, the masking levels of that particular subband and all the adjacent subbands can be estimated with the help of an empirically derived 32×32 matrix, which is described in the DCC standard. The algorithm assumes that the 32 frequencies of this matrix are positioned on the edges of the subband spectra, the most conservative approach.

Every block of 12 samples is converted to a floating-point notation; the mantissa determines resolution and the exponent controls dynamic range. As in MPEG-1 Layer I, the scale factor is determined and coded as a 6-bit exponent; it is valid for 12 samples within a block. The algorithm assigns each sample a mantissa with a variable length of 2 to 15 bits, depending on the ratio of the maximum signal to the masking threshold, plus an additional 4 bits for allocation information detailing the length of a mantissa.

4.3.2 Sony ATRAC

The Adaptive TRansform Acoustic Coding (ATRAC) algorithm was developed by Sony to support 74 min of recording and playing time on a 64-mm MiniDisc [Tsutsui, et al., 1992]. It supports coding of 44.1 kHz two-channel audio at a rate of 256 Kbit/s. The key to ATRAC's efficiency is that psychoacoustic principles are applied to both the bit allocation and the time-frequency mapping. The encoder (Figure 4.4) begins with two stages of quadrature mirror filters (QMFs) to divide the audio signal into three subbands that cover the ranges of 0–5.5 kHz, 5.5–11.0 kHz, and 11.0–22.0 kHz. These subbands are then transformed from the time domain to the frequency domain using the modified discrete cosine transform (MDCT). In addition, the transform block size adapts to signal characteristics to ensure dynamic tradeoffs between time and frequency resolution. The default transform block size is 11.6 ms, but in case of predicted pre-echoes the block size is switched to 1.45 ms in the high-frequency band and to 2.9 ms in the low- and mid-frequency bands. Following the time-frequency analysis, transform coefficients are grouped nonuniformly into 52 block floating units (BFUs) in accordance with the ear's critical band partitions. Transform coefficients are quantized using two parameters: word length and scale factor. The scale factor defines the full-scale range of the quantization and the word length defines the resolution within that scale. Each of the 52 BFUs has the same word length and scale factor, reflecting the psychoacoustic similarity within each critical band.

The bit allocation algorithm determines the word length with the aim of keeping the quantization noise below the masking threshold. One suggested algorithm makes combined use of fixed and variable bits. The algorithm assigns each BFU variable bits according to the logarithm of the transform coefficients. Fixed bits are mainly allocated to low-frequency BFU regions; this reflects the ear's decreasing sensitivity toward higher frequencies. The total bit allocation $b_{tot}(k)$ is the weighted sum of the fixed bit $b_{fix}(k)$ and the variable bit $b_{var}(k)$. Thus, for each BFU k, $b_{tot}(k) = T\,b_{var}(k) + (1-T)\,b_{fix}(k)$. The weight T describes the tonality of the signal, taking a

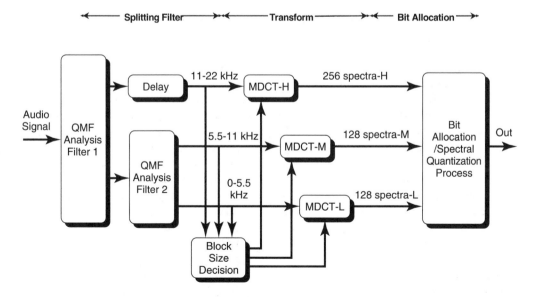

FIGURE 4.4
ATRAC audio encoder.

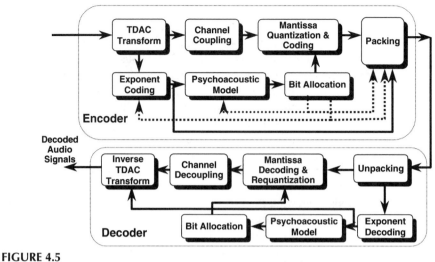

FIGURE 4.5
Dolby AC-3 encoder and decoder.

value close to 1 for pure tones and a value close to 0 for white noise. To ensure a fixed data rate, an offset b_{off} is subtracted from $b_{tot}(k)$ to yield the final bit allocation $b(k)$= integer $[b_{tot}(k)-b_{off}]$. As a result, the ATRAC encoder output contains MDCT block size mode, word length, and scale factor for each BFU, and quantized spectral coefficients.

4.3.3   Dolby AC-3

As illustrated in Figure 4.5, the Dolby AC-3 encoder first employs an MDCT to transform the audio signals from the time domain to frequency domain. Then, adjacent transform coefficients are grouped into nonuniform subbands that approximate the critical bands of the human auditory system. Transform coefficients within one subband are converted to a floating-point representation, with one or more mantissas per exponent. The exponents are encoded by a suitable strategy according to the required time and frequency resolution and fed into the psychoacoustic model. Then, the psychoacoustic model calculates the perceptual resolution according to the encoded exponents and the proper perceptual parameters. Finally, both the perceptual resolution and the available bits are used to decide the mantissa quantization. One distinctive feature of Dolby AC-3 is the intimate relationship among exponent coding, psychoacoustic models, and the bit allocation. This relationship can be described most conveniently by the hybrid backward/forward bit allocation. The encoded exponents provide an estimate of the spectral envelope, which, in turn, is used in the psychoacoustic model to determine the mantissa quantization. While most audio encoders need to transmit side information about the mantissa quantization, the AC-3 decoder can automatically derive the quantizer information from the decoded exponents. The basic problem with this approach is that the exponents are subject to limited time-frequency resolution and hence fail to provide a detailed psychoacoustic analysis. Further improvement can be realized by computing an ideal bit allocation based on full knowledge of the input signal and then transmitting additional side information that indicates the difference between the ideal bit allocation and the core bit allocation derived from the exponents.

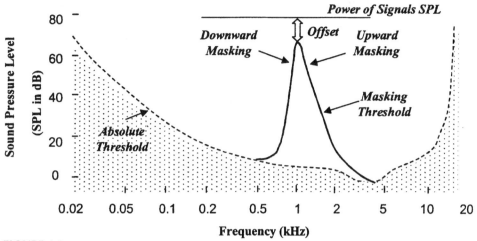

FIGURE 4.6
Masking threshold of a masker centered at 1 kHz.

4.4   ARCHITECTURAL OVERVIEW

The principles of perceptual coding can be considered according to eight aspects: T/F mapping, quantization and coding, psychoacoustic model, channel correlation and irrelevancy, long-term correlation, pre-echo control, and bit allocation. This section provides an overview of these standards through examination of the eight aspects.

4.4.1   Psychoacoustic Modeling

Most perceptual coders rely, at least to some extent, on psychoacoustic models to reduce the subjective impairments of quantization noise. The encoder analyzes the incoming audio signals to identify perceptually important information by incorporating several psychoacoustic principles of the human ear [Zwicker and Fasti, 1990]. One is the critical-band spectral analysis, which accounts for the ear's poorer discrimination in higher-frequency regions than in lower-frequency regions. Investigations have indicated that a good choice of spectral resolution is around 20 Hz, which has been implemented in MPEG-2 AAC and MPEG-4. The phenomenon of masking is another effect that occurs whenever a strong signal (masker) makes a spectral or temporal neighborhood of weaker signals inaudible. To illustrate this, Figure 4.6 shows an example of the masking threshold produced by a masker centered at 1 kHz. The absolute threshold (dashed line) is also included to indicate the minimum audible intensity level in quiet surroundings. Notice that the slope of the masking curve is less steep on the high-frequency side; that is, higher frequencies are more easily masked. The offset between masker and masking threshold is varied with respect to the tonality of the masker; it has a smaller value for a noiselike masker (5 dB) than a tonelike masker (25dB).

The encoder performs the psychoacoustic analysis based on either a side-chain FFT analysis (in MPEG) or the output of the filterbank (in AC-3 and PASC). MPEG provides two examples of psychoacoustic models, the first of which we now describe. The calculation starts with a precise spectral analysis on 512 (Layer I) or 1024 (Layer II) input samples to generate the magnitude spectrum. The spectral lines are then examined to discriminate between tonelike and noiselike maskers by taking the local maximum of magnitude spectrum as an indicator of

Table 4.3 CMFBs Used in Current Audio Coding Standards

CMFBs in Standards	Overlap Factor k	Number of Bands N	Frequency Resolution at 48 kHz (Hz)	Sidelobe Attenuation (dB)
MPEG Layers I and II	16	32	750	96
MPEG second hybrid level of Layer III	2	18	41.66	23
MPEG-2 AAC, MPEG-4 T/F coding	2	1024	23.40	19
Dolby AC-3	2	256	93.75	18

tonality. Among all the labeled maskers, only those above the absolute threshold are retained for further calculation. Using rules known from psychoacoustics, the individual masking thresholds for the relevant maskers are then calculated, dependent on frequency position, loudness level, and the nature of tonality. Finally, we obtain the global masking threshold from the upward and downward slopes of the individual masking thresholds of tonal and nontonal maskers and from the absolute threshold in quiet. The psychoacoustic model used in AC-3 is specially designed; it does not provide a means to differentiate the masking effects produced by either the tonal or the noise masker.

4.4.2   Time-Frequency Mapping

Since psychoacoustic interpretation is mainly described in frequency domain, the time-frequency mapping is incorporated into the encoder for further signal analysis. The time-frequency mapping can be implemented either through PQMF [ISO, 1992], time-domain aliasing cancellation (TDAC) filters [Prince, et al., 1987], or the modified discrete cosine transform [ISO, 1992]. All of them can be referred to as cosine modulated filterbanks (CMFBs) [Liu et al., 1998; Shlien, 1997]. The process of CMFBs consists of two steps: the window-and-overlapping addition (WOA) followed by the modulated cosine transform (MCT). The WOA performs a windowing multiplication and addition with overlapping audio blocks. Its complexity is $O(k)$ per audio sample, where k is the overlapping factor of audio blocks. In general, the sidelobe attenuation of the filterbank increases with the factor. For example, the factor k is 16 for MPEG-1 Layer II and 2 for AC-3.

The complexity of MCT is $O(2N)$ per audio sample, where N is the number of bands. The range of N is from 18 for MPEG-1 Layer III to 2048 for the MPEG-2 advanced audio coding. Table 4.3 compares the properties of CMFBs used in audio coding standards. Due to the high complexity of the MCT, fast algorithms have been developed following the similar concepts behind the FFT. As listed in Table 4.4, the MCTs used in current audio standards can be classified into three different types: time-domain aliasing cancellation (TDAC), variants of the TDAC filterbank, and the polyphase filterbank.

4.4.3   Quantization

For perceptual audio coding, quantization involves representing the outputs of the filterbank by a finite number of levels with the aim of minimizing the subjective impairments of quantization

Table 4.4 Comparison of Filterbank Properties

Classes	MCT Transform Pair	CMFBs in Standards
Polyphase Filterbank	$$X_k = \sum_{i=0}^{N-1} x_i \cos\left(\frac{\pi}{N}\left(i - \frac{N}{4}\right)(2k+1)\right)$$ $$x_i = \sum_{i=0}^{N/2-1} X_k \cos\left(\frac{\pi}{2N}\left(i + \frac{N}{4}\right)(2k+1)\right)$$ for k = 0, 1 ..., N/2 − 1 and i = 0, 1, ..., N − 1	MPEG Layers I and II (N=64), MPEG Layer III first hybrid level (N=64)
TDAC Filterbank	$$X_k = \sum_{i=0}^{N-1} x_i \cos\left(\frac{\pi}{2N}\left(2i + 1 + \frac{N}{2}\right)(2k+1)\right)$$ $$x_i = \sum_{k=0}^{N/2-1} X_k \cos\left(\frac{\pi}{2N}\left(2i + 1 + \frac{N}{2}\right)(2k+1)\right)$$ for k = 0, 1 ..., N/2 − 1 and i = 0, 1, ..., N − 1	MPEG-2—AAC (N=4096), MPEG-4-T/F Coding (N=4096), MPEG Layer III second hybrid level (N=36), AC-3 Long Transform (N=512)
TDAC-Variant Filterbank	$$X_k = \sum_{i=0}^{N-1} x_i \cos\left(\frac{\pi}{2N}(2i+1)(2k+1)\right)$$ $$x_i = \sum_{k=0}^{N/2-1} X_k \cos\left(\frac{\pi}{2N}(2i+1)(2k+1)\right)$$ for k = 0, 1 ..., N/2 − 1 and i = 0, 1, ..., N − 1	AC-3 Short Transform 1 (N=256)
	$$X_k = \sum_{i=0}^{N-1} x_i \cos\left(\frac{\pi}{2N}(2i+1+N)(2k+1)\right)$$ $$x_i = \sum_{k=0}^{N/2-1} X_k \cos\left(\frac{\pi}{2N}(2i+1+N)(2k+1)\right)$$ for k = 0, 1 ..., N/2 − 1 and i = 0, 1, ..., N − 1	AC-3 Short Transform 2 (N=256)

noise. The characteristics of a quantizer can be specified by means of the step size and the range, as shown in Figure 4.7. While a uniform quantizer has the same step size throughout the input range, a nonuniform quantizer does not. For a uniform quantizer, the range and the step size determine the number of levels required for the quantization. In contrast to a uniform quantizer, the quantization noise of a nonuniform quantizer is varied with respect to the input value. Such a design is more relevant to the human auditory system in the sense that the ear's ability to decipher two sounds with different volumes decreases with the sound pressure levels.

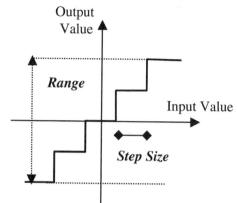

FIGURE 4.7
Quantizer characteristics.

Table 4.5 Quantization Schemes Used in Audio Coding Standards

Standards	Quantization Types	Range Adaptation with Time at 48 kHz (ms)	Range Adaptation with Frequency at 48 kHz (Hz)
MPEG Layer I	Uniform	8	750
MPEG Layer II	Uniform	8–24	750
MPEG Layer III	Nonuniform	24	No
Dolby AC-3	Uniform	5.3–32	93.75–375
MPEG-2 AAC	Nonuniform	21.33	No

For uniform quantization with fixed bit rate, the range directly affects the quantization error. Hence, an accurate estimation of the range leads to a good control of the quantization noise. On the other hand, for nonuniform quantization, the quantization noise depends more on the input values than on the ranges. In the current audio standards, uniform quantizers are used in MPEG Layer I, Layer II, and Dolby AC-3. For MPEG Layers I and II, the scale factor helps to determine the range of a quantizer. For Dolby AC-3, the exponent, which accounts for the range of a uniform quantizer, is adaptive with the time and frequency. Table 4.5 lists the quantization schemes used in the audio coding standards. For MPEG Layer III, MPEG-2 AAC, and MPEG-4 T/F coding, nonuniform quantizers are used and their ranges are not adaptive with the frequency. Until now, we have only considered the scalar quantization situation where one sample is quantized at a time. On the other hand, vector quantization (VQ) involves representing a block of input samples at a time. Twin-VQ has been adopted in MPEG-4 T/F coding as an alternative to scalar quantization for higher coding efficiency.

4.4.4   Variable-Length Coding

In MPEG Layer III and MPEG-2 AAC, variable-length coding is used to take advantage of different occurring probabilities of quantizer outputs. However, variable-length coding brings two implementation problems. The first problem is the decoder complexity. The decoder has to decode the packing values bit-by-bit and symbol-by-symbol, leading to much higher complexi-

Table 4.6 Multichannel Coding for the Audio Standards

Standards	Stereo Correlation and Irrelevancy	Multichannel Transform
MPEG-1 Layer I	Intensity coding	No
MPEG-1 Layer II	Intensity coding	No
MPEG-1 Layer III	M/S and intensity coding	No
MPEG-2 Layer I	Intensity coding	Matrixing
MPEG-2 Layer II	Intensity coding	Matrixing
MPEG-2 Layer III	M/S and intensity coding	Matrixing
Dolby AC-3	Coupling and matrixing	M/S and coupling
MPEG-2 AAC	M/S and intensity coding	M/S and intensity coding
MPEG-4 T/F coding	M/S and intensity coding	M/S and intensity coding

ty than would be the case with fixed-length coding. A parallel or multiple look-up table has been adopted to solve this problem. The second problem is on the bit allocation, which is discussed in Subsection 4.4.8.

4.4.5 Multichannel Correlation and Irrelevancy

Since the signals from different channels are usually recorded from the same sound source, it is a natural speculation that correlation exists among these channels. Furthermore, there is stereo irrelevancy for the multichannel signals. In particular, localization of the stereophonic image within a critical band for frequencies above 2 kHz is based more on the temporal envelope of the audio signal than on its temporal fine structure. Recognizing this, audio coding standards have developed techniques for dealing with the multichannel correlation and irrelevancy.

For stereo audio, there are two types of coding techniques: middle/side (M/S) coding and intensity coding. M/S coding removes the correlation by transforming the left and right channels into the sum and difference signals. In intensity coding mode the encoder combines some high-frequency parts of two-channel signals into a single summed signal by exploiting the stereo irrelevancy. For channel numbers greater than two, the correlation in the channels can be reduced by using an N × N matrix, where N is the channel number. In addition, multiple channels in high-frequency regions can be combined to form a single channel using the so-called coupling technique. MPEG-2 AAC separates all the channels in pairs and then applies the M/S and intensity coding. Table 4.6 lists the multichannel coding schemes used in various audio standards.

4.4.6 Long-Term Correlation

When dealing with stationary signals, an audio encoder can benefit from the correlation between adjacent frames to achieve further coding gain. In the current audio coding standards, MPEG-2 AAC and MPEG-4 T/F coding are two typical examples that make use of a second-order backward prediction tool to exploit the interframe correlation. The importance of the prediction tool increases with the sampling rates.

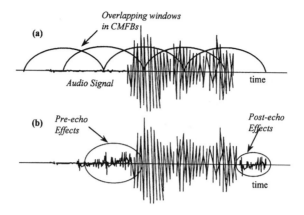

FIGURE 4.8
Pre-echo effects. (a) Original signal. (b) Reconstructed signal with pre-echo and postecho.

4.4.7 Pre-echo Control

Pre-echoes occur when a region of low power is followed by a signal with a sharp attack, as shown in Figure 4.8a. Using a Cosine Modulated Filterbank (CMFB) with its window length covering these two regions, the inverse transform will spread quantization noise evenly throughout the whole region. Hence, the quantization noise in the region of low power will be too large to be masked. This is called the pre-echo effect, as shown in Figure 4.8b. Pre-echoes can be masked by the premasking phenomenon of the human auditory system [Zwicker and Fasti, 1990] only if the window length is sufficiently small (about 1–4 ms). Similarly, the postecho effect can be found in the situation where a signal with a sharp attack is followed by a region of low power. Fortunately, the postecho effect can be masked because the duration within which postmasking applies is in the order of 50–200 ms.

A solution to the pre-echo problem is to switch between window sizes of different lengths. This approach has been implemented in MPEG Layer III, Dolby AC-3, MPEG-2 AAC, and MPEG-4 T/F coding. Table 4.7 lists the long and the short windows used in various audio coding standards. In addition to block size switching, MPEG-2 AAC and MPEG-4 T/F coding make use of the temporal noise shaping tool to control the pre-echo effect. This tool performs a forward prediction in the frequency domain and hence leads to a temporal shaping in the time domain. Through this shaping, the small signal and large signal can be shaped to similar amplitudes which, in turn, helps to reduce the pre-echo effects.

Table 4.7 Length of the Long and Short Windows

Standards	Length of Long Window at 48 kHz Sampling Frequency (ms)	Length of Short Window at 48 kHz Sampling Frequency (ms)
MPEG Layer III	48	16
Dolby AC-3	10.67	2.66
MPEG-2 AAC	42.67	2.66
MPEG-4 T/F Coding	42.67	2.66

4.4.8 Bit Allocation

The purpose of bit allocation is to apportion the total number of bits available for the quantization of filterbank outputs to achieve the best audio quality. The quality and the bit rate are two fundamental requirements for bit allocation. For MPEG Layers I and II, control over the quality and the bit rate is carried out by a uniform quantizer. Hence the purpose of bit allocation is simply to apportion the total number of bits available for the quantization of the subband signals to minimize the audibility of the quantization noise.

For MPEG Layer III and MPEG-2 AAC, control over the quality and the bit rate is difficult. This is mainly due to the fact that they both use a nonuniform quantizer whose quantization noise varies with respect to the input values. In other words, it fails to control the quality by assigning quantizer parameters according to the perceptually allowable noise. In addition, the bit-rate control issue can be examined from the variable-length coding used in MPEG Layer III and MPEG-2 AAC. Variable-length coding relies heavily on the quantizer outputs and hence cannot be derived from the quantizer parameters alone. For this reason, the bit allocation is one of the main tasks leading to the high encoder complexity in MPEG Layer III and MPEG-2 AAC.

For Dolby AC-3, it is also difficult to determine the bit allocation. As mentioned previously, AC-3 adapts its range according to the specified exponent strategy. However, there are 3072 possible strategies for the six blocks in a frame. Each of these strategies affects the temporal resolution and the spectral resolution of the quantizer's ranges. Encoded exponents are also used in the psychoacoustic model; this is a special feature of the hybrid coding in Dolby AC-3. Hence the intimate relationship among the exponents, the psychoacoustic model, and the quantization has led to high complexity in bit allocation. The bit allocation issue in Dolby AC-3 has been analyzed in one paper [Liu et al., 1998].

4.5 CONCLUSIONS

With the standardization efforts of MPEG, audio coding has been a historic success in the area of multimedia and consumer applications. A variety of compression algorithms has been provided ranging from MPEG-1 Layer I, which allows for an efficient decoder implementation, to MPEG-2 AAC, which provides high audio quality at a rate of 384 Kbit/s for five full-bandwidth channels. All of the important compression schemes build on the proven frequency-domain coding techniques in conjunction with block companding and perceptual bit-allocation strategies. They are designed not just for using statistical correlation to remove redundancies but also to eliminate the perceptual irrelevancy by applying psychoacoustic principles. While current audio coders still have room for improvement in terms of bit rates and quality, the main focus of current and future work has been switched to offer new functionalities such as scalability and editability, thereby opening the way to new applications.

4.6 DEFINITIONS OF KEY TERMS

Critical band: Psychoacoustic measure in the spectral domain that corresponds to the frequency selectivity of the human ear
Intensity coding: A coding method of exploiting stereo irrelevance or redundancy in stereophonic audio programs based on retaining at high frequencies only the energy envelope of the right and left channels
MDCT: Modified Discrete Cosine Transform that corresponds to the Time Domain Aliasing Cancellation Filterbank

Polyphase filterbank: A set of equal bandwidth filters with special phase interrelationships, allowing for an efficient implementation of the filterbank

4.7   REFERENCES

ISO/IEC JTC1/SC29/WG11. 1992. Information technology—Coding of moving pictures and associated audio for digital storage media up to about 1.5 Mb/s—IS 11172 (Part 3, Audio).

ISO/IEC JTC1/SC29/WG11. 1994. Information technology—Generic coding of moving pictures and associated audio information—IS 13818 (Part 3, Audio).

ISO/IEC JTC1/SC29/WG11. 1997. Information technology—Generic coding of moving pictures and associated audio information—IS 13818 (Part 7, Advanced audio coding).

ISO/IEC JTC1/SC29/WG11. 1999. Information technology—Coding of audiovisual objects—ISO/IEC.D 4496 (Part 3, Audio).

Liu, C. M. and Lee, W. J. 1998. A unified fast algorithm for the current audio coding standards. Proc. Audio Engineering Society Conv., preprint 4729. Amsterdam.

Liu, C. M., Lee, S. W., and Lee, W. C. 1998. Bit allocation method for Dolby AC-3 encoder. *IEEE Trans. Consumer Electronics* 44(3):883–887.

Lokhoff, G. C. P. 1992. Precision Adaptive Subband Coding (PASC) for the Digital Compact Cassette (DCC). *IEEE Trans. Consumer Electronics* 38(4):784–789.

Prince, J. P., Johnson, A. W., and Bradley, A. B. 1987. Subband/transform coding using filterbank design based on time domain aliasing cancellation. *Proc. ICASSP* 2161–2164. Dallas.

Shlien, S. 1997. The modulated lapped transform, its time-varying forms, and its application to audio coding standards. *IEEE Trans. on Speech and Audio Process.* 5(4):359–366.

Todd, C. C., Davidson, G. A., Davis, M. F., Fielder, L. D., Link, B. D., and Vernon, S. 1994. AC-3: Flexible perceptual coding for audio transmission and storage. Proc. Audio Engineering Society 96th Conv. preprint 3796. Amsterdam.

Tsutsui, K., Suzuki, H., Shimoyoshi, O., Sonohara, M., Akagiri, K., and Heddle, R. M. 1992. ATRAC: Adaptive Transform Acoustic Coding for MiniDisc. Proc. Audio Engineering Society 93rd Conv., preprint 3456. San Francisco, CA.

Zwicker, E. and Fasti, H. 1990. *Psychoacoustics: Facts and Models,* Springer-Verlag, Berlin.

4.8   BIBLIOGRAPHY

Additional technical information can be found in the following works.

Brandenburg, K. and Bosi, M. 1997. Overview of MPEG audio: Current and future standards for low-bit-rate audio coding. *J. Audio Eng. Soc.* 45(1/2):4–19.

Noll, P. 1997. MPEG digital audio coding. *IEEE Signal Processing Mag.* 14(5):59–81.

Pan, D. 1995. A tutorial on MPEG/Audio compression, *IEEE Multimedia Mag.* 2(2): 60–74.

Pohlmann, K.G. 1995. *Principles of Digital Audio,* 3rd ed. McGraw-Hill, New York.

CHAPTER 5

Still Image Compression Standards

MICHAEL W. HOFFMAN
KHALID SAYOOD

5.1 INTRODUCTION

Compression is one of the technologies that enabled the multimedia revolution to occur. However, for the technology to be fully effective there has to be some degree of standardization so that equipment designed by different vendors can talk to each other. The process through which the first image compression standard, popularly known as Joint Photographic Experts Group (JPEG), was developed was highly successful and has become a model for the development of other compression standards. The JPEG standard is effectively two standards, one for lossy compression of still images, and another for lossless compression of still images. A third standard for compressing binary images, popularly known as Joint Bilevel Image Group (JBIG) for the committee that created it, completes the trio of current international standards for compressing images. There are also de facto standards such as Graphics Interchange Format (GIF) and Portable Network Graphics (PNG) that include compression of graphic images. However, these are not the subject of this chapter. While JPEG2000 and JBIG are current international standards, there are new international standards on the horizon at the writing of this chapter. These are the JPEG-LS standard, and a new standard from the JBIG committee popularly referred to as JBIG2. Unlike the JPEG and JBIG standards, JPEG2000, JPEG-LS, and JBIG2 try to include to some extent all three types of compression represented by the earlier standards. Thus, JPEG-LS has a binary mode for compression of bilevel images and a "near-lossless" mode that is essentially a lossy mode with pixel-level distortion constraints. The JPEG2000 standard has both a mode for bilevel images and a mode for lossless compression. The JBIG2 algorithm allows for lossy compression of bilevel images. However, each standard has a primary focus, with lossless compression being the focus of JPEG-LS, lossy compression the primary focus of JPEG2000, and lossless compression of bilevel images the primary focus of JBIG2.

In the following sections we will describe the current standards in some detail and the proposed standards in as much detail as is available at the current time. We will describe the lossy portion of JPEG and the current status of the JPEG2000 standard in Section 5.2. The lossless part of JPEG and JPEG-LS are the subject of Section 5.3, and the JBIG and JBIG2 standards are described in Section 5.4.

5.2   LOSSY COMPRESSION

Any compression scheme can be viewed as a sequence of two steps. The first step involves removal of redundancy based on implicit assumptions about the structure in the data, and the second step is the assignment of binary codewords to the information deemed nonredundant. The lossy compression portion of JPEG and JPEG2000 relies on a blockwise spectral decomposition of the image. The current JPEG standard uses Discrete Cosine Transform (DCT) and the JPEG2000 standard uses wavelets.

5.2.1   JPEG

JPEG refers to a wide variety of possible image compression approaches that have been collected into a single standard. In this section we attempt to describe JPEG in a somewhat general but comprehensible way. A very complete description of the JPEG standard has been presented in Pennabaker and Mitchell [1993]. As mentioned in the preceding, the JPEG standard has both lossless and lossy components. In addition, the entropy coding employed by JPEG can be either Huffman coding or binary arithmetic coding. Figures 5.1 and 5.2 present very general image compression models that help describe the JPEG standard. In Figure 5.1 the compression process is broken into two basic functions: modeling the image data and entropy coding the description provided by a particular model. As the figure indicates, the modeling and entropy coding are separate. Hence whether Huffman or arithmetic entropy codes are used is irrelevant to the modeling. Any standard application-specific or image-specific coding tables can be used for entropy coding. The reverse process is illustrated in Figure 5.2.

The modes of operation for JPEG are depicted in Figure 5.3. Two basic functional modes exist: nonhierarchical and hierarchical. Within the nonhierarchical modes are the sequential lossless and the lossy DCT-based sequential and progressive modes. The sequential modes progress through an image segment in a strict left-to-right, top-to-bottom pass. The progressive modes allow several refinements through an image segment, with increasing quality after each refinement. The hierarchical mode allows combinations of nonhierarchical modes, progressive

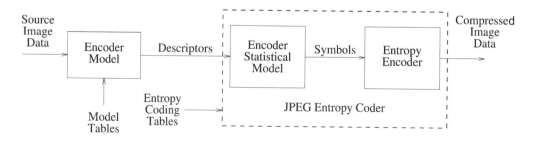

FIGURE 5.1

General JPEG encoder models. (After Pennabaker and Mitchell, 1993, Figure 5-2).

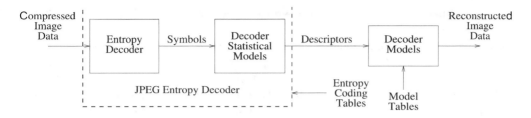

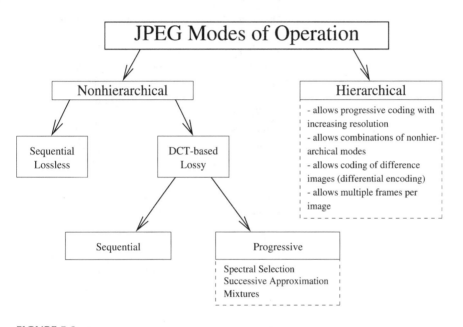

FIGURE 5.2
General JPEG decoder models. (After Pennabaker and Mitchell, 1993, Figure 5-3.)

FIGURE 5.3
JPEG modes of operation.

coding with increasing resolution, coding of difference images, and multiple frames per image (the nonhierarchical modes allow only a single frame per image). The sequential lossless mode will be discussed in Section 5.3.1. In the following sections we will discuss the nonhierarchical DCT-based lossy modes in more detail.

5.2.1.1 DCT-Based Image Compression

The basis for JPEG's lossy compression is the two-dimensional DCT. An image is broken into 8×8 blocks on which the transform is computed. The transform allows different two-dimensional frequency components to be coded separately. Image compression is obtained through quantization of these DCT coefficients to a relatively small set of finite values. These values (or some representation of them) are entropy coded and stored as a compressed version of the image.

The DCT-based compression is effective because the human visual system is relatively less sensitive to higher (spatial) frequency image components and the quantization is uneven for the various DCT frequencies. In addition, many images have very few significant high-frequency components, so many of the quantized DCT coefficients are zero and can be efficiently coded. Figure 5.4 illustrates a typical JPEG quantization table. The DC coefficient (top left) is encoded differentially from block to block and the quantization step size is given for the *difference* between the current block's DC coefficient and the previous block's quantized DC coefficient. The other 63 coefficients in the transform are referred to as the AC coefficients. These are quantized directly by the respective step size values in the table. All quantization is done assuming a midtread quantizer to allow for zero values in the quantizer output. The scaling in the JPEG quantization tables assumes that a 1 is equal to the least significant bit in a $P+3$ bit binary word when the original pixels have P bits of precision. This results from the DCT scaling providing a multiplicative factor of 8. JPEG users are free to define and use their own quantization tables (and hence transmit or store these with images) or use the tables provided by JPEG. Note that in the lossy DCT-based JPEG modes the allowable values of P are 8 or 12, with a value of 128 or 2048 being subtracted from the raw image data for 8- or 12-bit precision, respectively, prior to processing. Adaptive quantization has been added as a JPEG extension. The adaptive quantization allows either selection of a new quantization table or a modification (e.g., scaling) of an existing quantization table.

Additional compression of the AC coefficients is obtained because these coefficients are scanned in a zigzag fashion from lowest frequency to highest frequency, as illustrated in Figure 5.5. The zigzag scan in combination with the quantization table and midtread quantizer design results in long runs of zero coefficients for many image blocks. This substantially enhances the compression achieved with the subsequent entropy coding. In addition, this ordering is exploited by some of JPEG's progressive transmission schemes.

5.2.1.2 Progressive Transmission
Both sequential and progressive transmission are possible using JPEG lossy compression. In sequential transmission, the entire image is compressed using 8×8 blocks, scanning the image

DC

Horizontal Frequency $\longrightarrow$

Vertical Frequency

16	11	10	16	24	40	51	61
12	12	14	19	26	58	60	55
14	13	16	24	40	57	69	56
14	17	22	29	51	87	80	62
18	22	37	56	68	109	103	77
24	35	55	64	81	104	113	92
49	64	78	87	103	121	120	101
72	92	95	98	112	100	103	99

FIGURE 5.4
An example quantization table for the luminance DCT coefficients.

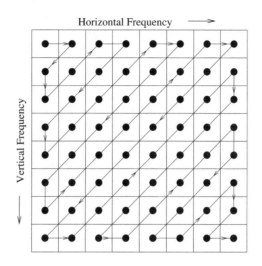

Horizontal Frequency →

Vertical Frequency

FIGURE 5.5
Zigzag scan order for 8 × 8 DCT coefficients.

from left to right and top to bottom. The full precision-compressed representation is provided as the image is decoded and reconstructed block by block. With progressive transmission all blocks within a section of the image are compressed with one level of precision that is enhanced by subsequent refinements to the original lower quality reproductions. Note that for nonhierarchical JPEG modes the *pixel sampling resolution* for progressive reconstructions is the same at each refinement. This is not necessarily the case for hierarchical JPEG modes.

Within JPEG, two means of achieving progressive transmission are provided: spectral selection and successive approximation. In spectral selection the DC coefficients and various bands of AC coefficients are coded separately and different reconstructions are made by adding the different bands of AC coefficients to the reconstructed image. The user is free to define which AC components belong in which bands. Extended runs of zeros are available for progressive transmissions that are substantially longer than those possible within a single DCT block—since the runs can progress through an array of 8 × 8 blocks.

In successive approximation varying precision is obtained for the DC and AC coefficients via point transforms (i.e., ignoring lower-order bits). The refinements occur by adding in more bits of precision to the DCT coefficients as the progression continues. Mixtures of spectral selection and successive approximation progressive approaches are allowed as well [Pennabaker and Mitchell, 1993].

5.2.1.3 General Syntax and Data Ordering

Figure 5.6 illustrates the general syntax for JPEG encoded data streams. Hierarchical modes of JPEG use multiple frames for a given image. As illustrated in the figure, nonhierarchical modes of JPEG can employ only a single frame for each image. A frame can include tables (e.g., quantization or Huffman coding), a frame header (required), and at least one scan segment. After the first of many scan segments the Define Number of Lines (DNL) segment is permitted. Headers and other controls in JPEG are indicated by markers. Markers are 16-bit, byte-aligned values that start with hex value FF. In order that the coded data not be confused with markers, any FF byte within coded data is padded with a zero byte. No such padding is done within marker segments, since these have known length. The scan segment is the main unit of JPEG compressed

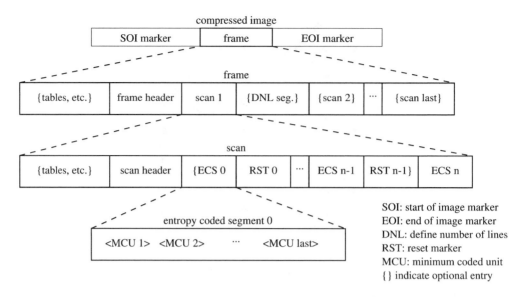

FIGURE 5.6
Syntax for nonhierarchical JPEG data.

image data. Additional tables can be defined before the required scan header. The scan is then comprised of a number (≥ 1) of entropy coded segments (ECS) and reset markers (RST_n). The reset markers allow for recovery from transmission or decoding errors. Within each entropy coded segment is a set of minimum coded units (MCU).

JPEG allows multiple components for images. A common example of an image with multiple components is a color image. Typically these data can be stored as a single luminance component and two chrominance components. For display, these components may be converted to Red-Green-Blue (RGB) or some other representation. Chrominance components are often subsampled relative to the luminance component, since the human visual system is less sensitive to high-frequency variations in chrominance and these occur infrequently in typical images. JPEG allows different quantization tables for each component in a scan.

JPEG allows multiple components of an image to be interleaved or noninterleaved. For a non-interleaved segment of an image a single scan of each of the components within the segment is made following the left-to-right, top-to-bottom scanning of 8×8 blocks. For large images these segments can require substantial memory for storing the various decoded components before (perhaps) converting them to RGB format for display. As a result, interleaving of components is permitted within a scan.

JPEG defines sampling factors for the horizontal and vertical dimensions of each scan component. For each component, Cs_i, these are specified as H_i and V_i for the horizontal and vertical dimensions, respectively. Within the segment the various components are broken into H_i by V_i sections (of 8×8 blocks) and ordered left-to-right, top-to-bottom (within a section) from the first scan component to the last. At this point the individual portion of the image with multiple components can be reconstructed with a minimum of buffering. This process is roughly illustrated for a four-component image in Figure 5.7. The data in the four components are arranged into a single scan (as opposed to the four scans required without interleaving) and ordered as shown in the figure with MCU 1–MCU 4.

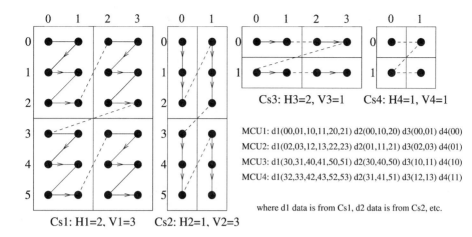

FIGURE 5.7
Order for data interleaving in a single scan.

5.2.1.4 Entropy Coding

At this point, the data are ready for entropy coding. As mentioned above, either binary arithmetic coding or Huffman coding can be used within JPEG. Details on these entropy coding schemes can be found elsewhere [Pennebaker, 1993; Sayood, 1996].

We will roughly describe the Huffman coding technique in this section. Two separate schemes are used for coding the DC and AC coefficients, respectively. For the DC coefficients, the difference between the current DC value and that of the previous DC value for that same component is encoded. A Huffman codeword exists for each of 16 possible sets of DC differences. Each set of differences implies an additional number of bits to completely specify the difference. Figure 5.8 illustrates the number of additional bits and the residual DPCM difference for DC coefficients. For example, if the code corresponding to 3 additional bits was sent, then the difference would be one of [–7, . . ., –4,4, . . .,7]. A subsequent transmission of 011 would indicate the difference was –4. The design of the Huffman codes exploits the likelihood of small DC differences from block to block. Note that using 8 bits of precision would limit the potential additional bits to 11, whereas, 12 bits of precision allows up to 15 additional bits.

The AC coefficients are ordered via a zigzag scan, as described previously. The run length encoding specifies two things: the length of the run of zeros and the number of additional bits needed to specify the value of the AC coefficient that ended the run of zeros. The additional bits are then transmitted in a similar fashion to what we described for the DC DPCM differences. Four bits are used for the run length and 4 bits are used to specify the additional bits—this 8-bit word is encoded via Huffman coding tables. A special End of Block (EOB) symbol indicates that the run of zeros goes to the end of the block. An additional symbol ZRL indicates runs of zeros that are larger than 15. These are followed by an additional run description.

In the case of progressive transmission, the process is similar, except that much longer runs of zeros are possible in bands than can occur in blocks. Thus, the End of Block (EOB) is interpreted as End of Band and the range of runs is specified with additional bits and can range from 1 to 32,767. In addition, point transforms of coefficient values make long runs of zeros much more likely in the early passes of a successive approximation image.

Required Additional Bits	DPCM Differences
0	0
1	−1, 1
2	−3, −2, 2, 3
3	−7, . . ., −4, 4, . . ., 7
4	−15, . . ., −8, 8, . . ., 15
5	−31, . . ., −16, 16, . . ., 31
6	−63, . . ., −32, 32, . . ., 63
7	−127, . . ., −64, 64, . . ., 127
8	−255, . . ., −128, 128, . . ., 255
9	−511, . . ., −256, 256, . . ., 511
10	−1023, . . ., −512, 512, . . ., 1023
11	−2047, . . ., −1024, 1024, . . ., 2047
12	−4095, . . ., −2048, 2048, . . ., 4095
13	−8191, . . ., −4096, 4096, . . ., 8191
14	−16383, . . ., −8192, 8192, . . ., 16383
15	−32767, . . ., −16384, 16384, . . ., 32767

FIGURE 5.8
Extra bits and possible differences for DC coefficients.

5.2.2 JPEG2000

The purpose of the JPEG2000 standard is to correct some perceived deficiencies in the IS 10918-1 standard as well as to incorporate advances in the state of the art that have occurred since the development of the old standard. The deficiencies that are corrected by the new standard include:

- poor subjective performance at rates below 0.25 bits per pixel (bpp)
- lack of the ability to provide lossy and lossless compression in the same codestream
- lack of robustness to bit errors in the compressed image
- poor performance with computer-generated imagery
- poor performance with compound documents (which include both text and images).

In order to be backward compatible the JPEG2000 standard includes a DCT mode that is identical to the current baseline JPEG standard. However, the default mode uses a wavelet decomposition. Subband decomposition entails dividing a sequence into subsequences with different spectral properties. In two-band decomposition we pass the signal through a low-pass and a high-pass filter where the transfer functions of the low-pass and high-pass filters are such that the outputs of the filters can be recombined to obtain the original sequence. As the bandwidth of the filter outputs is considerably less than the bandwidth of the original sequence, we can subsample the output of the filters without violating the Nyquist condition.

There are two approaches to the subband decomposition of two-dimensional (2D) signals: using 2D filters or using one-dimensional filters first on the rows then on the columns (or vice versa). Most compression strategies, including the JPEG2000 verification model, use the second approach. Furthermore, the rows or columns of an image are of finite length. Therefore, at the boundaries we reflect the pixel value across the boundary pixel, as shown in Figure 5.9.

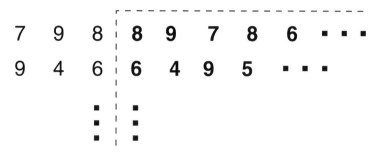

FIGURE 5.9
Pixel values reflected across the left boundary of an image.

In Figure 5.10. we show how an image can be decomposed using subband decomposition. We begin with an $N \times M$ image. We filter each row then down sample to obtain two $N \times M/2$ images. We then filter each column and subsample the filter output to obtain four $N/2 \times M/2$ images. Of the four subimages, the one obtained by low-pass filtering the rows and columns is referred to as the Low Low (LL) image, the one obtained by low-pass filtering the rows and high-pass filtering the columns is referred to as the Low High (LH) image, the one obtained by high-pass filtering the rows and low-pass filtering the columns is called the High Low (HL) image, while the subimage obtained by high-pass filtering the rows and columns is referred to as the High High (HH) image. This decomposition is sometimes represented as shown in Figure 5.11. Each of the subimages obtained in this fashion can then be filtered and subsampled to obtain four more subimages. This process can be continued until the desired subband structure is obtained. The structures currently being considered in the verification model are shown in Figure 5.12. In the

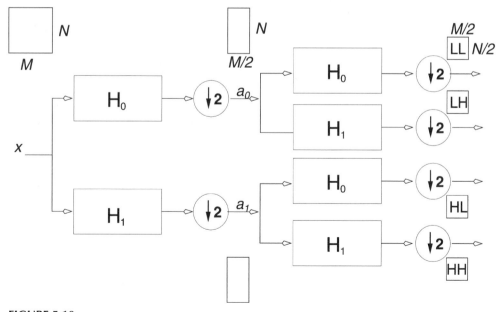

FIGURE 5.10
Subband decomposition of an $N \times M$ image.

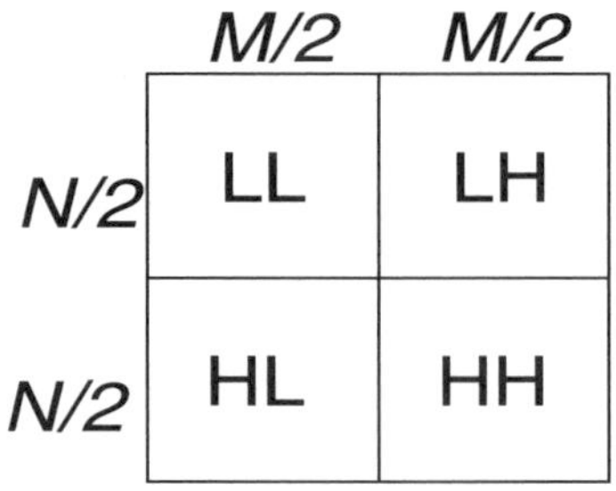

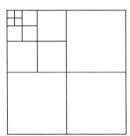

FIGURE 5.11
First-level decomposition.

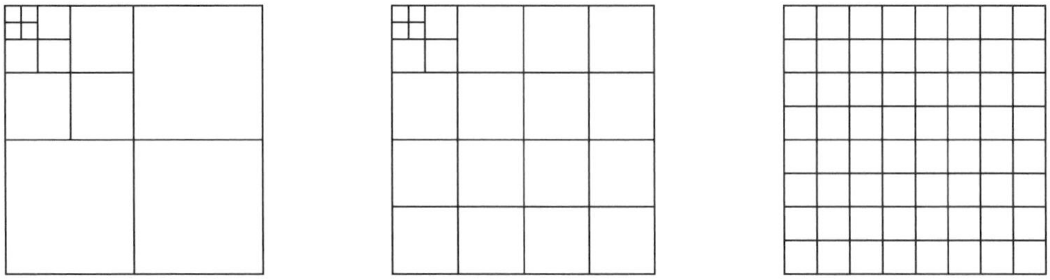

FIGURE 5.12
Subband structures in the verification model.

leftmost structure in Figure 5.12 the LL subimage has been decomposed after each decomposition into four subimages. This is one of the more popular decompositions.

The standard envisages the use of two type of filters. One type of wavelet filters results in integer-valued coefficients. These have been called *reversible* filters and are principally useful in lossless compression. The other type of filter is simply those filters that do not have this property—an obviously much broader class.

The JPEG committee envisaged a spatial organization consisting of tiles, blocks, and frames. *Tiling* refers to a partition of the original image into rectangular tiles. The dimensions of all tiles that do not lie along the right and lower boundaries of the image are restricted to be powers of 2. Furthermore, within a given image component all tiles are required to have the same dimensions. Each tile is compressed independently where compression includes the decomposition, quantization, and coding operations. However, the compressed representation of all tiles is put together in a single bit stream with a common header.

Partitioning into *blocks* is in the context of the coder. Each block is coded independently. However, they are not decomposed independently. Thus a subband from a decomposition can be divided into blocks for the purpose of coding. The blocks are constrained not to cross tile boundaries. Within a subband, all blocks that do not lie on the right or lower boundaries are required to have the same dimensions. A dimension of a block cannot exceed 256.

The frames are supposed to be similar to tiles, and have the same size requirements as tiles. However, unlike the tiles, the frames are allowed some weak dependencies.

The coding scheme being considered for JPEG2000 is based on a scheme originally proposed by Taubman [1994], and Taubman and Zakhor [1994]. It is known as Embedded Block Coding with Optimized Truncation (EBCOT). As its name indicates, EBCOT is a block coding scheme that generates an embedded bit stream. The embedding and blocking interact in an interesting manner. The bit stream is organized in a succession of *layers* with each layer contributing more bits toward the reconstruction. Within each layer each block is coded with a variable number of bits (which could be zero). The number of bits used to represent each block (referred to as *truncation points*) can be obtained using rate-distortion optimization procedures. The suggested data structure is a quad-tree-like data structure in which subblocks are organized into 2×2 *quads*. The next level consists of a collection of 2×2 quads. The coding consists of a sequence of significance map encoding followed by refinement steps. A structure (block or collection of blocks) is said to be significant if any of the constituent coefficients is significant. A coefficient c_{ij} is said to be significant at level p if $|c_{ij}| \geq 2^p$.

The JPEG2000 standard also includes the capacity for region of interest (ROI) coding in which a user-specified region can be coded with higher fidelity than the rest of the image. If the integer wavelet filters are used the coding can be lossless.

5.3 LOSSLESS COMPRESSION

Both the lossless compression schemes described in this section use prediction for redundancy removal.

5.3.1 JPEG

The current lossless JPEG standard is something of an add-on to the much better known lossy JPEG standard. The JPEG standard is a predictive coding technique that has eight different modes from which the user can select. The first mode is no prediction. In this mode the pixel values are coded using a Huffman or arithmetic code. The next seven modes are given in the following list. Three of the seven modes use one-dimensional predictors and four use 2D prediction schemes. Here, $I(i, j)$ is the (i, j)th pixel of the original image, and $\hat{I}(i, j)$ is the predicted value for the (i, j)th pixel.

$$\textbf{1} \quad \hat{I}(i, j) = I(i-1, j) \tag{1}$$

$$\textbf{2} \quad \hat{I}(i, j) = I(i, j-1) \tag{2}$$

$$\textbf{3} \quad \hat{I}(i, j) = I(i-1, j-1) \tag{3}$$

$$\textbf{4} \quad \hat{I}(i, j) = I(i, j-1) + I(i-1, j) - I(i-1, j-1) \tag{4}$$

$$\textbf{5} \quad \hat{I}(i, j) = I(i, j-1) + (I(i-1, j) - I(i-1, j-1))/2 \tag{5}$$

$$\textbf{6} \quad \hat{I}(i, j) = I(i-1, j) + (I(i, j-1) - I(i-1, j-1))/2 \tag{6}$$

$$\textbf{7} \quad \hat{I}(i, j) = (I(i, j-1) + I(i-1, j))/2 \tag{7}$$

Different images can have different structures that can be best exploited by one of these eight modes of prediction. If compression is performed in a non-real-time environment, (e.g., for the purposes of archiving), all eight modes of prediction can be tried and the one that gives the most compression is used.

5.3.2 JPEG-LS

JPEG-LS provides lossless and near-lossless modes of operation. The near-lossless mode allows users to specify a bound (referred to as NEAR) on the error introduced by the compression

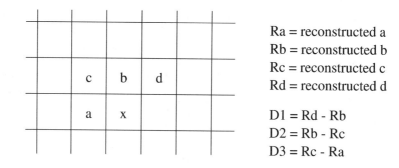

FIGURE 5.13
JPEG-LS context modeling and prediction neighborhoods.

algorithm. JPEG-LS exploits local structure and repetitive contexts within images to achieve efficient lossless (and near-lossless) compression. The context modeling for JPEG-LS is more elaborate than that performed by the original JPEG lossless standard described in Section 3.1. Figure 5.13 depicts the relevant neighborhood of pixels used to define the context. A set of gradients is formed by taking differences between various reconstructed pixels (these are available to both the encoder and decoder for the lossless and near-lossless modes). For the first line in an image the reconstructed values of b, c, and d are set to zero (as is the reconstructed value of a for the first pixel). For the first (or last) pixel in subsequent lines, the reconstructed value of a (or d) is set to the value of Rb, while c is set to the value assigned to Ra in the previous line. The gradients are computed using the reconstructed pixel values as indicated in the figure.

The context is defined by quantizing the gradients. A set of known nonnegative thresholds is defined as T1, T2, and T3. These thresholds are used to quantize each of the three gradients to one of nine levels. Figure 5.14 describes this quantization. By using NEAR as either 0 or the near-lossless allowable distortion, this quantization of gradients can be applied to either lossless or near-lossless modes of operations. Note that if the first nonzero context is negative, SIGN is defined as −1 and the quantized gradients are stored as the negative of their actual values. This process results in the contexts for the three gradients being stored as one of 365 possibilities in the range [0, . . .,364].

If the three gradients are all equal to zero (or less than NEAR for near-lossless compression), then JPEG-LS enters a "run mode" in which runs of the same (or nearly the same) value as the previous reconstructed pixel are encoded. Otherwise, "regular mode processing" is used. In regular mode processing, a median "edge-detecting" predictor is used. This prediction is then cor-

```
if (Di <= -T3) Qi = -4
else if (Di <= -T2) Qi = -3;
else if (Di <= -T1) Qi = -2;
else if (Di <= -NEAR) Qi = -1;
else if (Di <= NEAR) Qi = 0;
else if (Di <= T1) Qi = 1;
else if (Di <= T2) Qi = 2;
else if (Di <= T3) Qi = 3;
else Qi = 4;
```

where i=1,2,3 for Di and Qi
   T1, T2, and flight5.htmT3 are nonnegative thresholds
   NEAR is near lossless threshold (or 0 if lossless)

Context is (Q1,Q2,Q3) {or (-Q1,-Q2,-Q3) if first nonzero element is negative (with SIGN = -1) }

365 possible contexts

FIGURE 5.14
JPEG-LS gradient quantization and context definitions.

rected based on the context. The prediction error is then encoded to achieve lossless or near-lossless prediction—with the value of NEAR being used to quantize the prediction error for the near-lossless case. A bias correction for each context is maintained to remove the accumulated bias from the context correction term error. In addition, modulo arithmetic is used to map the prediction error to an allowable set of positive values. A context-dependent Golomb code (a special-case Huffman code matched to geometric distributions) is used to encode the remaining error for each sample.

In run mode processing, as long as the successive image pixels are equal to the Reconstructed pixel Ra (or within NEAR of Ra for the near-lossless case), a run continues. The run is terminated by either the end of the current image line or a run interruption sample (i.e., one that violates the match condition to Ra defining the run). The coding for the run mode is broken into two sections: run coding and run interruption coding. Run lengths in powers of two are encoded. In a "hit" situation, the run length is either that power of 2 or an end of line has occurred. In a "miss" situation, the location of the termination of the run is transmitted. Run interruption samples can be one of two classes: those where the interruption sample has neighboring pixels Ra and Rb with values that are more than NEAR apart and those where this absolute difference is less than NEAR. Different predictors and a slightly different updating of context information is used in these cases—with the entropy coding being identical to the regular mode processing.

For multiple component images (up to four components), line interleaving and sample interleaving are allowed. A single set of context counters is used across all components, but the pixel values used to compute gradients and predictors are component specific. The run length encoding variables are component dependent. In the line interleaved mode each component Ci has Vi lines coded (Vi defined in frame header). Only the first line of the first component is byte aligned. The last line of the last component is padded with zeros until byte aligned. In the sample interleaved mode all components must have the same dimension and they are placed one sample per component in the order of the components. Run length coding is used, *but the run must exist in every component.* The all zeros gradient quantized context may be used due to the fact that all of the components may not meet the condition for a run. Byte alignment is enforced only on the first sample from the first component and the last sample from the last component of the image.

5.4 BILEVEL IMAGE COMPRESSION

The final compression standard we look at is the bilevel image compression standard International Telecommunications Union Telecommunication Standardization Sector (ITU-T) Recommendation T.82 [also International Organization for Standardization/International Electrotechnical Commission (ISO/IEC) 11544], popularly known as JBIG, and its update released in final committee draft in December 1999, which we will refer to as JBIG2. The JBIG standard was proposed in 1993 as a replacement for the Modified Huffman (MH) and Modified Modified READ (MMR) algorithm used in the Group 3 and Group 4 facsimile standards. The JBIG2 standard is being proposed to replace the JBIG algorithm. Our description of JBIG2 is based on the working draft WD 14492 released by the JBIG committee on August 21, 1998.

5.4.1 JBIG

The JBIG standard allows for a progressive encoding of bilevel images. In progressive coding a low-resolution version of the image is initially transmitted, followed by refinements that allow the decoder to increase the resolution of the reconstructed image. Refinements are received until the decoded image is identical to the original image.

The resolution reduction starts with the original image from which a sequence of reduced resolution images is obtained. The nth reduced resolution image is obtained from the $(n-1)$th reduced resolution image. Each resolution reduction step results in a reduction by four of the number of pixels. The encoding proceeds in the other direction. The lowest resolution image is encoded first, followed by higher-resolution images until the original image is encoded losslessly. The process is shown pictorially in Figure 5.15.

Resolution reduction involves replacing 2×2 blocks of pixels in the higher-resolution layers with a single pixel in the lower-resolution layer. In gray-level images in this type of resolution reduction, the value of the pixel in the lower-resolution image is the average value of the four pixels in the higher-resolution layer. For bilevel images this approach does not work well as it results in the obliteration of thin lines in bilevel documents. Instead, the JBIG committee has proposed a somewhat more sophisticated approach that uses the pixel values in the higher-resolution layer as well as the already encoded pixels in the lower-resolution layer to determine the value of a pixel in the lower-resolution layer. To describe the resolution reduction rule we need to introduce a pictorial representation of the different resolution layers. In Figure 5.16 the squares represent the higher resolution layer and the circles represent the lower-resolution layer.

Suppose that we want to find the value of the lower resolution pixel marked X in Figure 5.16. According to the rule suggested in JBIG, the values of the 12 numbered pixels in Figure 5.16

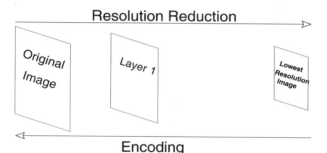

FIGURE 5.15
Resolution reduction and coding.

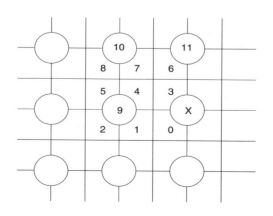

FIGURE 5.16
Pictorial representation of higher- and lower-resolution layers.

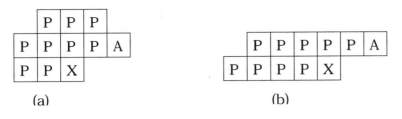

FIGURE 5.17
(a) Three-line and (b) two-line neighborhoods.

form a 12-bit index with the pixel labeled "0" forming the least significant bit, and the pixel label "11" forming the most significant bit. This 12-bit number forms an index into a lookup table.

While the resolution reduction starts with the original image and proceeds to the lowest-resolution image, the coding proceeds in the opposite direction. The lowest-resolution image is encoded first.

The coding method used in this standard is adaptive arithmetic coding. The particular table to be used by the arithmetic coder depends on the context of the pixel being encoded, as well as on whether prediction is being used. The recommendation provides two ways for developing the context for the encoding of the lowest-resolution images. These are shown in Figure 5.17. Information on which context is being used is transmitted to the decoder in the header. The pixel to be coded is marked **X**, while the pixels to be used for the context are marked **P** or **A**. The **A** and **P** pixels are previously encoded pixels and are available to both encoder and decoder. The **A** pixel, referred to in the standard as the adaptive template, can be thought of as a floating member of the neighborhood. Its placement can be changed by the encoder depending on the characteristics of the image being encoded. These pixels are principally used to capture periodic structure in the image. This is especially useful when encoding images rendering greyscale using halftoning.

In most images there is considerable correlation between lines. The encoding makes use of this correlation by comparing the current line with the previous line. If the line n does not differ at any pixel location from the line above line n, the marker $LNTP_n$ is set to 0. This procedure is called *typical prediction (bottom)* (there is a different typical prediction rule used to encode the information required for layers other than the lowest-resolution layer). The information about whether the current line can be obtained via typical prediction is sent to the decoder using a differential encoding scheme. A marker $SLNTP_n$ is generated for each row using

$$SLNTP_n = !(LNTP_n \oplus LNTP_{n-1})$$

For the top line of an image $LNTP_{n-1}$ is taken to be 1. The marker $SLNTP_n$ is treated as a virtual pixel to the left of the line being encoded, and is encoded by the arithmetic coder as another pixel. Typical prediction can allow the encoder to skip a significant number of pixels, depending on the type of document. If some pixels in the current line are different from the corresponding pixels in the line above, $LNTP_n$ is assigned a value of 1.

The arithmetic encoder uses the information regarding typical prediction and the value of the context to select a frequency table with which to arithmetically encode the pixels.

Given that the lower-resolution images are obtained from the higher-resolution images, it would be reasonable to use the information contained in the lower-resolution image to encode the information contained in the higher-resolution image. The already encoded pixels are used in two different ways, through typical prediction, and through *deterministic prediction*.

The typical prediction approach used during the encoding of the higher-resolution layers is different from the typical prediction strategy used in the lowest-resolution layer. When encoding a higher-resolution pixel the typical prediction strategy is to look at whether the low-resolution pixel associated with it is a "not typical" pixel. A *not-typical low-resolution pixel* is defined as one whose eight neighbors are the same color but one or more of the four high-resolution pixels associated with it have a different color. A low-resolution line is denoted "not typical" (the marker *LNTP* is set to 1) if it contains any nontypical pixels. The idea behind typical prediction is to provide efficient encoding of pixels in regions of solid color.

The idea behind deterministic prediction is that, with the pixels of a lower-resolution image and the already encoded pixels in the higher-resolution layer being encoded, a particular pixel may be determined. Recall that the lower-resolution pixels were obtained using a deterministic resolution reduction rule. Therefore for certain configurations of high- and low-resolution pixels, the pixel being encoded can take on only one possible value. If this is the case the value of that particular pixel does not need to be transmitted. The development of an "inverse map" from the resolution reduction tables to a table to be used for deterministic prediction is made much easier if we take into account the orientation of the high-resolution pixel with reference to the lower-resolution pixel. The four different orientations are referred to in the standard as the "phase" of the high-resolution image and are numbered as shown in Figure 5.18.

The index into the deterministic prediction table for different phases is generated using different sets of pixels. The pixels used for the generation of the indices are listed in Table 5.1. The numbers refer to the numbering scheme shown in Figure 5.19. Thus to obtain the deterministic prediction for pixel number 8 (in Figure 5.19) we would use the binary values of pixels 0–7 to construct an 8-bit index into the corresponding deterministic prediction table. The index in this case can take on 256 different values. For the case where the default resolution reduction tables from the standard are used, there are 20 cases (out of 256) in which pixel number 8 is exactly determined. This is listed as the "number of hits" in Table 5.1. For a pixel in the phase 1 orientation we get a 9-bit index, and of the 512 values that the index can take on (for the default tables), 108 configurations of pixels result in deterministic predictions. For the cases where a deterministic prediction is not available, the pixel is sent to the arithmetic coder to be encoded for transmission.

As was the case for the lowest-resolution layer, the adaptive arithmetic coder uses a different frequency table depending on the context of the pixel being encoded. Unlike the lowest-resolution layer, the arithmetic coder uses contexts formed by pixels from two layers, the layer to which the pixel belongs as well as the layer obtained via resolution reduction from the layer to which the pixel belongs. As each pixel in the higher-resolution layer can be at one of four different orientations to the associated pixel in the lower-resolution layer, the adaptive arithmetic coder uses

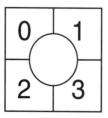

FIGURE 5.18
Numbering of the orientation of the higher-resolution pixels with respect to the associated lower-resolution pixel.

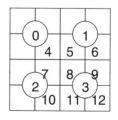

FIGURE 5.19
Numbering scheme used for deterministic prediction.

Table 5.1 Construction of Index for Deterministic Prediction (DP) for Different Orientations

Phase	Pixel to Be Predicted	Pixels Used to Construct Index into DP Table	Number of Hits (Using Default Resolution Reduction)
0	8	0,1, 2, 3, 4, 5, 6, 7	20
1	9	0,1, 2, 3, 4, 5, 6, 7, 8	108
2	11	0,1, 2, 3, 4, 5, 6, 7, 8, 9, 10	526
3	12	0,1, 2, 3, 4, 5, 6, 7, 8, 9, 10, 11	1044

four different contexts. These contexts are shown in Figure 5.20. As in the case of the contexts for the lowest-resolution layer, the pixels marked A can be moved around depending on the type of document being encoded.

The arithmetic encoder only encodes those pixels that are nontypical and for which there is no deterministic prediction. The arithmetic encoder operates on a two-symbol alphabet consisting of the MPS (most probable symbol) and LPS (least probable symbol). Depending on the context, a binary 0 or 1 can be either an MPS symbol or an LPS symbol. Details of the arithmetic coding can be found in Sayood (1996).

Finally, recall that JBIG was developed as a standard for compression of bilevel documents. A comparison of JBIG with the earlier Modified Huffman (MH), Modified READ (MR), and Modified Modified READ (MMR) algorithms used in Group 3 and Group 4 facsimile standards is shown in Table 5.2 [Arps, 1994]. The Modified READ algorithm was used with $K = 4$, while the JBIG algorithm was used with an adaptive three-line template.

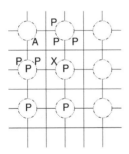

Phase 0

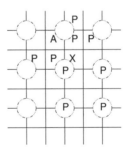

Phase 1

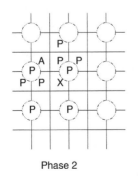

Phase 2

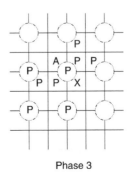

Phase 3

FIGURE 5.20
Contexts used for coding higher-resolution layers.

Table 5.2 Comparison of Binary Image Coding Schemes

Source Description	Original Size (pixels)	MH (bytes)	MR (bytes)	MMR (bytes)	JBIG (bytes)
Letter	4352×3072	20,605	14,290	8,531	6,682
Sparse text	4352×3072	26,155	16,676	9,956	7,696
Dense text	4352×3072	135,705	105,684	92,100	70,703

Source: From Arps and Truong (1994).

5.4.2 JBIG2

The JBIG standard provides a highly effective method for lossless compression of a generic bilevel image. The JBIG2 algorithm currently in working draft stage is an improvement on the JBIG standard in a number of different ways. The JBIG2 approach takes advantage of the properties of the source material. Noting that many bilevel images consist of text on some background, while another significant percentage of bilevel images are or contain halftone images, the JBIG2 approach provides techniques specific to the encoding of text and to the encoding of halftone images. Furthermore, the JBIG2 approach gives the user the option of using lossy compression, which significantly increases the amount of compression that can be obtained.

The working draft itself provides specifications only for the decoder, thus leaving a great deal of flexibility in the design of the encoder. The encoder is allowed to divide a page into a number of regions. There are three types of regions:

1. *Symbol regions:* regions that contain text data, that is, characters of a small size arranged in vertical or horizontal rows.
2. *Halftone regions:* regions containing photographic images that have been dithered to produce bilevel images.
3. *Generic regions:* essentially all portions of the page that do not fit into the previous two categories. These could include line drawings, large text, etc.

How the partitioning of a page takes place and how the encoder determines the nature of the regions are details not specified by the standard. The standard only specifies the operations of the decoder. This leaves a great deal of flexibility to manufacturers of implementations of JBIG2. The information required by the decoder in the JBIG2 encoded file is required to be organized in a sequence of segments. The different types of segments include a page information segment, symbol dictionary segments, symbol region segments, halftone dictionary segments, halftone region segments, and an end-of-page segment. Each segment should have a segment header, a data header, and segment data.

The page segment tells the decoder information about the page, including the size, resolution etc. Once a page buffer has been set up, the decoder decodes the various segments and places the different regions in the page buffer. We briefly describe the decoding of the various regions.

There are two procedures used for decoding the generic regions: the generic region decoding procedure and the generic refinement region decoding procedure. The generic region decoding procedure uses either the MMR technique from ITU-T recommendation T.6 or a variation of the technique used to encode the lowest-resolution layer in the JBIG recommendation described above. Just as in the lowest-resolution layer of JBIG, a flag (*TPON*) indicates whether typical prediction is being used. If it is, the *LNTP* bit is decoded and, depending on whether a particular line is "typical" or not, either the previous line is copied directly or the current line is decoded from the segment data.

The generic refinement decoding procedure is similar to the procedure used by the higher-resolution layers in JBIG. However, unlike the situation of the higher-resolution layer in JBIG, there is no lower-resolution layer available to the decoder. The role of the lower-resolution layer is played by a "reference" layer. The generation of the reference layer is not specified by the standard.

The symbol region decoding procedure is a dictionary-based decoding procedure. The segment data contains the location where a symbol is to be placed as well as the index to a symbol dictionary. The symbol dictionary is contained in a symbol dictionary segment and consists of a set of bitmaps. The dictionary itself is decoded using the different generic decoding procedures.

The halftone region decoding procedure is similar to the symbol region decoding procedure. The segment data contain the location of the halftone region and indices to a halftone dictionary. The dictionary is a set of fixed-size halftone patterns. As in the case of the symbol dictionary, the halftone dictionary is contained in a separate segment and is decoded using generic decoding procedures.

Because the JBIG2 standard provides specifications only for the decoder, it leaves open the possibility of lossy compression. Both the halftone encoding procedure and the symbol encoding procedure can be viewed as vector quantization techniques. This means that the symbols or halftone patterns in the dictionary need not losslessly match the symbols or halftone patterns in the original image. By allowing for nonexact matches the dictionaries can be kept small, resulting in higher compression. For the generic regions the encoder can subject the region to be encoded to lossy transformations prior to encoding, which allows for higher compression.

5.5 DEFINITIONS OF KEY TERMS

DCT: Discrete Cosine Transform. A means of decomposing a signal into frequency-dependent components. Particularly useful for image compression when extended to two dimensions.

JBIG: Joint Bilevel Image Group. Analogous body to JPEG that specified the standard for bilevel images.

JBIG2: Common name for upcoming bilevel image standard verification models, working drafts, etc.

JPEG: Joint Photographic Experts Group. The name of the committee that drafted the eponymous standard for still image compression.

JPEG-LS: An upcoming standard that focuses on lossless and near-lossless compression of still images.

JPEG2000: The current standard that emphasizes lossy compression of images. Wavelets are the default means of decomposing the images, but backward compatibility with JPEG is accommodated.

wavelets: A time-scale decomposition that allows very efficient energy compaction in images. They are particularly useful for achieving less than 0.25 bits per second (bps) without the artifacts present in DCT-based approaches used at those rates.

zigzag scan: See Figure 5.5 for a depiction of the zigzag scan pattern used for ordering two-dimensional image components after decomposition. This scan pattern allows for long runs of zero coefficients for typical images.

5.6   REFERENCES

R. B. Arps and T. K. Truong. Comparison of International Standards for Lossless Still Image Compression. *Proceedings of the IEEE,* 82:889–899, June 1994.

W. B. Pennebaker and J. L. Mitchell. *JPEG Still Image Data Compression Standard.* New York: Van Nostrand Reinhold, 1993.

K. Sayood. *Introduction to Data Compression.* San Francisco: Morgan Kaufmann, 1996.

D. Taubman. *Directionality and Scalability in Image and Video Compression.* Ph.D. thesis, University of California at Berkeley, May 1994.

D. Taubman and A. Zakhor. Multirate 3-D Subband Coding with Motion Compensation. *IEEE Transactions on Image Processing,* IP-3:572–588, September 1994.

5.7   BIBLIOGRAPHY

There are a number of sources for further exploration of the standards described in this chapter.

- The book *JPEG Still Image Data Compression Standard* by W. B. Pennebaker and J. L. Mitchell contains a detailed description of the JPEG standard. New York: Van Nostrand Reinhold, 1993.

- The November 1988 issue of the *IBM Journal of Research and Development* contains extensive descriptions of the arithmetic coder used in the JBIG and JPEG standards.

- An introduction to many of the standards described in this chapter can be found in *Introduction to Data Compression* by K. Sayood. San Francisco: Morgan Kaufmann, 1996.

- The JPEG standard is also described in *Digital Compression for Multimedia: Principles and Standards* by J. D. Gibson, T. Berger, T. Lookabaugh, D. Lindbergh, and R. Baker. San Francisco: Morgan Kaufmann.

- *Techniques and Standards for Image, Video and Audio Coding* by K. R. Rao and J. J. Hwang also contains a nice description of the JPEG standard. New Jersey: Prentice-Hall.

- A description of the new JPEG2000 standard appears in a paper entitled *Overview of JPEG2000* by C. Chrysafis, D. Taubman, and A. Drukarev, which was presented at the PICS 1999 conference. Other information is on the Web at: *www.jpeg.org.*

- The best source for detailed information about the JBIG standard is contained in ITU-T Recommendation *T.82 Information Technology—Coded Representation of Picture and Audio Information—Progressive Bi-Level Image Compression.*

CHAPTER 6

Multimedia Conferencing Standards

DAVID LINDBERGH

6.1 INTRODUCTION

The International Telecommunication Union Telecommunication Standardization Sector (ITU-T, known before 1993 as the CCITT) has produced a number of international standards ("Recommendations," in ITU parlance) for real-time digital multimedia communication, including video and data conferencing. This chapter covers the most important of these standards, the ITU-T H-series, including H.320 through H.324, and H.310, together with their associated video and audio codecs and component standards, as well as the ITU-T T.120 series for data/graphics conferencing and conference control. Audio and video codecs used by the ITU-T H-series are covered from a systems viewpoint, focusing on what the codecs do, not how they do it.

Table 6.1 summarizes the ITU-T H-series standards, their target networks, and the basic video, audio, multiplex, and control standards for each.

Table 6.1 ITU-T Multimedia Conferencing Standards (Basic Modes)

Standard	Network	Video	Audio	Multiplex	Control
H.320 (1990)	ISDN	H.261	G.711	H.221	H.242
H.321 (1995)	ATM/B-ISDN	Adapts H.320 to ATM/B-ISDN network			
H.322 (1995)	IsoEthernet	Adapts H.320 to IsoEthernet network			
H.323 (1996)	LANs/Internet	H.261	G.711	H.225.0	H.245
H.324 (1995)	PSTN	H.263	G.723.1	H.223	H.245
H.310 (1996)	ATM/B-ISDN	H.262	MPEG-1	H.222	H.245

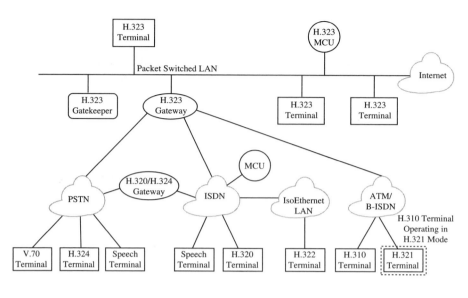

FIGURE 6.1
Interoperability of H-Series terminals.

All ITU-T H-series systems standards support real-time conversational two-way video and audio (limited to one stream of each in H.320, H.321, and H.322), with provisions for optional data channels for T.120 data/graphics conferencing and other purposes. Extensions allow multipoint operation (in which three or more sites can join in a group conference), and in some systems encryption, remote control of far-end cameras, and broadcast applications. Each standard specifies a common baseline mode to guarantee interoperability, but allows the use of other optional modes, both standard and nonstandard, to be automatically negotiated using the control protocol.

These systems fall into two generations. H.320, H.321, and H.322 are first-generation standards, based on H.320 for ISDN networks approved in 1990. H.321 and H.322 specify the adaptation of H.320 terminals for use on ATM and IsoEthernet networks, respectively.

H.323, H.324, and H.310 are the second-generation H-series system standards. Approved in 1995 and 1996, they benefit from industry's experience with H.320, avoiding the problems and limitations that were discovered. They all use the new H.245 control protocol and support a common set of improved media codecs. H.324, which like H.320, is intended for low-bit-rate circuit switched networks [initially analog public switched telephone network (PSTN), often known as plain old telephone service (POTS)], makes use of some H.320 extension standards, including H.233/H.234 encryption and H.224/H.281 far-end camera control.

All these H-series terminals can interoperate with each other through appropriate gateways and can participate in multipoint conferences, as illustrated in Figure 6.1.

6.2   H.320 FOR ISDN VIDEOCONFERENCING

The ITU-T H.320 standard, known during its development as "px64" for its use of bandwidth in 64-Kbit/s increments, covers videoconferencing and videotelephony over ISDN and switched-56 circuits at rates from 56 Kbit/s to 2 Mbit/s.

Like the other H-series systems, H.320 supports real-time conversational two-way video and audio (one channel each), with provisions for optional data channels. Extensions allow multi-point operation (in which three or more sites can join in a group conference), encryption, remote control of far-end cameras, and broadcast applications.

H.320 was developed during the late 1980s and approved by the CCITT (now ITU-T) in 1990. It was the first successful low-bit-rate video communications standard and remains the universally accepted standard for ISDN videoconferencing.

6.2.1   The H.320 Standards Suite

The H.320 document is a "systems standard" that calls out a number of other ITU-T standards for various parts of the system, as shown in Figure 6.2. The core components of H.320 are the following:

- *H.221 multiplex:* Mixes audio, video, data, and control information into a single bit stream. Uses synchronous time division multiplexing with 10-ms frames.

- *H.230/H.242 control:* Mode control commands and indications, capabilities exchange. Operates over a fixed 400-bits per second (bps) channel (BAS) in the H.221 multiplex.

- *H.231/H.243 multipoint:* Specifies central multipoint bridges and operation for multiway group conferences (optional in the H.320 standard, but universally implemented).

- *H.261 video coding:* Compresses color motion video into a low-rate bitstream. Quarter Common Intermediate Format (QCIF) (176 × 144) and Common Intermediate Format (CIF) (352 × 288) resolutions.

- *G.711 audio coding:* 8-kHz sample rate, 8-bit log-PCM (64 Kbit/s total) for toll-quality narrowband audio (3-kHz bandwidth).

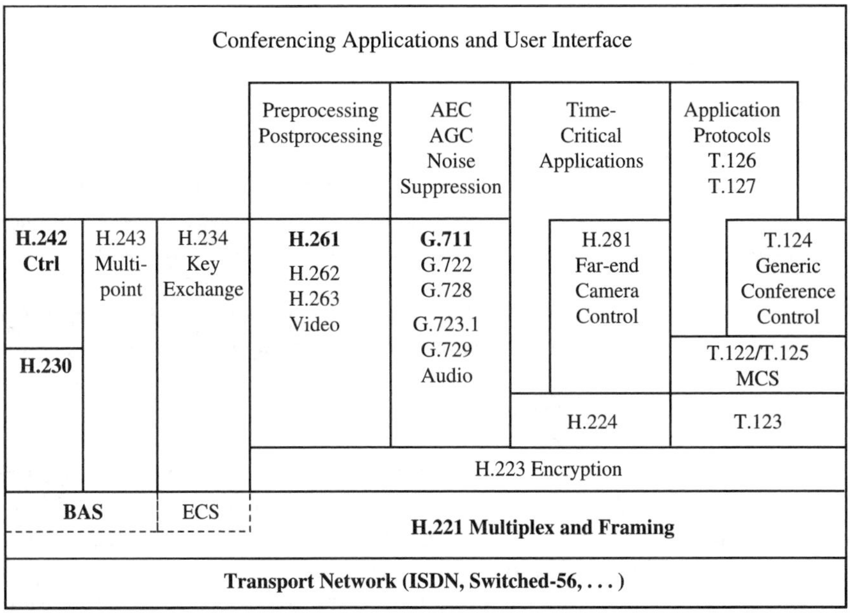

FIGURE 6.2
H.320 protocol stack.

Baseline H.320 components are shown in bold in Figure 6.2 and include QCIF resolution H.261 video, G.711 logarithmic-Pulse Code Modulation (PCM) audio, the H.221 multiplexer, and H.230/H.242 control. Improved standard modes, such as H.263 and Common Intermediate Format (CIF) resolution H.261 video and improved-quality or lower-bit-rate audio modes (leaving more bandwidth for video) can be negotiated using control protocol procedures, as can non-standard or proprietary modes.

In addition to the core components of H.320, optional standard extensions support remote control and pointing of far-end cameras (H.224/H.281), encryption according to H.233/H.234, and data conferencing using T.120 for sophisticated graphics and conference control. T.120 supports applications like Joint Photographic Experts Group (JPEG) still image transfer, shared document annotation, and personal computer (PC) application sharing. The H.331 standard (not shown) specifies how H.320 terminals can be used for low-bit-rate broadcast (send or receive only) applications.

6.2.2 Multiplex

The multiplexer component of a multimedia conferencing system mixes together the audio, video, data, and control streams into a single bit stream for transmission. In H.320, the H.221 time division multiplexer (TDM) is used for this purpose.

H.221 supports a total of eight independent media channels, not all of which are present in every call. The BAS and FAS channels carry H.320 system control and frame synchronization information and are always required. There is provision for one channel each of audio and video and three user data channels, LSD, HSD, and MLP. The optional ECS channel carries encryption control messages if encryption is used.

The multiplexing scheme uses repeating 10-ms frames of 80 bytes (640 bits) each. Various bit positions within each frame are allocated to the different channels in use. This system makes efficient use of the available bandwidth, except that the allocation of bits to different channels can change among only a small number of allowed configurations.

6.2.3 System Control Protocol

The control protocol operates between a pair of H.320 systems to govern the terminal's overall operational mode, including negotiation of common capabilities, selection of video and audio modes, opening and closing of channels, and transmission of miscellaneous commands and indications to the far end. H.242 and H.230 together define the basic H.320 control protocol and procedures.

All H.320 control messages are drawn from tables in the H.221 standard, with each 8-bit code assigned a particular meaning. Each table entry is called a codepoint. Longer or less frequently used messages can be sent as multibyte messages, using escape values in the initial table. The meaning of and procedures for using the messages are defined variously in H.221, H.230, H.242, and, for multipoint-related messages, H.243.

The messages are sent directly in the (net) 400-bit/s H.221 BAS channel without any packetization, headers, or CRC, and without the use of any acknowledgment or retransmission protocol. The low error rate of Integrated Services Digital Network (ISDN) channels, combined with the forward error correction applied to all BAS messages, ensures that control messages are nearly always received without error. Although this system has worked well for H.320 terminals, the potential for undetected errors in the control channel sometimes results in added procedural complexity, such as redundant transmission. The newer H.245 control protocol, which replaces H.230/H.242 control in the second-generation ITU-T conferencing standards, is based instead on a reliable link layer that retransmits errored messages automatically until a positive acknowledgment is received, thus allowing the control protocol to assume that messages are guaranteed to arrive correctly.

6.2.4   Audio Coding

6.2.4.1   G.711 Baseline Audio

The baseline audio mode for H.320 is the G.711 log-PCM codec, a simple 8-kHz sample rate logarithmic PCM scheme that has long been used as the primary voice telephony codec for digital telephone networks (long-distance voice telephone calls are today carried on digital networks, even if they originate from analog telephones).

G.711 is defined to use 8-bit samples, for a total bit rate of 64 Kbit/s, but for use with H.320 each sample is truncated to 6 or 7 bits, resulting in alternative bit rates of 48 or 56 Kbit/s. G.711 provides excellent toll-quality narrowband (3-kHz audio bandwidth) audio with insignificant codec delay (well under 1 ms) and very low implementation complexity.

To provide compatibility with normal G.711 voice telephone calls, all H.320 calls start by sending and receiving G.711 audio while performing initial synchronization and mode negotiation in the H.221 FAS and BAS channels. Unfortunately, G.711 specifies two alternative encoding laws, A-law and μ-law; both schemes were already in use in different parts of the world at the adoption of G.711, and the CCITT was unable to agree on a single law. As a result, H.320 systems must attempt to automatically detect the coding law in use by the far end at the start of each call or avoid using audio until H.320 control procedures can be used to establish another audio mode.

6.2.4.2   Lip Sync

Audio codecs generally involve less delay than video codecs. H.320 provides lip synchronization, in which the decoded audio output matches lip movements displayed in the video, by adding delay in the audio path of both the transmitter and receiver, keeping the audio and video signals roughly synchronized as they are transmitted. This makes it impossible for receivers to present audio with minimal delay if the user does not want lip sync. Second-generation systems support lip sync without adding audio delay at the transmitter, instead using timestamp or time-skew methods that let the receiver alone add all necessary audio delay if desired.

6.2.4.3   Optional Audio Modes

G.711 was chosen as the baseline H.320 audio mode for its low complexity and compatibility with ordinary telephone traffic, but it is quite inefficient in its use of bandwidth compared to optional H.320 audio modes. The data bandwidth saved by switching to an alternative audio mode can be used to send additional video bits, making a big difference to H.320 video quality, especially on common 2-B (128-Kbit/s) H.320 calls.

Table 6.2 summarizes the set of approved and planned ITU-T audio codecs used in H-series conferencing. [Note that narrowband codecs pass 200–3400 Hz audio, and wideband codecs pass 50–7000 Hz. In Table 6.2 and throughout this chapter, these are referred to as "3-kHz" and "7-kHz" audio bandwidth codecs, respectively. Also note that audio codec delay is highly dependent on implementation. The delay values in Table 6.2, and in the rest of this chapter, use a (3 * frame size) + look-ahead rule of thumb, which includes algorithmic delay, one frame time for processing, and one frame time for buffering. Gibson et al. (1998) provides a more complete discussion of codec delay and other noncodec factors contributing to total end-to-end delay.]

The most important and commonly supported optional H.320 audio modes are G.728, for 16-Kbit/s narrowband audio, and G.722, for 56-Kbit/s wideband (7-kHz audio bandwidth) audio.

Table 6.2 Audio Codecs Used in Multimedia Conferencing

Standard	Bit Rates Kbit/s	Audio Bandwidth (kHz)	Complexity (Fixed-Point) MIPS	Frame Size	Delay
G.711 (1977)	48, 56, 64	3	Near zero	125 μs	<<1 ms
G.728 (1992)	16	3	~35–40	625 μs	<2 ms
G.723.1 (1995)	5.3, 6.4	3	~18–20	30 ms	97.5 ms
G.729 (1995)	8	3	~18	10 ms	35 ms
G.729A (1996)	8		11	10 ms	35 ms
G.722 (1988)	48, 56, 64	7	~10	125 μs	<2 ms
G.722.1 (1999)	24, 32	7	~14	20 ms	60 ms

G.728 is the preferred and most commonly used narrowband audio mode for H.320. A low-delay code-excited linear prediction (LD-CELP) algorithm based on the usual 8-kHz narrowband sample rate and audio frames of five samples (0.625 ms), it uses 16 Kbit/s to provide excellent toll-quality audio with a total codec delay of about 1.875 ms.

Audio quality is even more important to successful videoconferencing than video quality. For this reason wideband audio (7-kHz audio bandwidth) is popular for use with H.320 conferencing systems. G.722 is used for wideband coding at 56 Kbit/s and, like G.711, has alternative bit rates of 64 and 48 Kbit/s. The newer G.722.1 wideband audio codec (approved 1999) offers 7 kHz wideband audio quality equivalent to G.722, at lower bit rates of 24 and 32 Kbit/s.

Recent work on improved narrowband audio codecs has produced ITU-T G.723.1, a 5.3/6.4-Kbit/s codec with about 100 ms of codec delay, and G.729, an 8-Kbit/s codec with about 35 ms of codec delay. Both codecs are options in H.320 but are not yet widely implemented. These codecs are described in Section 6.4.2.

6.2.5   Video Coding

Video is optional in the H.320 standard, but is included in essentially all products. The H.261 video codec, approved with H.320 in 1990, is the baseline video mode for H.320 systems.

The MPEG-2 video codec H.262, which supports a variety of picture formats and both interlaced and progressive scanning patterns, is an option in H.320, but since it is intended for use at high bit rates (many megabits per second), it is not widely implemented in H.320 systems.

The newer H.263 low-bit-rate video codec, approved in 1995 as part of the H.324 set, offers significantly improved video compression and additional features and is an option commonly implemented in H.320 systems. H.263 is described later in this chapter.

6.2.5.1   H.261 Video Codec

H.261 codes video frames using a DCT on blocks of size 8 × 8 pixels. An initial frame (called an INTRA frame) is coded and transmitted based on an input video picture. In typical video scenes, subsequent frames are often very similar to immediately preceding ones, except for small motions of objects in the scene. These can be coded efficiently (in the INTER mode) using motion compensation from the previous frame; the displacement of groups of pixels from their position in the previous frame (called motion vectors) is transmitted together with the DCT coded difference between the thus predicted and original images, rather than retransmitting the coded pixels themselves.

H.261 supports two video picture formats, CIF (common intermediate format) and QCIF (quarter CIF). Only QCIF is part of the H.320 baseline requirements; CIF format support is optional but widely implemented.

CIF has 352 × 288 luminance pixels. Color is sent at half this resolution in each dimension (176 × 144) because the human eye is less sensitive to color resolution than brightness resolution. This pixel array is mapped onto the 4:3 aspect ratio television screen and results in non-square individual pixels.

QCIF has one-quarter of the CIF resolution, 176 × 144 luminance pixels. It uses the same pixel aspect ratio and color sampling format as CIF.

A third H.261 image format, Annex D graphics, is used only for still-image transmission and not for motion video. Annex D mode doubles the CIF resolution in each dimension, for a 704 × 576 pixel still image. Annex D graphics are being replaced by more sophisticated still-image transfer methods in T.120, including the JPEG and JBIG formats used by T.126.

All H.261 video is noninterlaced, using a simple progressive scanning pattern. The video frame rate is based on the National Television Standards Committee (NTSC) standard rate of 30,000/1001 (29.97) Hz. The low video bit rates used in H.320 often make it impossible to code each 30-Hz frame with reasonable detail, so the effective frame rate can be reduced by skipping frames at the encoder, allowing more bits to be sent for each remaining frame.

H.261 receivers may be limited in the rate at which they can decode incoming video, so included in the H.320 video capabilities are minimum picture intervals (MPI) for each supported H.261 resolution. The MPI values are the minimum number of frame periods (0, 1, 2, or 3 frames) the H.261 decoder requires between each coded frame and correspond to maximum frame rates of 30, 15, 10, or 7.5 frames per second.

H.261 specifies a standard coded video syntax and decoding procedure, but most choices about encoding methods, such as allocation of bits to different parts of the picture, are left to the discretion of the implementor. The result is that the quality of H.261 video, even at a given bit rate, greatly depends on the cleverness of the encoder implementation and on the number of computing cycles available to the encoder in real time.

6.2.6   H.231 and H.243: Multipoint

Multipoint operation, in which three or more terminals can participate in a single joint conference, is a widely implemented option in H.320. H.231 and H.243 define multipoint operation for H.320.

H.320 is oriented around ISDN switched circuit network connections, which are inherently point-to-point in nature. Multipoint calls work by connecting all participating terminals to a central bridge device, called a *multipoint control unit* (MCU). Each connection is equivalent to an ordinary point-to-point call. The MCU receives audio and video from all connected terminals and sends selected audio and video out to all the terminals. The MCU can also include participants in the conference using regular voice-only telephones or other standards supported by the MCU.

The MCU has great flexibility in choosing what to send to each terminal, but usually it mixes the few loudest audio signals and sends this to all receivers. The video signal from the current speaker (based on automatic speech detection or manually selected in various ways) is normally sent to all receiving terminals. The speaker is sent video of the previous speaker and audio without her own speech. Many other methods of choosing who is seen and heard are possible, but this common implementation allows all sites to see and hear the current speaker and the speaker to view the reactions of the preceding speaker.

MCUs themselves may be connected together to create a larger joint conference or to save on communications costs—for example, with one MCU on each continent, with a single overseas connection between the MCUs. Figure 6.3 illustrates the connections in a multipoint call.

H.231 specifies the requirements for H.320 MCUs, in which each port of the MCU behaves much like an H.320 terminal. H.243 specifies procedures for setting up H.320 multipoint calls, including terminal addressing and choosing a single common audio and video mode for the conference.

In a multipoint call, T.120 data conferencing and conference control on the MLP data channel terminates at the MCU, where the T.120 protocol stack routes messages among the terminals. The LSD and HSD data channels operate in a broadcast mode; one terminal transmits at a time, and all others receive the data, relayed through the MCU. H.243 specifies procedures for passing the LSD and HSD tokens, which grant permission to transmit, between the terminals. The bit rate assigned to these channels in the H.221 multiplex must be the same for all terminals, or receivers may be unable to accept the bit rate being transmitted.

Similarly, video switching in the MCU requires that terminals be able to receive the exact video bit rate being transmitted by the source terminal, so video bit rates must match among all terminals in a multipoint conference. The H.320 video bit rate is determined by the bandwidth consumed by audio and data channels. H.243 provides procedures for an MCU to force all connected terminals into a selected communication mode (SCM), in which all terminals use the same video, audio, and data modes and bit rates to facilitate data and video switching in the MCU.

The audio and video processing that takes place in an MCU adds more end-to-end delay to the already undesirably long delays in videoconferencing, but this has not been a big problem in practice. Perhaps this is because group conferences are usually more formally structured than one-on-one conversations, with participants speaking one at a time, making the end-to-end delay less obvious.

H.243 also specifies various optional multipoint features, such as passwords for entry into a group conference and a chair control mode, in which the chairperson of a conference can con-

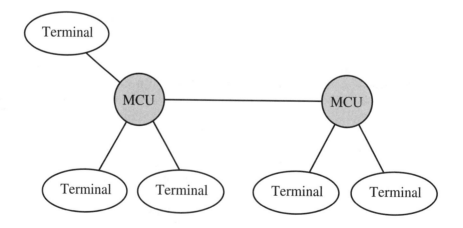

FIGURE 6.3
Connections in a multipoint call.

trol who is seen, drop terminals from the conference, and so on. These optional functions of H.243 are being replaced with more sophisticated equivalents from the T.120 standards.

A video mixing mode is described in H.243, where the MCU combines scaled-down video images from several terminals into a single output video image. This results in a "Hollywood Squares"-style output picture, with video from many different sites visible at all times in adjoining rectangular regions, without the need to switch away from any participant. This mode provides continuous presence for participants and is an indirect way to deal with the limitation of a single channel of video in H.320. Aside from the greatly reduced resolution, this method involves significantly larger end-to-end delay because the MCU must decode, frame synchronize, compose, and recode the input video streams. For these reasons, this method of supporting continuous presence operation is discouraged in the second-generation conferencing systems based on H.245. Instead, second-generation systems have the option of simultaneously decoding separate channels of video from each site.

6.2.7   H.233 and H.234: Encryption

Encryption is an option in H.320 and makes use of ITU-T H.233 and H.234, both of which are also used with H.324.

H.233 covers encryption procedures and algorithm selection; FEAL, B-CRYPT, and DES may be used as well as other ISO-registered algorithms and, in H.324, nonstandard algorithms. Key length is dependent on the algorithm in use and is practically unlimited. The interests of different national governments unfortunately made it impossible for the ITU to agree on a single baseline encryption algorithm.

H.234 covers key exchange and authentication using the ISO 8732, extended Diffie-Hellman, or RSA algorithms.

In H.320, encryption can be applied to any or all of the audio, video, and data channels in the H.221 multiplex. In a multipoint encrypted call, the link between each terminal and MCU is encrypted. The MCU decrypts and reencrypts all channels, acting as a "trusted entity."

6.2.8   H.331 Broadcast

Standard H.320 terminals can, with only small modifications, be used on networks that provide one-way data transmission, to provide essentially a low-bit-rate one-way digital television service. Applications include classroom broadcasts, with a teacher transmitting video and audio to many receiving students at different locations, large corporate or political broadcasts, and so on. Often low-bit-rate digital satellite transmission is used for such services, with one uplink site sending H.320 video and many receiving stations.

H.331 specifies how to use H.320 terminals in situations where there is no data path from receivers back to transmitters, so normal two-way control negotiation is impossible. In such situations, no MCU is needed because there is no coordination of modes or combining of video and audio, since there is only a single transmitter.

6.3   H.320 NETWORK ADAPTATION STANDARDS: H.321 AND H.322

Two standards broaden the use of H.320 to networks other than ISDN. They directly adapt standard H.320 terminals to other networks and offer simple interworking with ISDN-based terminals and features identical to H.320 on ISDN. H.321 covers adaptation of H.320 terminals to Asynchronous Transfer Mode (ATM)-based broadband ISDN (B-ISDN) networks. H.322 covers H.320 on IsoEthernet (ISLAN-16T) LANs.

6.3.1   H.321: Adaptation of H.320 to ATM and B-ISDN

H.321 specifies the use of standard H.320 terminals, intended for ISDN use, on ATM and B-ISDN networks. It uses the services of ATM Adaptation Layer 1 (AAL1), with both the segmentation and reassembly (SAR) and convergence sublayer (CS) functions of I.363, to provide data transport channels equivalent to ISDN channels.

6.3.2   H.322: Adaptation of H.320 to IsoEthernet

Like H.321, H.322 adapts standard ISDN H.320 terminals to a different network type. H.322 is described as intended for "local area networks which provide a guaranteed quality of service"—local area netowrks (LANs) that provide what are essentially virtual ISDN B channels. The only existing LAN that meets this criterion is the Institute of Electrical and Electronics Engineering (IEEE) 802.9a isochronous Ethernet, ISLAN-16T (IsoEthernet), which provides 96 virtual B channels in addition to 10-Mbit/s Ethernet. The H.322 standard itself, only two pages long, is almost null, stating that a standard H.320 terminal is adapted to work on the LAN via an unspecified LAN interface.

6.4   A NEW GENERATION: H.323, H.324, AND H.310

The success of the original H.320 standard sparked the development of many extensions, including H.233/H.234 for encryption, H.224/H.281 for real-time far-end camera control, and H.331 for broadcast applications, as described in the preceding sections. The H.321 and H.322 standards for adapting H.320 terminals to new networks were also a result of the popularity of H.320.

   At the same time, industry experience with H.320 since its 1990 approval revealed limitations in the standard, and improved techniques for video and audio compression appeared. A new generation of standards has been developed that avoid the problems in H.320 and take advantage of the latest compression technology: H.323 for packet switched networks, H.324 for low-bit-rate circuit switched networks, and H.310 for ATM-based broadband ISDN.

   Some of the major improvements over H.320 common to all the second-generation standards include:

- faster call start-up upon initial connection
- support for multiple channels of video, audio, and data
- simpler and more flexible assignment of bandwidth among the various channels
- separate transmit and receive capabilities
- means to express dependencies between capabilities
- explicit description of mode symmetry requirements
- receiver-driven mode request mechanism
- improved audio and video coding
- larger range of video modes and resolutions
- cleaner mechanisms for extension to future standard and nonstandard features

Many of these improvements come from the new, more flexible H.245 control protocol, which replaces the functions formerly provided by H.242, H.230, and part of H.221.

   H.323, H.324, and H.310 all support the improved compression standards G.723.1, G.729, and H.263 (which are now also optional in H.320), share the same basic system architecture, and fully support T.120 data, graphics, and control. Smooth H.320 interworking, as well as interworking with each other, was a prime design consideration, which is why H.261 video has been retained as a baseline video mode for all the new standards.

Like H.320, these new standards can be used in both point-to-point and multipoint calls and start with a baseline for guaranteed interoperability, with many additional features as options. H.323, H.324, and H.310 are all "systems standards," calling out H.245 and other component standards that together make up the complete conferencing system.

6.4.1   H.245 Control Protocol

The new H.245 multimedia system control protocol benefits from the experience gained with H.320 system control. The control model of H.245 is based on *logical channels,* independent unidirectional bit streams with defined content, identified by unique numbers arbitrarily chosen by the transmitter. There may be up to 65,535 logical channels.

H.245 operates in its own logical channel and carries end-to-end control messages governing system operation, including capabilities exchange, opening and closing of logical channels, mode preference requests, flow control messages, and general commands and indications. The H.245 structure allows future expansion to additional capabilities as well as manufacturer-defined nonstandard extensions to support additional features.

H.245 requires an underlying reliable link layer such as TCP/IP (for H.323), V.42 (for H.324), or Q.922 (for H.310). The H.245 protocol assumes that this link layer guarantees correct, in-order delivery of messages.

6.4.2   Audio and Video Codecs

All the second-generation standards use H.261 video as a baseline for backward compatibility with H.320. H.323 and H.310 require G.711 log-PCM audio. However, new, more efficient codecs have recently been developed, including the G.723.1 5.3/6.4-Kbit/s speech codec, the G.729 8-Kbit/s speech codec offering lower delay, the G.722.1 wideband audio codec (24 and 32 Kbit/s), and the H.263 video codec. These improved codecs, while generally optional, are widely implemented in second-generation systems.

G.723.1 and G.729 both provide for silence suppression, in which the average audio bit rate is reduced by not transmitting during silence or by sending smaller frames carrying background noise information. In typical conversations, both ends rarely speak at the same time, so this can save significant bandwidth for use by video or data channels. As with most Code-Excited Linear Prediction (CELP) codecs, G.723.1 and G.729 work best on speech and less well on other audio sources such as music.

6.4.2.1   *G.723.1—5.3/6.4-Kbit/s Audio*

G.723.1 was developed as part of the H.324 suite, originally for use on V.34 modems at very low bit rates. It uses only 5.3 or 6.4 Kbit/s and is estimated to require about 18–20 MIPS in a general-purpose DSP. Transmitters may use either of the two rates and can change rates for each transmitted frame, since the coder rate is sent as part of the syntax of each audio frame.

The codec provides near-toll-quality speech based on 8-kHz audio sampling. The 6.4-Kbit/s rate provides better audio quality than the 5.3-Kbit/s rate, especially in the presence of background noise. Coders can switch between the two rates on every 30-ms frame if desired. When coding each frame, G.723.1 looks ahead 60 samples (7.5 ms) into the following audio frame, resulting in a typical codec delay of 97.5 ms. This amount of delay is quite large for voice telephony applications (especially since codec delay is only one component of total end-to-end delay), but it is suitable for use in low-bit-rate video telephony, as even more delay is normally needed in the audio path to achieve lip synchronization with video.

6.4.2.2   G.729—8-Kbit/s Low-Delay Audio

G.729 is a high-quality, high-complexity 8-Kbit/s narrowband (8-kHz sample rate) speech codec using Conjugate-Structure Algebraic Code-Excited Linear Prediction (CS-ACELP). The codec produces 10-ms frames and uses 5 ms of look-ahead, for about 35 ms of total codec delay. While the data rate used by G.729 is higher than that of G.723.1, its considerably lower delay, higher voice quality, and improved ability to withstand sequential encoding and decoding ("tandem-ing") make it a better choice for applications that do not require lip synchronization or that do not use video at all.

Annex A of G.729 (sometimes called "G.729A") specifies a lower-complexity (about 11 fixed-point MIPS) encoding scheme that produces a bit stream fully compatible with the standard G.729 decoder and that shares the same frame size and delay. This provides a low-complexity codec that can interoperate with G.729, suitable for inexpensive DSPs and PC software implementation. However, both the full G.729 and the 6.4-Kbit/s mode of G.723.1 offer better audio quality than G.729A.

6.4.2.3   G.722.1—7 kHz Wideband Audio

The G.722.1 wideband audio codec, approved in 1999, provides 50-7000 Hz audio, closer to frequency modulation (FM) radio quality than to that of traditional telephones, at 24 and 32 Kbit/s. Wideband audio avoids the "tinny" quality of narrowband codecs, and is more relaxing and less stressful to listen to for long periods. G.722.1 is a transform codec, and is therefore capable of handling music and sound effects just as well as speech. It has a 20 ms frame length and 1 frame of look-ahead, for a total codec delay of 60 ms.

6.4.2.4   H.263 Video Codec

The H.263 video codec is optional in all the H-series standards except H.324, where both H.263 and H.261 are required in systems that support video. Approved in 1995 as part of the H.324 suite, H.263 is a general-purpose low-bit-rate video codec based on the same DCT and motion compensation techniques as H.261 and targeted at the same set of applications.

H.263 is the result of many small incremental improvements to the techniques used in H.261, which together add up to a very large improvement in video quality. At the low video bit rates typical in H.324 operation on V.34 modems (10–20 Kbit/s of video signal), H.263 video quality is considered equivalent to that of H.261 at up to double the bit rate, although the difference is less at the higher bit rates used by H.320 and H.323. H.263 is expected to gradually replace H.261 in all applications, except that H.261 support will continue to be needed for backward compatibility.

Video coding improvements in H.263 include half-pixel precision motion compensation (H.261 uses full pixel precision and a loop filter), improved variable-length coding, reduced overhead, and optional modes including unrestricted motion vectors (where MVs are allowed to point outside the picture), arithmetic coding instead of variable-length (Huffman) coding, an advanced motion prediction mode including Overlapped Block Motion Compensation (OBMC), and a PB-frames mode, which combines a bidirectionally predicted picture with a normal forward predicted picture.

H.263 supports the same CIF and QCIF picture formats used in H.261, as well as new formats, as shown in Table 6.3.

Like H.261, H.263 works with noninterlaced pictures at frame rates up to 30 frames per second (fps). Extensions enable H.263 to handle square pixels and picture formats of arbitrary size.

Table 6.3 H.263 Standard Video Picture Formats

Picture Format	Luminance Pixels	Video Decoder Requirements	
		H.261	H.263
SQCIF	128×96	Not defined	Required
QCIF	176×144	Required	Required
CIF	352×288	Optional	Optional
4CIF	704×576	Not defined	Optional
16CIF	1408×1152	Not defined	Optional

H.263 does not contain any equivalent to the Annex D still-image mode of H.261. This still-image transfer function is provided instead by the T.120 series standards, specifically the still-image transfer protocol of T.126.

6.4.3   H.323 for Packet Switched Networks

Originally titled "Visual Telephone Systems and Equipment for Local Area Networks Which Provide a Non-Guaranteed Quality of Service," ITU-T H.323 at its conception was intended exclusively for real-time multimedia conferencing on corporate LANs. As H.323 matured toward its 1996 approval, it became a generic conferencing standard for all types of packet switched networks (PSNs), including Novell IPX/SPX, Ethernet and token-ring corporate LANs, and, most importantly, Transmission Control Protocol/Internet Protocol (TCP/IP) networks such as corporate intranets and the public Internet. This expanded scope is reflected in H.323's revised 1998 title: "Packet Based Multimedia Communication Systems."

In packet switched networks, each packet is a variable-sized unit of data containing headers and payload, which is routed hop-by-hop to its destination. Each packet may follow a different path through the network, resulting in variable and unpredictable packet arrival times, potential packet loss, out-of-order arrival, and so on. By far the most popular type of packet network is IP, which is used on the Internet as well as on many corporate LANs.

Figure 6.4 illustrates the H.323 terminal, which can provide multiple channels of real-time, two-way audio, video, and data. The H.323 standard specifies H.245 for control; defines the use of audio codecs, video codecs, and T.120 data channels; and calls out H.225.0 for audio and video packetization as well as call setup. When each logical channel is opened using H.245 procedures, the responding terminal returns a network address and port number, which together form a *transport address* to which packets carrying that channel should be addressed. Most H.323 calls consume only a small fraction of a fast network's capacity, usually 100–500 Kbit/s for each call, with the remainder of the network bit rate available for other uses.

The multiplex for H.323 is the packet structure of the network itself, as each packet can contain a different type of content (audio, video, data, control, etc.) as specified by the packet header.

G.711 audio is mandatory in H.323 terminals, with G.723.1 and G.729 among the optional audio modes. Video is optional, but if provided, QCIF format H.261 is the baseline video mode, with H.263 the most important optional video mode.

6.4.3.1   *H.225.0—Media Packetization and Call Setup*
The H.225.0 standard specifies procedures for setting up H.323 calls on the network, as well as audio and video packetization for H.323.

Call setup makes use of a small subset of the Q.931 messages and procedures originally defined to implement ringing, busy, answer, and related call setup functions in ISDN telephony. Terminals can be addressed using their network address [such as an internet protocol (IP) address] or, more commonly, alias addresses such as telephone numbers or email addresses. Alias addresses are simply text strings that are automatically mapped into network addresses.

H.225.0 specifies that the output of the audio and video codecs is packetized and transported using the Real-Time Transport Protocol (RTP/RTCP) [Schulzrinne et al., 1996] defined by the Internet Engineering Task Force (IETF) for sending real-time data on the Internet. RTP uses unreliable channels, such as User Datagram Protocol/Internet Protocol (UDP/IP) or Novell Internet Packet Exchange (IPX), to carry video and audio streams. On unreliable channels, lost or corrupted packets are not retransmitted. This ensures that all audio and video packets are delivered with minimal delay and allows multicast operation, but it means that receivers must deal with missing packets through error concealment or other means. Timestamps are included on RTP packets to allow received audio and video to be synchronized.

The H.245 control channel, call setup messages, and data channels such as T.120 are carried on *reliable* channels such as TCP/IP, which use retransmission to guarantee correct, in-order delivery of all data packets. These data types require delivery of all packets and are less time-critical than audio and video, so they can tolerate the larger and more variable delays resulting from retransmission in the network.

6.4.3.2 *Modems and Internet Telephony*

On current IP networks, H.323 packets carry headers of 40–54 bytes. In order to have a reasonably high ratio of payload to header, and therefore reasonable line efficiency, packets must be very large. At network bit rates above 1 Mbit/s, such large, efficient packets can be sent quickly.

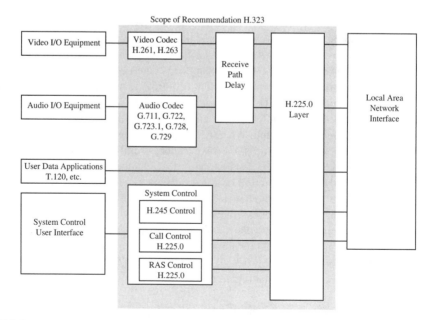

FIGURE 6.4
Block diagram of H.323 terminal.

At very low bit rates, these large packets can block transmission of small, urgent packets such as audio frames, thereby causing significant blocking delay. As a result, H.323 is not well suited to operation at very low bit rates, such as on V.34 modems that run at up to 33,600 bps. Alternatively, small packets can be used, but this can result in extremely poor efficiency. The IETF is addressing this problem with proposals for header compression [Bormann, 1997a; Casner and Jacobson, 1997] and mechanisms for urgent packets to interrupt large packets [Bormann, 1997b], but these techniques may take some years to be fully deployed in the public Internet. In the meantime, real-time Internet communication must trade off delay against efficiency when running at low bit rates.

For example, carriage of each 24-byte G.723.1 audio frame in a packet with 40 bytes of header results in the 6.4-Kbit/s audio codec occupying over 17 Kbit/s of network bandwidth. Reducing the header overhead to 5% would require grouping together 32 audio frames in each packet, representing 960 ms of audio, which would then take an additional 270 ms to transmit at a typical modem connect rate of 24,000 bps. Sending a video packet of the same size would block transmission of audio for 270 ms.

While this delay is not significant for one-way applications, it is a serious problem in two-way communications where data must be carried reasonably efficiently to achieve good video quality or net data throughput. If only an audio channel is to be transmitted, the high overhead associated with small, low-latency packets can be tolerated because it still results in a bit rate that can be carried by the modem.

This is how Internet telephony is possible today using H.323 over V.34 modems, even without header compression. The Voice on IP (VoIP) activity group of the International Multimedia Teleconferencing Consortium (IMTC) has selected H.323 as the standard for Internet telephony. When operating on V.34 modem links, the VoIP has recommended use of the G.723.1 audio codec, but H.323 itself suggests G.729 for low-bit-rate audio-only use and G.723.1 for low-bit-rate multimedia applications. As the only multivendor interoperable Internet telephony standard, H.323 is being supported by almost all manufacturers and service providers for Internet telephony. At this writing, it seems that G.723.1 will be the predominant low-bit-rate codec for H.323, but G.729 may eventually also be widely supported.

6.4.3.3   *Network Congestion and the Internet*

The other factor influencing H.323 performance is network congestion and delay.

Good-quality, low-latency H.323 performance requires networks that are lightly loaded or that are quality-controlled, such as Ethernet LANs with switched hubs and reservation protocols like RSVP [Braden et al., 1996]. Many corporate LANs and intranets are able to provide this, but the public Internet currently cannot.

H.323 can be used on the public Internet, and in some cases will work well, but network congestion problems can require very large delays (often well over 1 second each way) to buffer unpredictable packet arrival times, even when most packets arrive quickly.

Nevertheless, H.323 performance is limited only by the quality of service offered by the network.

Aside from the H.323 terminal, three additional major system components are described by H.323: gateways, gatekeepers, and multipoint controllers. Each of these components is optional in H.323 systems but provide important services.

6.4.3.4   *Gateways*

H.323 gateways provide interworking between H.323 terminals on the packet switched network and other network and terminal types, such as H.320 on ISDN, H.310 on ATM/B-ISDN, and H.324 and regular analog telephones on the PSTN.

Smooth, low-delay operation between H.323 and other H-series terminals should be possible through H.323 gateways, since all H-series conferencing systems support QCIF H.261 video in common, which ensures that time-consuming video transcoding is not needed.

Except for H.324, all H-series systems support G.711 log-PCM audio, but in many situations it will be desirable for gateways to use a more efficient audio codec when connecting an H.323 terminal to an H-series terminal on the ISDN or PSTN. Transcoding the G.711 audio to another codec saves limited bandwidth for use by video. This transcoding normally takes less time than is already needed to delay audio to synchronize with video, so if lip sync is being used, no additional net delay need be added by audio transcoding.

6.4.3.5   Gatekeepers

Gatekeepers are logical entities that grant or deny permission for terminals to make calls, as a way of controlling network traffic. When permitting a call, a gatekeeper can place a limit on the amount of bandwidth to be used for the call. Decision rules for gatekeepers are not defined by the standard and are under the control of implementors and system administrators. Another function of the gatekeeper is to locate users by translating alias addresses into network addresses. Terminals can be identified both with a network address, appropriate to the particular network type in use, and with an alias address, such as a telephone number or email address. Alias addresses can be used to provide users with an easily remembered address, perhaps the same as their office telephone number, and can provide an address to be used for incoming calls from an H.320 or H.324 terminal, which may not be aware of network addressing schemes. Similarly, the gatekeeper can locate gateways and direct calls through an appropriate gateway when needed.

6.4.3.6   Multipoint and Multicast

Unlike H.320 and H.324 multipoint, which requires all terminals to connect to a central MCU, H.323 allows the terminals in a multipoint conference to directly send audio and video streams to other participating terminals using multicast network mechanisms. H.323 multipoint calls can also use a traditional centralized MCU or a combination of both techniques [Thom, 1996]. These three modes are called *decentralized multipoint, centralized multipoint,* and *hybrid multipoint,* in which some media types are centrally mixed or switched, and others are multicast.

6.4.4   H.324 for Low-Bit-Rate Circuit Switched Networks

H.324, approved in 1995, was the first of the second-generation ITU-T multimedia conferencing standards. H.324 defines multimedia terminals for low-bit-rate circuit switched networks (CSNs), initially packet switched telephone network (PSTN) analog phone lines (often known as "plain old telephone service," or POTS) using V.34 modems at up to 33,600 bps. H.324 has been extended to other CSN networks, including ISDN and wireless networks (digital cellular and satellite). In the future, H.324 may eventually replace H.320 on ISDN networks, just as Group 3 fax replaced Group 2, while still retaining backward compatibility for many years.

Unlike the packet switched networks covered by H.323, CSNs are characterized by direct point-to-point synchronous data links, operating at constant bit rates over long periods of time. End-to-end latency across the CSN link is fixed, and no routing is performed, so there is no need for packet addressing or other overhead to cope with unpredictable arrival times or out-of-order delivery. Above all, H.324 was designed to provide the best performance possible (video and audio quality, delay, and so on) at low bit rates.

H.324 is a "toolkit standard," which allows implementors to choose the elements needed in a given application. Figure 6.5 illustrates the major elements of the H.324 system, which can support real-time video, audio, and data, or any combination. The mandatory components are the

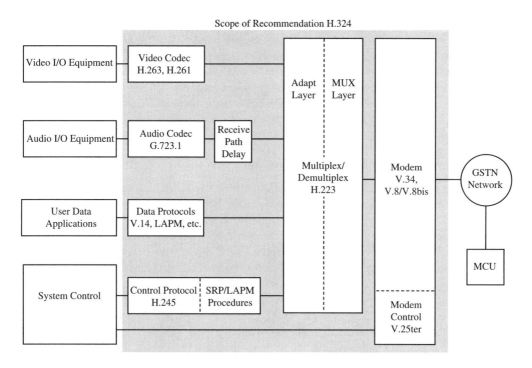

FIGURE 6.5
H.324 block diagram. (GSTN = General Switched Telephone Network; LAPM = Link Access Procedure for Modems; MCU = Multipoint Control Unit; SRP = Source Routing Protocol.)

V.34 modem (for PSTN use), H.223 multiplexer, and H.245 control protocol. Video, audio, and data streams are all optional, and several of each kind may be used simultaneously. H.324 enables a wide variety of interoperable terminal devices, including PC-based multimedia video-conferencing systems, inexpensive voice/data modems, encrypted telephones, World Wide Web browsers with live video, remote security cameras, and standalone videophones.

6.4.4.1   Modem
When operating on PSTN analog phone lines (POTS), H.324 uses the V.34 modem, which has a maximum speed of 33,600 bps. Typical telephone connections are too noisy to support this maximum rate, so the usual long-distance V.34 connect rate is in the range of 20–26 Kbit/s.

H.324 uses the modem directly as a synchronous data pump. V.42bis data compression and V.42 retransmission are not used at the modem level, although these same protocols can be used within an individual H.223 logical channel to support data applications. Since most PCs have only asynchronous RS-232 interfaces, PC-based H.324 terminals need a synchronous interface to the modem or another method to pass the synchronous bit stream to the modem.

The error burst behavior of V.34 modems is important to H.324 system design. V.34 provides bit rates between 2400 and 33,600 bps, in steps of 2400 bps. On a given telephone connection, each step changes the bit error rate by more than an order of magnitude. Most modems are designed for V.42 data transfer, so modem DSP is tuned for a target error rate such that retransmissions will consume, on average, less than 2400 bps. Since in H.324 errored audio is not retransmitted, it may be desirable to run the modem at the next lower 2400-bps rate step to avoid excessive audio dropouts or errors in decoded video.

6.4.4.2 Extensions

H.324 has been extended to operate on other networks besides the PSTN. When operating on these networks, the V.34 modem is replaced by another network interface. H.324/I specifies how to operate H.324 on ISDN networks, while H.324/M extends H.324 to operate on wireless mobile networks by adding variable amounts of forward-error-correction to cope with high bit-error rates. H.226 provides a low-delay channel aggregation method to combine multiple channels for a larger total bit rate. These extensions are specified annexes to the H.324 standard.

6.4.5 H.310 for ATM and B-ISDN Networks

H.310 defines a native ATM/B-ISDN standard for videoconferencing, as well as one-way video for applications like video surveillance. Broadband ISDN (B-ISDN) is based on asynchronous transfer mode (ATM) and operates at rates of up to several hundred megabits per second.

Like the other second-generation H-series standards, H.310 uses the H.245 control protocol, supports T.120, and was designed with interworking with other H-series terminals in mind.

H.310 includes as a subset the H.321 ATM network adaptation standard for H.320, described earlier. This is included primarily for interworking with H.320 and H.321 terminals. The native H.310 mode is based on MPEG-2 systems and video coding. Terminals are required to support MPEG-2 video coding (H.262 ‖ ISO/IEC 13818-2) as well as H.261 video and G.711 audio for two-way conversational use. H.263 is optional. MPEG-1 audio (ISO/IEC 11172-3 Layer II) is mandatory for unidirectional terminals and optional for conversational terminals [Okubo et al., 1997].

H.310 terminals can operate at a wide variety of data rates, but all terminals must support the common rates of 6.144 and 9.216 Mbit/s, corresponding to the MPEG-2 Main Profile at Main Level for medium- and high-quality services, respectively. Multipoint operation is possible in H.310 using a centralized multipoint control unit (MCU), as with H.320.

6.5 T.120 FOR DATA CONFERENCING AND CONFERENCE CONTROL

The ITU-T T.120 series standardizes data and graphics conferencing as well as high-level conference control. T.120 supports point-to-point and multipoint video, audio, data, and graphics conferencing with a sophisticated, flexible, and powerful set of features, including support of nonstandardized application protocols. T.120 multipoint data transport is based on a hierarchy of T.120 MCUs that route data to its proper destination.

Early in its development, T.120 was called Multi-Layer Protocol (MLP), after the data channel MLP in H.221 intended to carry T.120.

T.120 is network independent, so terminals using H.320 on ISDN, H.323 on packet networks, H.324 on PSTN, H.310 on ATM, voice/data modems, and so on, can all join in the same T.120 conference. Figure 6.6 illustrates the T.120 protocol stack.

6.6 SUMMARY

This chapter has discussed the two generations of ITU-T H-series standards for real-time multimedia conferencing and the T.120 standard for data/graphics conferencing and multimedia conference control.

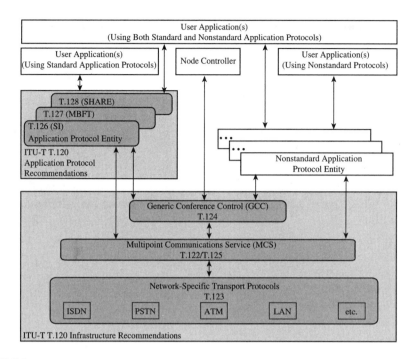

FIGURE 6.6
ITU-T.120 system model.

Multimedia and videoconferencing systems have so far achieved practical and commercial success in specific applications such as business meetings, telecommuting, and education. Consumer desire for video telephony is still mostly speculative, but at a minimum, issues of cost, end-to-end delay, video quality, and ease of use will have to be addressed before multimedia communication can hope to supplant traditional telephone service.

The standards described in this chapter, together with efforts by manufacturers and industry organizations like International Multimedia Teleconferencing Consortium (IMTC), seem likely to address the ease-of-use and interoperability issues. The cost of the necessary hardware is steadily declining and will reach consumer levels very soon. But major improvements in end-to-end delay and video quality may require higher bit rates than those available to consumers thus far. Happily, new high-bit-rate access technologies such as cable modems and xDSL seem to be coming quickly. These, possibly combined with effective Internet resource allocation mechanisms, may allow a real revolution in the communication tools we use every day.

6.7 REFERENCES

This chapter was adapted from J. Gibson, T. Berger, D. Lindbergh, and R. Baker, *Digital Compression for Multimedia* (San Francisco: MKP, 1998).

Bormann, C. 1997a (date accessed). The multi-class extension to multi-link PPP. Internet draft, May. ftp://ietf.org/internet-drafts/draft-ietf-issll-isslow-mcml-02.txt

Bormann, C. 1997b (date accessed). PPP in a real-time oriented HDLC-like framing. Internet draft, July. ftp://ietf.org/internet-drafts/draft-ietf-issll-isslow-rtf-01.txt

Bormann, C. 1997c (date accessed). Providing integrated services over low-bitrate links. Internet draft, May. ftp://ietf.org/internet-drafts/draft-ietf-issll-isslow-02.txt

Braden, R., L. Zhang, S. Berson, S. Herzog, and S. Jamin. 1997 (date accessed). Resource reservation protocol (RSVP)-Version 1 functional specification. Internet draft, June. ftp://ds.internic.net/internet-drafts/draft-ietf-rsvp-spec-16.txt

Casner, S., and V. Jacobson. 1997 (date accessed). Compressing IP/UDP/RTP headers for low-speed serial links. Internet draft, July. ftp://ietf.org/internet-drafts/draft-ietf-avt-crtp-03.txt

Okubo, S., S. Dunstan, G. Morrison, M. Nilsson, H. Radha, D. Skran, and G. Thom. 1997. ITU-T standardization of audiovisual communication systems in ATM and LAN environments. *IEEE Journal on Selected Areas in Communications,* August.

Schulzrinne, H., S. Casner, R. Frederick, and V. Jacobson. 1996 (date accessed). RTP: A transport protocol for real-time applications. IETF RFC 1889, January. ftp://ds.internic.net/rfc/rfc1889.txt

Thom, G. 1996. H.323: The multimedia communications standard for local area networks. *IEEE Comm. Magazine* 34(12):52–56.

CHAPTER 7

MPEG-1 and -2 Compression

TOM LOOKABAUGH

7.1 INTRODUCTION

The Moving Pictures Experts Group (MPEG), an international standards committee, has produced a set of standards for audio and video compression that are good examples of sophisticated compression strategies and that are also commercially important. Starting from an initial focus on video compression at approximately 1.2 Mbit/s on compact disc read-only memory (CD-ROM), the MPEG standards have expanded to become the lingua franca of high quality audio-video compression in the 1990s.

The focus of this chapter is the completed work embodied in the MPEG-1 and -2 standards. As of this writing, MPEG continues to work to produce new standards in several important areas, including advanced audio compression (AAC), digital storage media command and control, MPEG-4, and MPEG-7. This chapter is not an exhaustive restatement of the MPEG-1 and MPEG-2 standards (which run to several hundred pages); rather, it describes the major methods by which MPEG achieves compression and organizes information, to illuminate the potential effectiveness and applications of the standards.

7.2 THE MPEG MODEL

A key to understanding MPEG is understanding both the problems that MPEG set out to address in developing MPEG-1 and -2 (although it is likely MPEG will be applied in many unanticipated places as well) and the fundamental models that underlie the algorithms and that are used to foster interoperability.

7.2.1   Key Applications and Problems

Some of the most important applications to drive the development of MPEG include disk-storage-based multimedia, broadcast of digital video, switched digital video, high-definition television, and networked multimedia.

MPEG-1 had its genesis as the solution to a very specific compression problem: how to best compress an audio-video source to fit into the data rate of a medium (CD-ROM) originally designed to handle uncompressed audio alone. At the time MPEG started, this was considered a difficult goal. Using an uncompressed video rate for 8-bit active video samples [Comité Consultatif International des Radiocommunications or International Telecommunications Union—Radio (CCIR)-601 chroma sampling)] of approximately 210 Mbit/s, this requires a rather aggressive 200:1 compression ratio to achieve the slightly greater than 1 Mbit/s or so available after forward error correction and compressed audio on a typical CD-ROM.

Aside from the substantial compression requirement, another requirement of many videos on CD-ROM applications is a reasonable random access capability, that is, the ability to start displaying compressed material at any point in a sequence with predictable and small delay. This is a key attribute, for example, of many interactive game and training materials.

More recently, a new optical disk format with much higher capacity than CD-ROM has been developed. Called digital versatile disc (DVD) (renamed from the original "digital video disc" to include data-storage applications), its higher rates, combined with the use of variable-rate encoding, can provide video quality that surpasses that currently available to consumers through other video media [(e.g., Video Home System (VHS) tape, laserdisc, cable and off-air analog television], as well as cinematic audio and a number of unique access and format features.

7.2.2   Strategy for Standardization

A key purpose of a standard is to facilitate interoperability. A good standard achieves this while maximizing support for current and as-yet-unknown applications and the development of a variety of implementations. For MPEG, this is achieved by focusing standardization on two related questions:

- What is a legal MPEG bit stream?
- What should be done with it to generate displayable audio-video material?

The first question is answered by a careful specification of the legal syntax of an MPEG bit stream, that is, rules that must be followed in constructing a bit stream. The second is answered by explaining how an idealized decoder would process a legal bit stream to produce decoded audio and video.

This particular strategy for standardization allows a great deal of latitude in the design of encoding systems because the range between truly efficient bit streams (in terms of quality preserved versus bits spent) and inefficient but legal bit streams can be large. There is less latitude in decoder design, since the decoder must not deviate from the idealized decoder in terms of the bit streams it can process, but there is still room for different and clever implementations designed to reduce cost in specific applications, improve robustness to slightly damaged bit streams, and provide different levels of quality in post-MPEG processing designed to prepare a signal for display (e.g., interpolation, digital-to-analog conversion, and composite modulation).

The actual details of what constitutes a legal bit stream are largely conveyed through the specification of the syntax and semantics of such a bit stream.

7.3 MPEG VIDEO

The MPEG video algorithm is a highly refined version of a popular and effective class of video compression algorithms called Motion-Compensated Discrete Cosine Transform (MC-DCT) algorithms. While not universally used for video compression, these algorithms have served as a basis for many proprietary and standard video compression solutions developed in the 1980s and 1990s. The algorithms use the same basic building blocks, including:

- temporal prediction: to exploit redundancy between video pictures
- frequency domain decomposition: the use of the DCT to decompose spatial blocks of image data to exploit statistical and perceptual spatial redundancy
- quantization: selective reduction in precision with which information is transmitted to reduce bit rate while minimizing loss of perceptual quality
- variable-length coding: to exploit statistical redundancy in the symbol sequence resulting from quantization as well as in various types of side information

These basic building blocks provide the bulk of the compression efficiency achieved by the MPEG video algorithm. They are enhanced, however, by a number of detailed special techniques designed to absolutely maximize efficiency and flexibility.

7.3.1 The Basic Algorithm

Figure 7.1 shows the basic encoding and decoding algorithms. The incoming video sequence is preprocessed (interpolated, filtered), then motion estimation is used to help form an effective predictor for the current picture from previously transmitted pictures. The motion vectors from

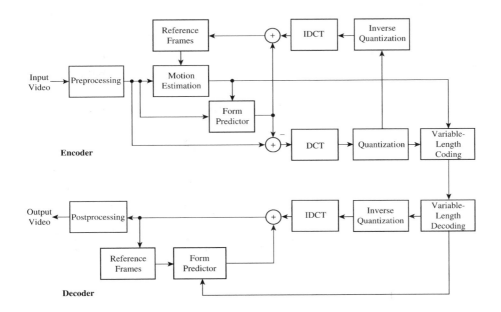

FIGURE 7.1
Block diagram of MPEG encoding and decoding.

motion estimation are sent as side information if used. The predictor for each block is subtracted, and the resulting prediction residual undergoes a Discrete Cosine Transform (DCT). The DCT coefficients are quantized, and the quantized coefficients are variable-length-coded for transmission. The quantized coefficients also undergo reconstruction, inverse DCT, and combination with the predictor, just as they will in the decoder, before forming reference pictures for future motion estimation and prediction.

The decoder decodes the variable-length codes, performs reconstruction of DCT coefficients, inverse DCT, formation of the predictor from previous reconstructed pictures, and summing to form the current reconstructed picture (which may itself serve to predict future received pictures). Postprocessing interpolates and filters the resulting video pictures for display.

7.3.1.1 Representation of Video

The video that MPEG expects to process is composed of a sequence of frames or fields of luma and chroma.

Frame-Based Representation MPEG-1 is restricted to representing video as a sequence of frames. Each frame consists of three rectangular arrays of pixels, one for the luma (Y, black and white) component, and one each for the chroma (Cr and Cb, color difference) components. The luma and chroma definitions are taken from the CCIR-601 standard for representation of uncompressed digital video.

The chroma arrays in MPEG-1 are subsampled by a factor of two both vertically and horizontally relative to the luma array. While MPEG does not specify exactly how the subsampling is to be performed, it does make clear that the decoder will assume subsampling was designed so as to spatially locate the subsampled pixels according to Figure 7.2 and will perform its interpolation of chroma samples accordingly.

Typically, MPEG-2 expects chroma subsampling to be consistent with CCIR-601 prescribed horizontal subsampling. Spatially, this implies the chroma subsampling pattern shown in Figure 7.3, termed 4:2:0 sampling.

Field-Based Representation MPEG-2 is optimized for a wider class of video representations, including, most importantly, field-based sequences. *Fields* are created by dividing each frame into a set of two interlaced fields, with odd lines from the frame belonging to one field and even lines to the other. The fields are transmitted in interlaced video one after the other, separated by

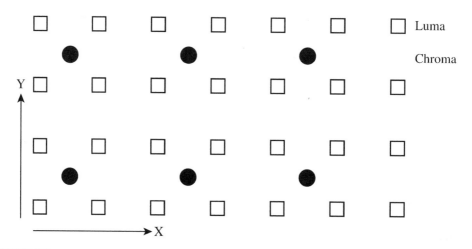

FIGURE 7.2
Relationship between luma and chroma subsampling for MPEG-1.

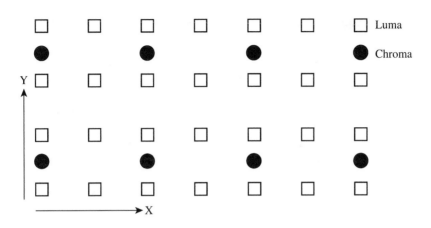

FIGURE 7.3
Relationship between luma and chroma subsampling for MPEG-2.

half a frame time. Interlacing of video is in fact a simple form of compression by subsampling. It exploits the fact that the human visual system is least sensitive to scene content that has both high spatial and temporal frequencies (such as a fast-moving item with much detail). An interlaced source cannot represent such scenes effectively, but can build up the full detail of a frame (within two field times) and can also update low-resolution items that are changing every field time; these latter types of material are the most visually important.

For field-based sequences, MPEG-2 expects the chroma associated with each field to be vertically subsampled within the field, yet maintain an expected alignment consistent with frame-based sequences. This leads to a vertical resampling pattern as shown in Figure 7.4.

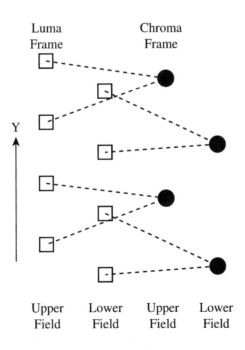

FIGURE 7.4
Relationship between luma and chroma samples vertically and in time for field-based video material.

7.3.2   Temporal Prediction

Temporal prediction exploits the intrinsic redundancy between video pictures. For instance, the background of a scene may be unchanged, with only disturbances in the foreground. In such a case, application of the classic Differential Pulse Code Modulation (DPCM) compression strategy can provide substantial compression, with the prediction being applied to entire video pictures. The typical application of this strategy uses the last picture to predict the current picture and is called *picture differencing*.

We can go beyond picture differencing, however, by noting that when there are changes in the current picture, many times they are caused by the motion of objects. Another common occurrence is motion of the entire picture caused by panning of a camera. Such observations motivate the inclusion of a motion model in the prediction to further reduce the amount of spatial information that must be encoded as a prediction residual. Motion compensation, then, is the application of a motion model to prediction.

7.3.2.1   Motion Compensation

The motion model used by MPEG is blockwise translation. For simplicity, the blocks are chosen to be a single fixed size: 16×16 pixels in the luma component. Since for typical 4:2:0 chroma subsampling, the chroma components are vertically subsampled by a factor of 2 in both dimensions relative to the luma, each chroma component includes an 8×8 block of pixels corresponding to the same spatial region as a 16×16 block from the luma picture. The collection of the 16×16 region from the luma and the two corresponding 8×8 regions from the chroma components is called a *macroblock*.

Motion compensation consists of taking a macroblock from the current picture and determining a spatial offset in the reference picture at which a good prediction of the current macroblock can be found. The offset is called a *motion vector*. The mechanics of motion compensation are to use the vector to extract the predicting block from the reference picture, subtract it, and pass the difference on for further compression. The vector chosen for a macroblock is applied directly to determine the luma predictor, but is scaled by half in both dimensions (corresponding to their subsampled size) before being applied to find the chroma predictors.

Motion estimation consists of finding the best vector to be used in predicting the current macroblock. This is typically the most computationally expensive activity in an MPEG encoder, but can have a substantial impact on compression efficiency.

The effectiveness of motion compensation can be enhanced by allowing the search for an effective prediction region in the reference picture to include not only positions at integral pixel offsets but also fractional pixel offsets. For a fractional pixel offset, the predicting macroblock is constructed by linearly interpolating pixel values relative to the nearest actual pixels. An example of a 1/2, 1/2 pixel offset prediction block is shown in Figure 7.5. The dark circle locations have pixel values x calculated as

$$x = (a + b + c + d) / 4 \qquad\qquad (1)$$

where *a, b, c,* and *d* are the closest pixels in the original reference picture.

MPEG allows half-pixel interpolation vertically, horizontally, and both (the previous example being both).

7.3.2.2   Picture and Macroblock Prediction Types

One of the key requirements of MPEG is reasonable support for random access. Random access into an arbitrary motion-compensated sequence is difficult for the same reason that starting

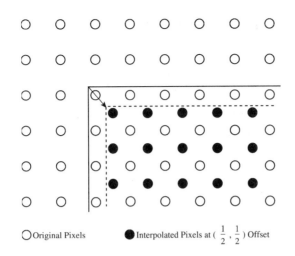

○ Original Pixels   ● Interpolated Pixels at ($\frac{1}{2}$, $\frac{1}{2}$) Offset

FIGURE 7.5
Example of half-pixel interpolation in motion compensation.

decoding in a delta modulation or Differential Pulse Code Modulation (DPCM) sequence is difficult: the received sequence consists of a lot of information about changes from one picture to
another, but not necessarily the starting point or original reference. This is a dangerous situation
in any case from the point of view of dealing with transmission imperfections; the effect of such
imperfections on the decoded images can persist indefinitely. These are usually managed
through one of two main strategies: refresh and leaky prediction.

Refresh involves periodically sending an entire picture (picture-based refresh) or portion of a
picture (region-based refresh) without any prediction, allowing the decoder to resynchronize at
that point. Region-based refresh is achieved by guaranteeing that a particular subset of macroblocks are encoded without prediction and rotating this subset through subsequent pictures
used in prediction until all macroblocks have been forced to be encoded without prediction. This
mechanism reduces the burst of bits required for a picture compressed entirely without prediction, but does not provide a guaranteed random-access entry point or a deterministic time to
complete refresh (since nonrefreshed material may be shifted into a just-refreshed area by
motion compensation before the refresh cycle is complete).

Leaky prediction involves slightly reducing the effectiveness of the prediction in order that
the decoder will progressively "forget" any perturbations in the decoded sequence caused by
transmission errors. Both encoder and decoder predictors are multiplied by an agreed upon leak
factor less than 1. Since this reduces the accuracy of the predictor, it causes an increase in the
bits required to transmit the compressed residual. However, any discrepancy between decoder
and encoder predictor memories is reduced during each iteration of prediction, resulting in an
exponential decay in the difference.

Only picture-based refresh aids in random access, however, as it allows a predictable location
to begin decoding and a bounded time to decode to any target picture.

Given the likelihood that streams would be generated with the regular use of nonpredicted
pictures, the MPEG committee decided to investigate some new prediction strategies that might
further improve compression efficiency and developed the concept of forward and backward
prediction. We first define an *anchor picture* as one that can be used for prediction. Then we
define four different prediction strategies that can be used for a particular macroblock, as shown
in Table 7.1. Each picture has a declared type that limits the type of macroblock allowed in that
picture, as shown in Table 7.2.

Table 7.1   Allowed Prediction Modes by Macroblock Type

Macroblock Type	Prediction
Nonpredicted macroblock	None
Backward-predicted macroblock	References temporally nearest subsequent anchor picture
Forward-predicted macroblock	References temporally nearest previous anchor picture
Bidirectionally predicted macroblock	Averages predictions from temporally nearest previous and subsequent anchor pictures

Table 7.2   Allowed Macroblock Types and Anchor Picture Definition by Picture Type

Picture Type	Anchor Picture	Macroblock Types
I	Yes	Nonpredicted
P	Yes	Nonpredicted
		Forward predicted
B	No	Nonpredicted
		Forward predicted
		Backward predicted
		Bidirectionally predicted

Figure 7.6 uses arrows to show which pictures can be referenced for motion estimation in a typical sequence of pictures. In fact, while not required by the standard, the set of sequences in which there is a fixed spacing between I-pictures and between anchor pictures (I- and P-pictures) is widely used, so widely that a pair of parameters is commonly used to describe these spacings: N is the number of pictures from one I-picture (inclusive) to the next (exclusive), and M is the number of pictures from one anchor picture (inclusive) to the next (exclusive). So, the pattern shown in Figure 7.6 would be called an $N = 9$, $M = 3$ pattern.

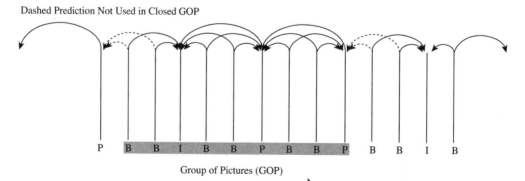

FIGURE 7.6
Typical group of pictures and allowed predictors for pictures.

There are several implications to this set of possible predictions:

- Using bidirectionally predicted blocks allows effective prediction of uncovered background, areas of the current picture that were not visible in the past but are visible in the future.
- Bidirectional prediction can provide for interpolation equivalent to an even finer degree than the half-pixel interpolation for motion compensation already allowed.
- Bidirectional prediction can reduce noise in the predictor.
- Bidirectional prediction increases motion estimation complexity in two ways: First, it obviously requires twice as much work to perform motion estimation in two different anchor pictures. More significantly, though, since anchor pictures are now possibly separated by several picture times, we need a larger overall motion estimation range in order to track objects of a given velocity.
- Since B-pictures cannot themselves be used to predict other pictures, the quality of compression is solely determined by the visual acceptability of the result. For P- and I-pictures, since the reconstruction will also be used for other predictions, quality must reflect both perceptual fidelity and prediction efficiency. This allows a substantial reduction in the bits allocated to B-pictures relative to I- and P-pictures. A ratio of 5:3:1 in bits spent on I-, P-, and B-pictures is not uncommon. These ratios can and should vary dynamically, though.

7.3.2.3   *Display and Transmit Order*

Combining the use of B-pictures with a desire to economize on the use of memory in high-volume decoders leads to a difference in the order in which pictures are transmitted versus the order in which they are displayed. The reordering is intended to ensure that, regardless of the spacing between anchor pictures, a decoder need only have enough memory to store two anchor pictures and the scratch memory for the picture currently being decoded. This is achieved by transmitting the anchor picture's compressed version before those of the B-pictures that will precede it in display.

7.3.2.4   *Field and Frame Prediction*

While MPEG-1 does not consider the possibility of field-based video representations, MPEG-2 specifically does allow for increased compression efficiency on interlaced material. One way in which this is accommodated is by allowing either field- or frame-based prediction for each macroblock (often termed *adaptive field/frame compression*). In the case of field-based prediction, the macroblock is divided into two field macroblocks. The top field macroblock is predicted from the top fields in one or two anchor pictures using the modes appropriate for the picture type. The lower field macroblock is predicted from the lower fields in one or two anchor pictures. Note that the subdivision of chroma here corresponds to the chroma subsampling constraints for field representation described previously.

Field-based prediction can be highly effective in instances of substantial horizontal acceleration. In these cases, an object captured in two different fields appears to have serrated vertical edges when viewed as a frame, with the depth of the serration varying with time due to acceleration. Frame-based motion estimation is ineffective in this case and leaves a good deal of difficult-to-compress high-frequency serrations left over. Field-based prediction can perform quite well. On the other hand, for more benign motion cases, frame-based prediction can more effectively exploit vertical redundancy. Allowing a choice between the modes to occur on a macroblock basis gives the best of both worlds at the cost of a slight increase in overhead (the net effect being in favor of allowing the choice).

7.3.3   Frequency Domain Decomposition

After motion compensation has been performed, the residual information (or original picture information in the case of a nonpredicted block) is transformed using the DCT. To prepare the macroblock for transformation, six 8×8 blocks are extracted from the macroblock, four from the luma picture and one each from each of the chroma pictures. The luma blocks for MPEG-1 and the frame DCT mode of MPEG-2 are shown in Figure 7.7; the field DCT mode of MPEG-2 is shown in Figure 7.8.

The field DCT mode is useful for much the same reason that field motion compensation is useful: Material with a high degree of horizontal motion, especially acceleration, captured using

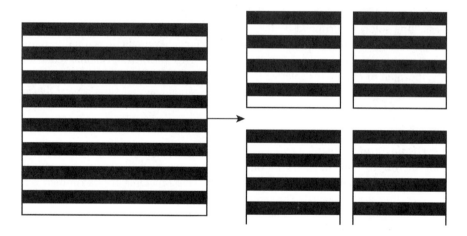

FIGURE 7.7
Mapping from 16×16 block to 8×8 blocks for frame-organized data.

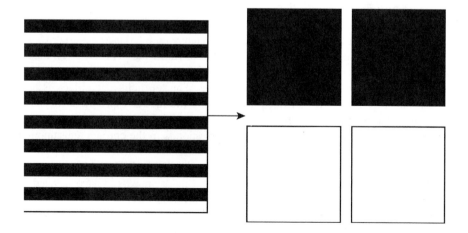

FIGURE 7.8
Mapping from 16×16 block to 8×8 blocks for field-organized data.

an interlaced video standard, tends to produce difficult-to-compress high frequencies when transformed using a frame-based DCT. The choice of frame versus field mode for motion compensation and DCT is not bound together, and there is a small benefit to be had by making the decision independently.

7.3.3.1 *Discrete Cosine Transform*

The MPEG standard does not specify exactly how to perform the 8×8 DCT and Inverse Discrete Cosine Transform (IDCT) required. This is too limiting given that many different software and hardware implementations of MPEG are expected. Instead, MPEG calls for the IDCT as implemented to be able to pass a test developed as part of the ITU-T standardization of H.261, a videoconferencing standard. In this test, Institute of Electrical and Electronics Engineers (IEEE) P1180/D2, the proposed IDCT inverse transforms a particular set of blocks, and the results are compared to an idealized set of results. Providing the proposed IDCT has less than a specified number of discrepancies at each of several magnitudes, it is an acceptable IDCT.

Allowing similar but not identical IDCTs in different implementations will cause mismatch in the encoders and decoders of different manufacturers. Since the encoder includes an IDCT for the purpose of reproducing the pictures that the decoder generates, the fact that an encoder and decoder have slightly different DCTs will introduce a gradually increasing difference between these pictures, resulting in inappropriate coding decisions on the encoder's part. This phenomenon is managed by limiting to 132 the number of times a specific macroblock location may be coded as predicted in P-pictures before that macroblock location must be coded without prediction. Moreover, MPEG-2 added some specific quantization rules that were found to further mitigate the problem of mismatch.

7.3.4 Quantization

After transformation, the DCT coefficients typically have 12 bits or more of precision. On the face of it, this is not a very auspicious compression, since the data started with 8 bits per pixel. Quantization is the key step that exploits the efforts provided by motion compensation and the DCT to reduce the bits required, since the information has now been organized so that many coefficients are essentially irrelevant to the final reproduction quality and only a few need be treated with care.

7.3.4.1 *Quantizer Step Size*

MPEG applies a uniform step-size quantizer to each coefficient. However, the step size of the quantizer may vary from coefficient to coefficient and macroblock to macroblock. In fact, the quantizer step size is determined at the decoder by the following equation:

$$ss = qf\ [m,n] \times qs \qquad (2)$$

The factor $qf\ [m,n]$ is dependent on the location of the coefficient within a block. The factor qs is the base quantizer step size. This allows emphasis on more perceptually important lower frequencies. MPEG provides two default weighting matrices $qf\ [m,n]$ for use on predicted blocks and nonpredicted blocks, shown in Figures 7.9 and 7.10. Note that for nonpredicted blocks, the DCT DC coefficient, the upper left-hand coefficient that is proportional to the average value of the space domain block, is not quantized using a weighting matrix value; hence that location in the weighting matrix has an * in Figure 7.9. This unusual treatment reflects the fact that the eye is highly sensitive to errors in the DC level of nonpredicted blocks and tends to quickly recognize these as blocking or tiling distortion in the reconstructed picture.

*	16	19	22	26	27	29	34
16	16	22	24	27	29	34	37
19	22	26	27	29	34	34	38
22	22	26	27	29	34	37	40
22	26	27	29	32	35	40	48
26	27	29	32	35	40	48	58
26	27	29	34	38	46	56	69
27	29	35	38	46	56	69	83

16	16	16	16	16	16	16	16
16	16	16	16	16	16	16	16
16	16	16	16	16	16	16	16
16	16	16	16	16	16	16	16
16	16	16	16	16	16	16	16
16	16	16	16	16	16	16	16
16	16	16	16	16	16	16	16
16	16	16	16	16	16	16	16

FIGURE 7.9
Intraframe quantizer weighting matrix.

FIGURE 7.10
Interframe quantizer weighting matrix.

7.3.5   Variable-Length Coding

Both the quantized coefficients and several different types of side information (macroblock prediction type, motion vectors, etc.) exhibit statistical concentration so that the overall average bit rate can be lowered by using variable-length coding. Note that the use of variable-length coding will necessitate buffering when MPEG is to operate over a fixed-rate channel.

7.3.5.1   Modified Huffman Codes

The type of variable-length code used throughout MPEG is a modified Huffman code. Huffman coding provides optimal variable-length coding for the chosen alphabet. However, for large alphabets, storing, encoding, and decoding the Huffman code can be expensive. Hence, the table size is generally restricted to a subset consisting of the most probable symbols. An additional codeword is added to the code, termed "ESCAPE." When an input symbol is observed that does not belong to the chosen high-probability symbol set, it does not have its own Huffman codeword; instead, the ESCAPE codeword is issued followed by the explicit (i.e., standard indexed value) input symbol. A second modification to the MPEG Huffman codes for video is that the all-zero codeword is not used to avoid emulation of a start code via a legitimate sequence of other codes.

7.3.5.2   Two-Dimension to One-Dimension Conversion

The quantized DCT coefficients form the bulk of the material that needs to be variable-length encoded. Empirically, though, after quantization, the 8×8 blocks of DCT coefficients tend to exhibit a substantial number of zeros, particularly in the higher frequencies, as in Figure 7.11.

MPEG exploits this phenomenon by first converting the 8×8 array into a one-dimensional array and then applying coding that can exploit long runs of zeros. Both MPEG-1 and MPEG-2 use a zigzag scan for cases, such as in Figure 7.11, where there are a large number of zeros in

	Increasing Horizontal Frequency →						
	−42	12	3	−1	0	0	0	1
Increasing	−19	14	−8	0	2	0	0	0
Vertical	16	8	−9	3	0	0	0	0
Frequency	3	−2	0	0	1	0	0	0
↓	3	0	0	1	−1	0	0	0
	0	0	0	0	0	0	0	0
	0	0	0	0	0	0	0	0
	0	0	0	0	0	0	0	0

FIGURE 7.11
Example of an 8×8 array of quantized DCT coefficients.

the lower right-hand quadrant. MPEG-2 provides an alternative, vertical scan, which is often more effective, especially when a field-based DCT has been applied. The scan path of zigzag and vertical scans is shown in Figure 7.12.

The decoder will perform the inverse scanning as part of its reconstruction process.

7.3.5.3   *Runlength Amplitude Coding*
Once a 64-symbol vector has been created by scanning the 8×8 array for each block, runlength amplitude coding is used. First, the DC quantized coefficient (containing the block average) is treated specially, receiving its own dedicated Huffman code, because the DC quantized coefficient tends to have a unique statistical characteristic relative to other quantized coefficients. Moreover, since there is some redundancy between adjacent DC quantized coefficients in nonpredicted blocks, only the difference between these is Huffman coded. The remaining quantized coefficients are parsed into a sequence of runs, where a run is defined as zero or more 0s followed by a single nonzero value. This becomes a new symbol for Huffman coding, consisting of a runlength-amplitude pair.

7.3.5.4   *Side Information Coding*
There are a variety of different types of side information, designed in general to reduce the number of blocks that actually have to undergo runlength-amplitude coding and to reduce the amount of information in those that do. The most important of these are:

Macroblock address
Macroblock type
Coded pattern
Motion vectors

7.3.6   Rate Control

One of the most important encoder design issues is how to include an effective rate control system into a compressor. As mentioned earlier, once variable-length coding has been applied, some

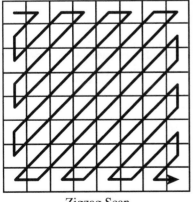
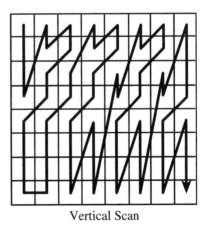

Zigzag Scan   Vertical Scan

FIGURE 7.12
Coefficient scan pattern for zigzag and vertical scans.

buffering is required to absorb instantaneous fluctuations in bit rate if the stream is to be delivered over a fixed-rate channel. Such buffering must be available at both encoder and decoder. A tenet of MPEG is that once the decoder buffer size has been specified, it must be possible to decode a legal bit stream in that buffer without overflowing or underflowing (except for the low-delay mode of MPEG-2, which does allow underflows).

7.3.6.1   Fixed-Rate Channels

For fixed-rate channels, with a little effort, the requirement on the decoder buffer can be made equivalent to a requirement on the encoder buffer as follows:

- If the encoder and decoder buffer are of the same size,
- and the encoder compresses units at the pace at which they arrive,
- and the decoder decompresses units at the pace at which they are to be displayed,
- and the decoding starts when the decoder buffer's fullness equals the buffer size minus the encoder's buffer fullness,
- and the encoder ensures that the encoder buffer never overflows (by throttling information when necessary) and never underflows (by using stuffing to send filler bits when necessary),
- then the decoder buffer will never underflow or overflow.

7.3.6.2   Variable-Rate Channels

MPEG supports variable-rate channels. The overriding constraint is that the decoder's buffer not overflow or underflow during decoding. Variable-rate channels come in two major classes: slave and master. A slave variable-rate channel has the rate selected by the MPEG system. A master variable-rate channel tells the MPEG system what the current rate is. Variable-rate channels are often also constrained by a maximum and minimum allowable instantaneous bit rate.

7.3.7   Constrained Parameters, Levels, and Profiles

The MPEG video algorithm provides an enormous flexibility in terms of the image sizes, bit rates, and other key parameters that can be supported. It is unreasonable to expect simple applications to bear the cost of decoding at the substantially higher complexity of the worst-case use of the standard, particularly if the simpler applications will be built in highly cost-sensitive high-volume applications. Hence, MPEG has defined compliance points within the standard.

For MPEG-1, the *constrained parameters* case provides the following set of specific restrictions that are reasonable for the kind of moderate-resolution, multimedia applications in the 1–3 Mbit/s range for which MPEG-1 was optimized.

- Horizontal size less than or equal to 720 pixels
- Vertical size less than or equal to 576 pixels
- Total number of macroblocks/picture less than or equal to 396
- Total number of macroblocks per second less than or equal to $396 \times 25 = 330 \times 30$
- Picture rate less than or equal to 30 pictures per second
- Bit rate less than or equal to 1.86 Mbit/s
- Decoder buffer less than or equal to 376,832 bits

Although the first two constraints seem to allow large image sizes (up to the full resolution of broadcast digital CCIR-601), the third and fourth constraints are more limiting in this regard. For instance, the Standard Intermediate Format (SIF) sizes of 352H × 240V (NTSC countries) and 352H × 288 (PAL countries) just satisfy these constraints; a CCIR-601 resolution would not. Much of the commodity silicon targeted at MPEG-1 applications is designed to the constrained parameters level.

MPEG-2 is designed to address a much larger number of potential applications, and in order to be cost-effective in each, it has a much larger set of compliance points. These are indexed by *levels,* which set a rough limit on processing power based on image size, and *profiles,* which restrict the algorithmic features available. Table 7.3 shows the grid of compliance points for MPEG-2.

The compliance point addressed by a particular system is stated as "*x* profile at *y* level." The most popular compliance point is, not surprisingly, main profile at main level (or MP ML, for short). Following are brief discussions of each of the levels:

- Low: This sets constraints on image size of approximately half CCIR-601 vertically and horizontally, namely, 352H × 288V.
- Main: CCIR-601 size images, 720 × 576V × 25Hz or 720H × 480V × 29.97Hz.
- High 1440: HDTV up to 1440H × 1152V.
- High: HDTV up to 1920H × 1152V.

Following are brief descriptions of the profiles:

- Simple: No B-pictures. This allows a reduction in the memory required at decoders of as much as 8 Mbits, at some cost in bit-rate efficiency.
- Main: The material in this chapter describes most of the aspects of the main profile.
- SNR: Adds a type of scalability based on transmitting two bit streams. The first, when decoded, will provide a reasonable reproduction. The second provides additional information to refine the reconstructed DCT coefficients, so that when both are decoded together, an excellent reconstruction results. Such scalability allows simpler and more complex decoders to share the same bit stream.
- Spatial: Adds a different type of scalability, in which the first stream provides a low-resolution reproduction and the combination of both streams provides a full-resolution reproduction.
- High: Allows SNR, spatial, and an additional type of scalability based on the first stream providing a low picture-rate reconstruction and both streams providing full picture rate, as well as a number of other high-complexity options.

Table 7.3 Defined Levels and Profiles for MPEG-2

	Profile				
Level	Simple	Main	SNR	Spatial	High
High		X			X
High 1440		X		X	X
Main	X	X	X		X
Low		X	X		

7.4   SUMMARY

The MPEG standard suite is a comprehensive set of video and audio decompression algorithm specifications, as well as specifications of how the compressed bit streams can be multiplexed together and how the resulting decoded media can be synchronized. MPEG is used in a number of substantial commercial applications, including playback of media from disk storage, digital broadcast of audio-video programming over a variety of channels, point-to-point switched connections for delivery of digital audio-video material, high-definition television, and networked multimedia, as well as in a large number of specialized commercial and noncommercial applications.

MPEG compression relies on a number of fundamental techniques described in this chapter and in other specific applications. For video, these include temporal prediction from video frame to video frame based on a simple model of motion, the application of the DCT to prediction residuals, quantization to select the extent to which information about individual DCT coefficients will be retained, and the application of both runlength and Huffman lossless variable-rate codes to extract remaining statistical redundancy in the quantized DCT coefficients.

CHAPTER 8

MPEG-4 and MPEG-7

JERRY D. GIBSON

8.1 INTRODUCTION

The activities of the Moving Pictures Experts Group (MPEG), a subcommittee of the International Standards Organization (ISO), have led to the establishment of two influential standards, widely known as MPEG-1 and MPEG-2. MPEG-1 and MPEG-2 both began by targeting major killer applications, and have led to the development of several commercial products and services, including the digital versatile disk (DVD), high-definition television (HDTV), digital video cameras that record in MPEG-1 format, video on compact discs (CDs), MP3 audio, and digital audio broadcasting (DAB). The MPEG-1 and MPEG-2 standards were the first to combine audio and video capabilities, and by all measures they have been extremely successful [1].

It has always been true that a compression standard set for one application is ported to other applications very quickly. Witness the wireline voice compression standard at 32 Kbit/s, G.726 (formerly G.721) that found applications in low-power wireless telephone systems [2], and the Joint Photographic Experts Group (JPEG) still image coding standard that was quickly used for video over the Internet as "motion JPEG." With such experiences as background and with the rapid development of the Internet and the wireless industry, expectations for standards have been raised so that the broadest possible set of applications is now sought. This is the environment in which MPEG-4 came into being. The evolution of digital audio-visual standards is continuing with the development of MPEG-7. This chapter describes highlights of the MPEG-4 standard and the developing MPEG-7 standard.

8.2   MPEG-4

Scanning the applications of MPEG-1 and MPEG-2 listed in Section 8.1, it is clear that MPEG standards are employed in communications, computing, and entertainment, and it is expected that these three areas will continue to converge in the future. Indeed, it is this expected convergence that is driving many international standards setting bodies. Thus, a goal of MPEG-4 was to develop functionalities to support applications from this convergence. An excellent starting point for our discussions is a comparison of the general reference models for the several MPEG standards. The general reference models for MPEG-1 and MPEG-2 are shown in Figures 8.1 and 8.2, respectively. While these figures appear very simple, it is important to note that the goals of MPEG-1 and MPEG-2 were not considered trivial at the time. Indeed, the target bit rates and quality sought by MPEG-1 and MPEG-2 were considered ambitious, and there was a real demand for the functionality implied by the "Interaction" box in Figure 8.2. Recognizing this demand and realizing the limitations of the MPEG-2 interactive capabilities, the Multimedia and Hypermedia Experts Group (MHEG) created a standard called MHEG-5, which supports extended functionalities that can work with MPEG-2 and other compression methods.

MHEG-5 is not a compression standard, but it provides the capability of composing a scene with text strings, still images, and graphic animations in addition to audio-visual streams. These parts can be provided by the author, and there is some capability for the user to modify the evolution of the scene, through text input and selection menus. The reference model for MHEG-5 is shown in Figure 8.3 and can be contrasted with the MPEG-1 and MPEG-2 compression-dominated standards.

MPEG-4 takes these ideas a step further by allowing more interaction by the user, and also includes some new video compression methods that are more object based; it also offers extended capabilities and choices for audio and voice encoding. Furthermore, MPEG-4 allows audio and video information that may be natural or synthetic, or even a combination of the two. Figure 8.4 specifies the general reference model for MPEG-4, where it is clearly quite a different standard than, say MPEG-1 and -2, and where it is obviously an extension of the MHEG-5 enabled MPEG structure.

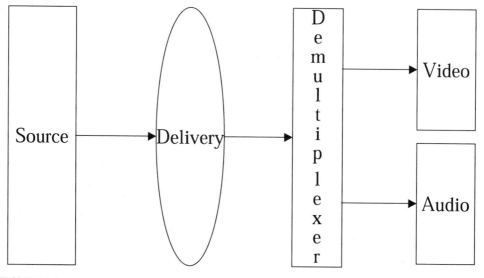

FIGURE 8.1
General reference model for MPEG-1.

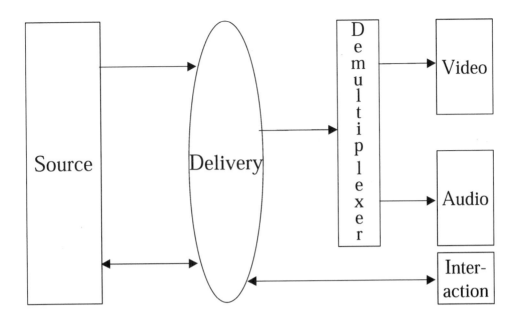

FIGURE 8.2
General reference model for MPEG-2.

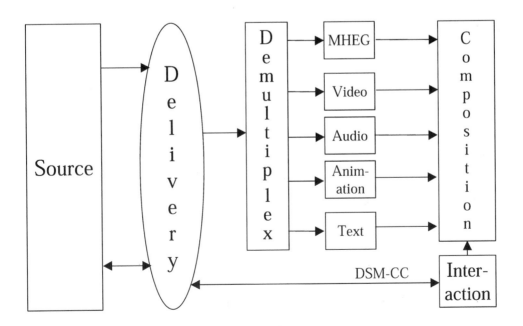

FIGURE 8.3
MHEG-5 enabled MPEG-2 reference model.

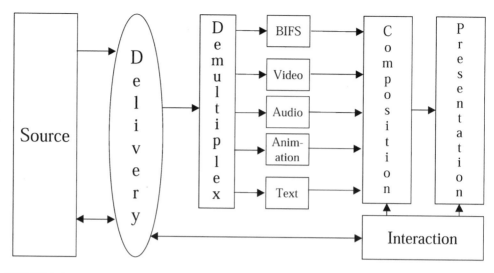

FIGURE 8.4
General reference model for MPEG-4.

8.2.1   MPEG-4 Systems Model

A major difference between MPEG-4 and the MPEG-1 and -2 standards is that MPEG-4 breaks a scene down into components called audio-visual objects, codes these objects, and then reconstructs the scene from these objects. Figure 8.5 is a representation of the MPEG-4 encoding and decoding process [3]. A number of key ideas are represented by this figure. The Binary Format for Scenes (BIFS) data stream specifies the scene composition, such as the spatial and temporal

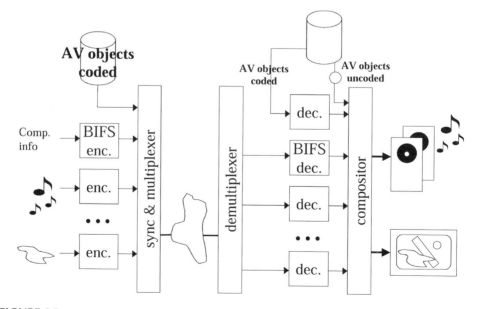

FIGURE 8.5
The MPEG-4 scene decomposition/object-based architecture.

locations of objects in scenes. There are different types of objects to be coded, including natur-al images, natural audio, voice, synthetic audio and video, textures, and physical objects, and thus there are several encoders shown in the figure. Thus, the scene is decomposed, the objects are compressed separately, and this information is multiplexed with the BIFS composition infor-mation for transmission to the decoder. At the decoder, the BIFS information is decoded, the compressed versions of the scene components are decoded, and the scene is recomposed.

There are a number of advantages to utilizing a scene decomposition approach in conjunc-tion with separate coding of the components [4]. The most obvious advantage is that one com-pression method does not have to compress a complicated scene that includes people, arbitrarily shaped objects, and (maybe) text, which often produces visual artifacts. Once decomposition is achieved, each component can be compressed with an encoder that is much better matched to the specific source to be encoded. Another advantage is that the data stream can be scaled based upon content. That is, the data stream can be modified by removing or adapting content, depend-ing upon bandwidth or complexity requirements. This is really a new idea, and can be very pow-erful. For example, if there is a need to lower the transmitted bit rate, rather than discard bits that affect the quality of the entire scene, it may be possible to discard an unimportant object in the scene that provides a sufficient bit-rate reduction without reducing the quality of the delivered scene as determined by the particular application. Another advantage of using scene decompo-sition and object-based compression is that the user can be allowed to access various objects in the scene and change the scene content.

The hierarchical description of a scene is an essential element of MPEG-4 functionality. Figure 8.6 shows a possible breakdown of a video scene into several possible layers. The video session consists of a Visual Object Sequence (VS), as shown at the top of the figure. Within the scene there will be Video Objects (VO), which can then be encoded in the Video Object Layer (VOL). The Video Object Plane (VOP) consists of time samples of a video object. The layer above the VOP in the figure is the Group of Video Object Planes (GOV), and this layer provides the opportunity for the VOPs to be coded independently of each other, thus allowing random access in the bit stream.

Figure 8.7 provides a less abstract view of how a scene might be decomposed. Here we see the scene broken down into multiple audio and video objects, including a person, the back-ground, furniture, and an audio-visual presentation. Thus, there are objects and other compo-nents of the scene, such as the background, that one may expect to remain in the scene and to not change for some amount of time. These components are coded separately, and if they do not change, they need not be encoded and transmitted again until they do change. Notice that the person object is further broken down into a sprite and a voice. The *sprite*, which represents the video of the person excised from the rest of the image, can be coded separately, and the voice can be sent to the appropriate voice coding method. Of course, the sprite and voice information is likely to be changing constantly, and thus it must be continually coded and transmitted, but the background image may not change appreciably over relatively long periods of time, and thus need not be retransmitted very often. The audio-visual presentation could contain high quality audio, and in this case, the audio is coded by one of the coders designed specifically for this task. It should be clear from this description and Figure 8.7 that the object-based approach offers the possibility of much better compression because a particular compression algorithm does not have to handle such a wide variety of inputs.

Just as in MPEG-1 and MPEG-2, a major component of the MPEG-4 standard is the speci-fication of a System Decoder Model (SDM). The architecture of the MPEG-4 Systems Decoder Model is shown in Figure 8.8, along with the network interfaces [4]. The composition and separate encoding of the several audio-visual objects is clearly represented here, but one also sees some network layer interfaces. Delivery Multimedia Integration Framework (DMIF)

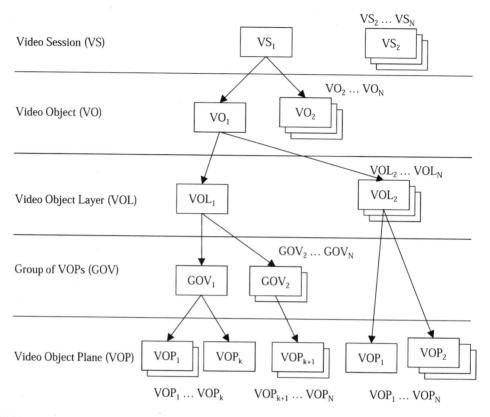

FIGURE 8.6
MPEG-4 video hierarchical description of a scene.

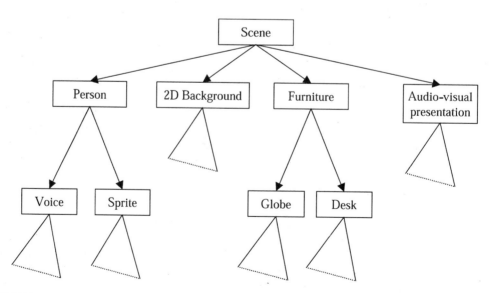

FIGURE 8.7
Hierarchical breakdown of a typical scene.

applications interface shown is intended to isolate MPEG-4 from specifics of the delivery layer, thus allowing MPEG-4 content to be delivered using a variety of delivery systems. The delivery systems shown in Figure 8.8 include Uuser Datagram Protocol (UDP) over the Internet Protocol (IP), Public Switched Telephone Network (PSTN), Asynchronous Transfer Mode (ATM), MPEG-2 streams, or a DAB multiplexer, in addition to MPEG-4 files. Also shown in Figure 8.8 is a FlexMux tool. The FlexMux tool provides the possibility of interfacing with systems that do not have means for multiplexing information, like Global System for Mobile Communications (GSM), or for interfacing with systems whose multiplexing method may not match MPEG-4

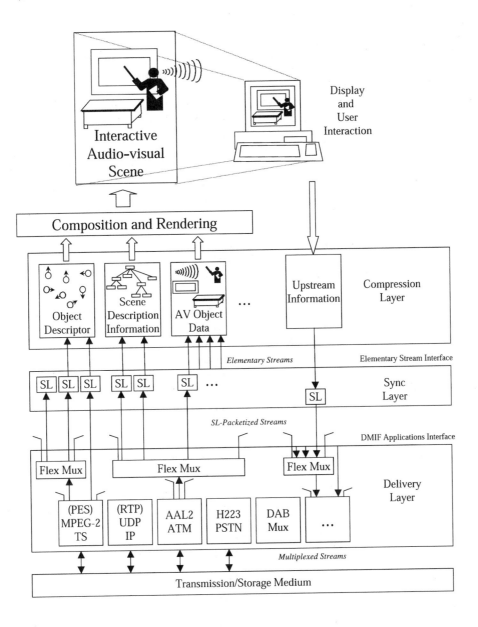

FIGURE 8.8

MPEG-4 systems decoder architecture. (AAL2 = Adaptation Layer Type 2; PES = Packetized Elementary Stream; RTP = Real-Time Transfer Protocol; SL = Sync Layer; TS = Transport Stream.)

utilization needs [4]. Perhaps the main impact of Figure 8.8 is that it highlights the wide range of functionalities that MPEG-4 attempts to address [7].

8.2.2 Natural Video Coding

In addition to being object based and allowing interaction with the objects, MPEG-4 compression also targets a wide range of applications (video streaming, digital television, and mobile multimedia) and provides interface points for most existing compression standards. Table 8.1 lists the video profiles and level definitions for MPEG-4. Video formats can be progressive or interlaced, and the chrominance subsampling is 4:2:0, which means that the chrominance components are sampled at half the luminance rate. While MPEG-4 video coding has been optimized for the three bit-rate ranges, under 64 Kbit/s, 64–384 Kbit/s, and 384 Kbit/s–4 Mbit/s, higher rates are supported, as shown in Table 8.1. There is also a baseline coding layer that is bit-stream compatible with H.263.

Since there is a desire to use MPEG-4 for mobile multimedia, error resilience is an important functionalilty. A common approach to error resilience is to insert resynchronization markers so that in case of an error, the received data can be ignored until the next resynch marker. MPEG-4 uses resynchronization markers, but with approximately a constant number of compressed bits in between, which is different from MPEG-2 and H.263. Another technique to obtain robustness is to separate coded data that has different error sensitivities. In MPEG-4, this idea is employed by separating the texture and motion compensation data streams. Thus, the more error-sensitive motion compensation information can be allocated additional error protection as needed without incurring the overhead for the entire data stream. Reversible variable-length codes are also employed so that decoding can take place in the reverse direction when a resync marker is received. Since header information is so critical, there is an option for error protecting the header information. More discussion of MPEG-4 over wireless channels can be found in [5].

Another critical functionality in MPEG-4 is scalability. Both spatial and temporal scalability options are available in MPEG-4 for video compression. For spatial scalability, the input VOPs are downsampled to yield the base layer VOPs, which are then passed through the encoder. The reconstructed base layer values are next upsampled and subtracted from the input VOP to generate an error signal that is encoded as the enhancement layer. The enhancement layer for temporal scalability is designated as either Type I or Type II. Only a portion of the base layer is enhanced in the Type I case. This would happen, for example, if only the head and shoulders of

Table 8.1 Subset of MPEG-4 Video Profile and Level Definitions

Profile and Level		Typical Screen Size	Bit Rate	Maximum Number of Subjects	Total Memory (Macroblock Units)
Simple profile	L1	QCIF	64 Kbit/s	4	198
	L2	CIF	128 Kbit/s	4	792
	L3	CIF	384 Kbit/s	4	792
Core profile	L1	QCIF	384 Kbit/s	4	594
	L2	CIF	2 Mbit/s	16	2,376
Main profile	L2	CIF	2 Mbit/s	16	2,376
	L3	ITU-R 601	15 Mbit/s	32	9,720
	L4	$1,920 \times 1,088$	38.4 Mbit/s	32	48,960

a moving person in a video sequence was enhanced. For Type II, enhancement is provided for the whole base layer.

8.2.3   Audio and Speech Coding

The MPEG-4 natural audio coding tools probably fit the definition of a toolbox better than in the case of the video coding tools. Several audio and speech encoding methods that are structurally incompatible are included in the standard and the choice among them depends on the audio object to be coded and the user preferences. The MPEG-4 audio coding tool is backward compatible with MPEG-2 Advanced Audio Coding (AAC), with supplements for bit-rate scalability and a long-term prediction mode. The targeted per-channel rates are 16 Kbit/s for better than AM-radio quality and 64 Kbit/s for transparent quality. At rates less than 16 Kbits/s per channel, the scale factors and spectral information are encoded using an interleaved, weighted vector quantization (VQ) method called Twin VQ.

The natural speech coding tools in MPEG-4 cover both 4 kHz and 7 kHz input bandwidths and a wide range of bit rates. There are two basic speech coders, a Code-Excited Linear Predictive Coder (CELP) and a Harmonic Vector Excitation Coder (HVXC). Tables 8.2 and 8.3 summarize the parameters and operating ranges for these coders [6]. The extra functionalities for the MPEG-4 speech coding tools include bit-rate scalable and bandwidth scalable coding to allow for multicast transmission. The available bit rates from HVXC are only 2 and 4 Kbit/s, but the CELP coder has bit-rate scalability at an increment as small as 200 bit/s. In fact, from Table 8.3 it is seen that for telephone bandwidth inputs there are 28 bit rates for the CELP coder in the range from 3.85 to 12.2 Kbit/s, and for wideband speech inputs (50 Hz–7 kHz), there are 30 bit rates from 10.9 up to 23.8 Kbit/s. Of course, there are variations in quality and coding delay as the rates are changed, since shorter frame sizes are used for higher rates and longer coding delays are needed to improve speech quality at the lower rates. The telephone bandwidth CELP coder is bandwidth scalable to the wider band inputs.

Table 8.2 Specifications for the HVXC Coder

Sampling frequency	8 kHz
Bandwidth	300–3,400 Hz
Bit rate	2,000 and 4,000 bit/s
Frame size	20 ms
Delay	33.5–56 ms
Features	Multi-bit-rate coding
	Bit-rate scalability

Table 8.3 Specifications for the CELP Coder

Sampling frequency	8 kHz	16 kHz
Bandwidth	300–3,400 Hz	50–7,000 Hz
Bit rate	3,850–12,200 bit/s (28 bit rates)	10,900–23,800 bit/s (30 bit rates)
Frame size	10–40 ms	10–20 ms
Delay	15–45 ms	15–26.75 ms
Features	Multi-bit-rate coding	
	Bit-rate scalability	
	Bandwidth scalability	

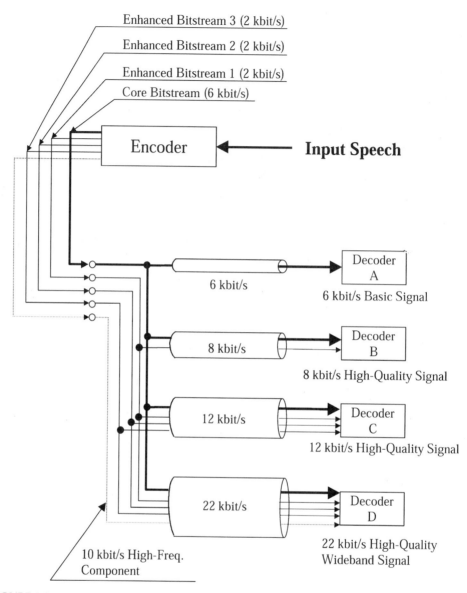

FIGURE 8.9
Scalabilities in MPEG-4/CELP.

An example of bit-stream scalability for CELP is shown in Figure 8.9 [6]. In this figure, the baseline coder is at 6 Kbit/s and enhancement layers at 2 Kbit/s per layer are provided. Using one enhancement layer gives an 8 Kbit/s decoded output, and using all three enhancement layers produces a 12 Kbit/s output. The dashed line indicates the possibility of bandwidth scalability by including a 10 Kbit/s enhancement data stream on top of the 12 Kbit/s already available. Bandwidth scalability bit streams on top of the baseline coders have a more restricted set of possible data rates, and these are shown in Table 8.4. The quality of the speech coders in MPEG-4 compares very favorably with other speech coding standards at the same rates, but without the functionalities offered by MPEG-4 [7].

Table 8.4 Bandwidth Scalable Bit Streams

Core Bit Stream (bit/s)	Enhancement Bit Stream (bit/s)
3,850–4,650	9,200, 10,400, 11,600, 12,400
4,900–5,500	9,467, 10,667, 11,867, 12,667
5,700–10,700	10,000, 11,200, 12,400, 13,200
11,000–12,200	11,600, 12,800, 14,000, 14,800

8.3   MPEG-7

The MPEG-7 standard under development continues the general trend evident in MPEG-4, where the standards are as concerned about functionality as they are about compression. MPEG-7 is designated Multimedia Content Description Interface and is concerned with the interpretation of information in such a way that it can be used or searched by devices or computers. Possible applications include content-based searches or queries, video and audio summarization, and accelerated browsing of Internet sites. The diagram in Figure 8.10 shows a broad view of possible applications for the MPEG-7 standard. The standard intends to specify a set of descriptors, a set of description schemes, and a description definition language, and these are shown in Figure 8.10. MPEG-7 implies a high level of "understanding" by the computing devices in the terminals and networks.

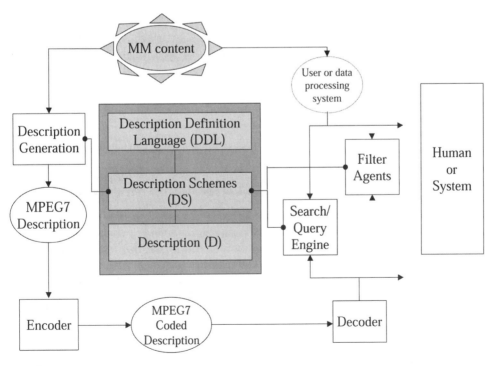

FIGURE 8.10
MPEG-7 applications overview. (MM = multimedia.)

There are several key terms in the figure from MPEG-7 terminology that need a little elaboration. These definitions are [8]:

Feature: A Feature is a distinctive characteristic of the data that signifies something to someone.

Descriptor: A Descriptor (D) is a representation of a Feature.

Descriptor Value: A Descriptor Value is an instantiation of a Descriptor for a given data set.

Description Scheme: A Description Scheme (DS) specifies the structure and semantics of the relationships between its components, which may be both Descriptors and Description Schemes.

Description: A Description consists of a DS and the set of Descriptor Values that describe the data.

Coded Description: A Coded Description is a Description that has been encoded to fulfill requirements such as compression efficiency, error resilience, or random access, for example.

Description Definition Language: The Description Definition Language (DDL) is a language that allows the creation of new Description Schemes and possibly new Descriptors.

The Description generation step will require algorithms for feature extraction, but the specific form of these algorithms is outside the scope of the MPEG-7 standard. This is logical since it is desirable to take advantage of future developments and to allow competitors to develop algorithms that distinguish their systems from others.

8.4   SUMMARY

As the MPEG standards move forward from the groundbreaking contributions of MPEG-1 and MPEG-2, the watchword is functionality. The goals of MPEG-4 and MPEG-7 are to add functionality to support a host of current and expected applications. These applications are motivated by the convergence of the computing, communications, and entertainment fields, and they are forming the basis for a new economy.

The toolbox approach is also highly evident in MPEG-4. In fact, this is so much so that portions of the standard appear to be a patchwork or simply a compendium of techniques. MPEG-4 is an extraordinary standard, however, and is certain to be a catalyst for a host of future applications.

8.5   REFERENCES

[1] L. Chiariglione, "Impact of MPEG Standards on the Multimedia Industry," *Proceedings of the IEEE,* Vol. 86, pp. 1222–1227, June 1998.

[2] D. C. Cox, "Wireless Personal Communications: A Perspective," Chapter 15 in *The Mobile Communications Handbook,* J. D. Gibson, ed., second edition, CRC Press, 1999.

[3] F. Pereira, "MPEG-4: Why, What, How and When?," *Image Communication,* Vol. 15, pp. 271–279, 2000.

[4] O. Avaro, A. Eleftheriadis, C. Herpel, G. Rajan, and L. Ward, "MPEG-4 Systems: Overview," *Image Communication,* Vol. 15, pp. 281–298, 2000.

[5] M. Budagavi, W. Rabiner Heinzelman, J. Webb, and R. Talluri, "Wireless MPEG-4 Video Communication on DSP Chips," *IEEE Signal Processing Magazine*, Vol. 17, pp. 36–53. January 2000.

[6] K. Brandenburg, O. Kunz, and A. Sugiyama, "MPEG-4 Natural Audio Coding," *Image Communication,* Vol. 15, pp. 423–444, 2000.

[7] ISO/IEC JTC1 SC29/WG11, N2725, "Overview of the MPEG-4 Standard," Seoul, South Korea, March 1999.

[8] ISO/IEC JTC1 SC29/WG11, N3158, "Overview of the MPEG-7 Standard," Maui, Hawaii, December 1999.

CHAPTER 9

ATM Network Technology

YOICHI MAEDA
KOICHI ASATANI

9.1   INTRODUCTION

Standardization activities on broadband ISDN (B-ISDN) services and Asynchronous Transfer Mode (ATM) networks commenced in the 1980s, and were targeted toward next-generation network technologies and services for high-speed and multimedia communication. General aspects and principles related to B-ISDN were standardized in 1990, and ATM networks are now considered the key service platform for providing high-speed computer communication and multimedia services. In accordance with this concept, activities have been expanded from standards in the public network area to implementation agreements on private networks and user terminals. This will accelerate standardization activities for ATM networks.

In the next section, the international standardization activities concerning ATM technology by International Telecommunication Union—Telecommunication Standardization (ITU-T) are overviewed. In the following sections, a technical overview for ATM standards is presented. First, the physical layer, ATM layer, and ATM Adaptation Layer (AAL) standard, which are used for the User Network Interface (UNI) and the Inter-Network Network Node Interface (NNI), are explained. In addition to these standards released by the ITU-T, the ATM Forum's physical layer standards that are mainly used for private UNIs and NNIs are explained. Furthermore, network aspects such as traffic control, performance, Operation and Maintenance (OAM), and signaling protocols are explained. The emphasis of this chapter will be on discussing relevant standards and on highlighting some key technological aspects concerning ATM networks.

9.2   OVERVIEW

9.2.1   Background

Studies on B-ISDN services started at the international standardization body Comité Consulatif International des Radiocommunications (CCITT) [now International Telecommunications Union—Telecommunications Section (ITU-T)]. Issues concerning B-ISDN were recognized as one of the major study items in CCITT SG XVIII (now ITU-T SG13) and a broadband task group (BBTG) was organized at the first ISDN experts meeting in January 1985. In the course of standardization activities, it was agreed to initiate the study of a unique digital transmission hierarchy that would be applicable for B-ISDN UNIs and NNIs. The Synchronous Digital Hierarchy (SDH) was completed as Recommendations G.707, G.708, and G.709, which are currently being merged into G.707. Parallel to these activities, during the BBTG activities focusing on basic technology to support B-ISDN services, Asynchronous Time Division (ATD) multiplexing technology was proposed by France in 1986. After enhancing the ATD technology to meet several requirements for B-ISDN, the Asynchronous Transfer Mode (ATM) was proposed and defined as a unique solution for B-ISDN services. These results were published as the guideline ITU-T Recommendation I.211 in 1988.

During the ninth study period (1989–1992) of ITU-T, basic ATM specifications (including cell size, header size, and ATM layer functions), adaptation layer functions, and the Operations, Administration, and Maintenance (OAM) principle were discussed. The fixed-size cell structure of a 5-octet cell header and a 48-octet information field, which simplifies protocol procedures, was agreed upon in June 1989. After the basic ATM specifications were settled, the ATM adaptation layer was developed. This layer enables segmentation and reassembly of higher-layer information into or from a size suitable for the 48-octet information field of an ATM cell under the framework of the service class and protocol type. These agreed-upon specifications were included in a set of 13 recommendations in November 1991 at the Matsuyama meeting. Based on the above basic recommendations, a stepwise approach was adopted for detailed protocol specifications. UNI specifications to support various media were discussed, especially those related to the ATM layer, the ATM adaptation layer applicable for data transfer, the OAM principle, and various traffic control recommendations, as release 1 capabilities. The set of detailed specification recommendations forms the minimum set of functions that support multimedia services. Parallel to these activities in SG XVIII, SG XI initiated studies on a signaling system to support B-ISDN services. These studies included a User Network Interface/Network Node Interface (UNI/NNI) signaling protocol above the ATM layer and signaling information elements and procedures. These signaling specifications will be produced to align with the release steps, and the first set of signaling recommendations will be produced while retaining compatibility with Q.931 for N-ISDN UNI signaling and Integrated Services Digital Network User Part (ISUP) [National Integrated Services Digital Network (N-ISDN User Part)] NNI signaling that was issued in September 1994. Subsequent activities focused on release 2 (1994–1995) capabilities that support the variable-bit-rate capability for voice and video, point-to-multipoint connection, and multiple Quality of Service (QoS). Release 3 (1996–) is intended to address enhanced connection configuration for distribution and broadcast services.

Although the ATM Forum and other consortia are not de-jure international standards organizations, their activities strongly influence the de-jure standards. The ATM Forum was organized in October 1991 with the aims of rapidly producing ATM implementation specifications with high levels of interoperability and of promoting ATM products and services. ITU-T was at that time developing detailed specifications for release 1, and further implementation agreements, which were not covered in ITU-T recommendations, were urgently required and were a primary goal of the ATM Forum activities. Furthermore, the ATM Forum is considering next-generation technologies for LANs and private communication systems. Applying ATM technology to Customer Premises Networks (CPNs) has several advantages in that we can realize multimedia services and seamless communication by using protocols commonly defined for both public and private networks. The success of the ATM Forum is mainly due to strong market needs and specific study items raised from the user side.

9.2.2 Basic ATM Concept

In the ITU-T Recommendations, B-ISDN is defined as the broadband aspects of ISDN, and it covers interfaces whose bit rates are higher than the primary rate (1.5 or 2 Mbit/s) and provides a variety of information-transfer capabilities applicable to multimedia services (including full-motion video). B-ISDN was developed to meet the following objectives:

- high-speed communication (>> 1.5 Mbit/s)
- flexible communication (flexible speeds, point-to-point, point-to-multipoint, and multi-point-to-multipoint connection configurations)

To meet these objectives, the following principles were recommended:

- the adoption of Asynchronous Transfer Mode (ATM) technology
- a 155.520-Mbit/s or 622.080-Mbit/s physical interface for the UNI as a minimum set
- the introduction of high-speed transmission systems with very low error rates (e.g., optical-fiber transmission systems)

All types of information, continuous or interrupted, are put into information blocks of fixed length. The information block together with its header is called a cell (an ATM cell) and cells are assigned to the information originating from a calling-party terminal. The cell-generation rate is proportional to the rate at which the information originates for any type of information (e.g., constant bit rate or variable bit rate). The ATM cell is composed of a 5-octet header field and a 48-octet information field, as shown in Figure 9.1. The ATM cell header includes Header Error Control (HEC), Cell Loss Priority (CLP), Payload Type (PT), Virtual Path/Virtual Channel Identifier (VPI/VCI), and Generic Flow Control (GFC) fields. The details of the header functions are described in the next section.

ATM can operate in a connection-oriented mode, in which a setup phase allows the network to reserve a route and necessary resources. Furthermore, an ATM cell has a small fixed length to reduce the buffer capacity in the switching nodes and to limit queuing delays by using high-speed, hardware-based switching mechanisms. ATM can satisfy the diverse loss, delay, and jitter requirements of various services, and supports higher reliability and faster recovery from failures in a high-speed communication environment.

9.2.3 ATM Network Protocol Structure

The protocol structure of an ATM network for B-ISDN is defined as an extension of that of N-ISDN, based on the Open Systems Interconnection (OSI) layering principle. Each layer has its specific functions, and a service of a given layer is enabled by using the service of its underlying layer. Figure 9.2 shows the ATM network protocol structure. The ATM layer provides a cell transfer function common to all services. On top of the ATM layer, the ATM Adaptation Layer

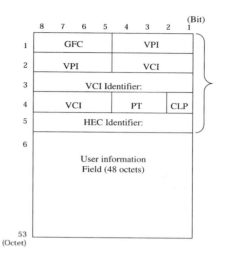

FIGURE 9.1
ATM cell format.

(AAL) provides service-dependent functions to the higher layers supported at the edges of the network (i.e., at the user terminals). Network functions for user information transfer are limited to the ATM layer and below (the user plane), to achieve high-speed and low-delay transfer within the network. That is, in the user plane, protocol handling inside a network is limited to no higher than the ATM layer. A network examines only the ATM cell header.

9.2.4   International Standardization and Recommendations

The basic principles and guidelines for B-ISDN study were described in I.121 (Blue Book, 1988). In the course of developing studies on B-ISDN details, the contents were split into sev-

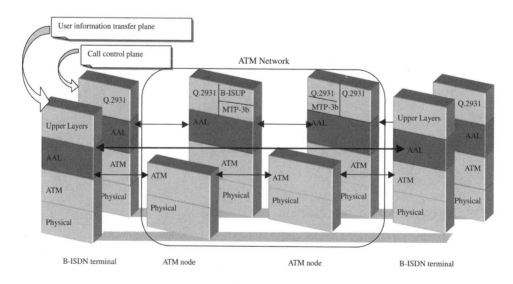

FIGURE 9.2
ATM network protocol structure.

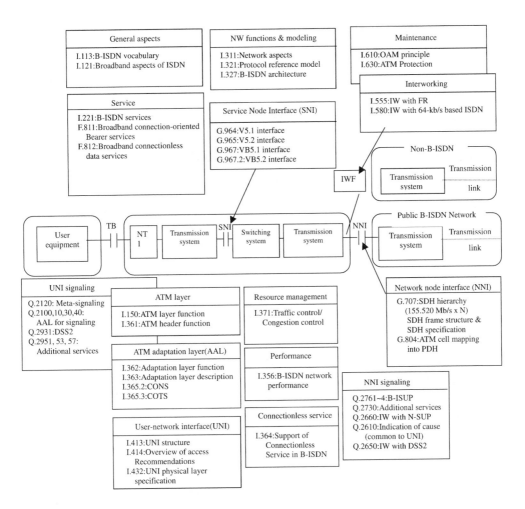

FIGURE 9.3
Key ITU-T recommendations for B-ISDN services.

eral new recommendations. The structure of key B-ISDN-related recommendations is shown in Figure 9.3. The I-series recommendations were produced to describe the aspects of B-ISDN. F-series recommendations for service definitions, Q-series recommendations for signaling protocols, G-series recommendations for digital line systems, and other series have been and will be enhanced to cover the B-ISDN aspects of each area.

9.3   PHYSICAL LAYER SPECIFICATIONS

The physical layer provides transmission resources to carry the ATM cell stream. It consists of the Physical Medium (PM) sublayer that specifies bit-stream transmission characteristics over optical fiber or metallic cable, and the Transmission Convergence (TC) sublayer that specifies, independently of the PM sublayer, transmission frames and ATM cell mapping rules for the transmission frame payload. Information crossing the boundary between the ATM layer and the physical layer is a flow of valid cells.

9.3.1   Basic Characteristics of the TC Sublayer

Two types of specifications are given for the TC in I.432.1. One is for Synchronous Digital Hierarchy (SDH)-based TC and the other is for cell-based TC. Currently, the SDH-based system is applied to actual ATM networks.

Every ATM cell has a 5-octet header that contains a 1-octet Cyclic Redundancy Check (CRC) to protect the remaining 4 octets in the header. This is called Header Error Control (HEC), and this function is also used for cell delineation. During a communication, if there is no user information to transmit, a nonuser cell is inserted into the connection to keep the cell flow synchronized over the SDH frame. Two nonuser cell types are defined. An Idle Cell is generated and terminated in the TC layer within the Physical layer. Therefore, an Idle Cell is invisible at the ATM layer. The other type, an Unassigned Cell, is generated and terminated by the ATM layer. The payload of an Unassigned Cell includes no effective user information except for a Generic Flow Control (GFC) field in the cell header. A more detailed ATM cell structure is discussed in Section 9.4.

9.3.2   Interface Bit Rates

Ultimately, 155.520 Mbit/s and 622.080 Mbit/s interfaces are the target for ATM user-network interfaces to accommodate future services. However, lower-bit-rate interfaces and various physical medium interfaces are expected to support ATM services in the short term at very low cost. ITU-T has defined 155.520 Mbit/s and 622.080 Mbit/s interfaces in I.432.2, 1.544 Mbit/s and 2.048 Mbit/s interfaces in I.432.3, a 51.840 Mbit/s interface in I.432.4, and a 25.600 Mbit/s one in I.432.5. In addition, the ATM Forum has defined additional interfaces for various transmission media such as twisted-pair cables and plastic-fiber cables. Physical interfaces available in accordance with the ITU-T and the ATM Forum are summarized in Table 9.1.

9.4   ATM LAYER SPECIFICATIONS

The ATM layer is independent of the underlying physical layer. An ATM cell stream mapped into the payload of the physical layer contains cell delineation information, so that it can be carried by both the SDH system and other existing transmission systems. This enables the same ATM cell stream to be exchanged between interfaces on different transmission systems and allows intercommunication between the systems. The ATM layer provides a cell multiplexing/demultiplexing function and a VC/VP routing function as defined in I.150.

The ATM layer specification is defined in I.361 and the ATM cell structure at the UNI is shown in Figure 9.1. The UNI ATM cell header has a 4-bit Generic Flow Control (GFC) field to support flow control between end-user terminals and a network terminal in Customer Premises Network (CPN). On the contrary, at the NNI, these 4 bits are used for the VPI field to support a large number of Virtual Path Connections (VPCs) at the NNI. The rest of the fields are the same for the UNI and NNI. The GFC protocol specified in the 1995 version of I.361 can only be applied to a star configuration for the end-user terminals. If GFC is not used or the equipment complies with the previous version of I.361, this field should be set to all 0. GFC for other configurations is under study. A Virtual Path/Virtual Channel Identifier (VPI/VCI) field is used as an ATM cell-routing identifier. Some special values are reserved, and these are used to identify operation-and-control channel components, such as meta-signaling channels, general broadcast signaling channels, and VP-OAM cells. The Payload Type Identifier (PTI) 3-bit field, which can represent eight patterns, indicates the contents of the 48-octet ATM-cell user-information field, such as user information, the OAM cell, or the resource management cell. The Cell Loss

Table 9.1 Physical Interface Specifications for ATM Interfaces

Specifications	Bit Rate (Mb/s)	Medium	Notes
ITU-T I.432.1	—	—	General specs. on B-ISDN UNI
ITU-T I.432.2	155.52 / 622.08	Optical	TB, SB (UNI)
ITU-T I.432.3	1.544 / 2.048	Coaxial	TB, SB (primary-rate UNI)
ITU-T I.432.4	51.284	Twisted pair	SB (UNI)
ITU-T I.432.5	25.6	Twisted pair	SB (UNI)
AF-PHY-15	155.52	Twisted pair	Private UNI
AF-PHY-16	1.544	G.703	DS1 public UNI
AF-PHY-17	Up to 155	Data path	Utopia Level 1 for data bus
AF-PHY-18	51.84	Twisted pair	Private UNI (same as I.432.4)
AF-PHY-29	6.312	G.703	J2 public UNI
AF-PHY-34	34.368	G.703	E3 public UNI
AF-PHY-39	155 / 622	Data path	Utopia Level 2 for PCI bus
AF-PHY-40	25.6	Twisted pair	Private UNI (same as I.432.5)
AF-PHY-43	155.52 / 622.08	Optical	Cell-based TC
AF-PHY-46	622.06	Optical	Private UNI
AF-PHY-47	155.52	Twisted pair	Private UNI (UTP3)
AF-PHY-53	155.52	Twisted pair	120ohm 155M TP
AF-PHY-54	44.736	G.703	DS3 public UNI
AF-PHY-62	155.52	MMF	155M MMF SWL private UNI
AF-PHY-63	Up to 622	Data path	WIRE for data bus
AF-PHY-64	2.048	G.703	E1 public UNI
AF-PHY-79	155.52	Plastic fiber	POF cable

Note: TB and SB are UNI reference points defined by ITU-T. TB and SB are similar to, respectively, the Public and Private UNIs defined by the ATM Forum.

Priority (CLP) bit field may be used for resource management mechanisms for priority or selective-cell-discard control.

9.5   ATM ADAPTATION LAYER (AAL) SPECIFICATIONS

The ATM Adaptation Layer (AAL) provides adaptation between the ATM layer and upper layers that use various communication media as defined in the I.362- and I.363-series Recommendations. The ATM layer provides a uniform and media-independent transfer capability. On the other hand, the AAL is a media- and application-dependent protocol, such as a voice, video, or data user-information protocol. The AAL is divided into two sublayers: a Segmentation and Reassembly (SAR) sublayer that provides user-information segmentation and reassembly of the ATM cell payload, and a Convergence Sublayer (CS) that provides a convergence function from user information to ATM cells.

9.6   NETWORK ASPECTS OF B-ISDN

9.6.1   Traffic Control

The traffic control architecture for ATM networks is designed to handle various types of multimedia traffic, as shown in Figure 9.4. At a UNI, the user should state the desired traffic

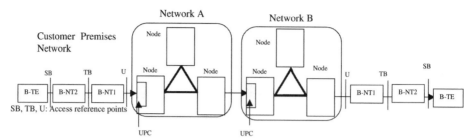

B-TE: Broadband Terminal Equipment
B-NT1: Broadband Network Termination 1
B-NT1: Broadband Network Termination 2

ATM-layer Traffic Parameter	
Traffic Descriptors	Peak Cell Rate
	CDV
	Average Cell Rate
	Burst Size
QoS Parameter (1.356)	Cell Loss
	Cell Misinsertion
	Cell Error
	Cell Delay
	Cell Delay Variation

ATM Network Traffic Control		
Connection Admission Control (CAC)		Connection admission control by source traffic Descriptors, required QoS, and network state information
Usage Parameter Control (UPC / NPC)		To monitor and control traffic in terms of traffic volume offered and validity of the ATM connection
Priority Control		Priority control by CLP bit Priority control by VCI
Congestion Control	Preventive	Taken by CAC,UPC, or Credit
	Reactive	Congestion control by traffic enforcement, flow control, etc.

FIGURE 9.4
Traffic control architecture in B-ISDN.

parameters, such as the cell rate and the required QoS for the communication. The parameters requested by the user are sent to the network, which uses them for connection-admission control to determine whether a virtual-channel/virtual-path connection request should be accepted or rejected. During the connection setup, negotiated traffic parameters at the user access and the network access are monitored to protect network resources from malicious as well as unintentional misbehavior that can affect the QoS of other already established connections. This mechanism is called Usage/Network Parameter Control (UPC/NPC). Furthermore, priority control, such as selective cell discarding or feedback control, is used to enhance the statistical multiplexing gain, if necessary.

The I.371 set of recommendations addresses functions and parameters for traffic and congestion control in the B-ISDN. A user-network and an internetwork traffic contract are defined in terms of a traffic descriptor that includes traffic parameters and associated tolerances of an ATM layer transfer capability and of QoS requirements. Relevant traffic parameters and generic conformance definitions for these parameters are specified. ATM transfer capabilities that make use of these traffic parameters to allow for different combinations of QoS objectives and multiplexing schemes, and specific conformance definitions that apply to these ATM transfer capabilities, are provided. In addition, traffic control and congestion control functions are further specified, among which are traffic parameter control functions at user-network and internetwork interfaces. Some specific traffic-control interworking configurations are described.

An ATM Transfer Capability (ATC) specifies a set of ATM layer parameters and procedures that is intended to support an ATM layer service model and a range of associated QoS classes. Each individual ATC is further specified in terms of a service model, a traffic descriptor, specific procedures if relevant, a conformance definition, and associated QoS commitments. Open-loop controlled ATCs [Deterministic Bit Rate (DBR), Statistical Bit Rate (SBR), and Guaranteed Frame Rate (GFR)] and closed-loop controlled ATCs [ATM Block Transfer (ABT) and Available Bit Rate (ABR)] are specified as follows.

9.6.1.1 Deterministic Bit Rate (DBR)
The DBR ATM transfer capability is intended to be used to meet the requirements of CBR traffic, and therefore to provide for QoS commitments in terms of the cell-loss ratio, cell-transfer delay, and cell-delay variation suitable for such traffic. However, DBR is not restricted to CBR applications and may be used in combination with looser QoS requirements, including unspecified requirements as indicated in Recommendation I.356.

9.6.1.2 Statistical Bit Rate (SBR)
The SBR ATM transfer capability uses the sustainable cell rate and intrinsic burst tolerance in addition to the peak cell rate, and is suitable for applications where there is prior knowledge of traffic characteristics beyond the peak cell rate, from which the network may obtain a statistical gain. QoS commitments are given in terms of the cell-loss ratio. There may or may not be QoS commitments concerning delay.

9.6.1.3 Guaranteed Frame Rate (GFR)
The GFR ATM Transfer Capability provides a minimum cell rate (MCR) for loss-tolerant, non-real-time applications with the expectation of transmitting data in excess of the MCR. It is assumed that the user-generated data cells are organized in the form of frames that are delineated at the ATM layer. The network does not provide feedback to the user concerning instantaneous available network resources. This ATC is decided to be added as a new ATC recommendation of I.371 series in 2000.

9.6.1.4 ATM Block Transfer (ABT)
The ABT ATM transfer capability is intended for applications that can change their instantaneous peak cell rate on a per-block basis. An ATM block is a group of cells delimited by Resource Management (RM) cells. ABT uses static parameters declared at connection setup and dynamic parameters renegotiable on an ATM-block basis via RM procedures using the RM cells.

9.6.1.5 Available Bit Rate (ABR)
The ABR ATM transfer capability is intended to support non-real-time elastic applications that can adapt to the instantaneous bandwidth available within the network. In such a case, the network may share the available resources between connections supporting such applications. ABR uses static parameters declared at connection setup and dynamic parameters renegotiable via RM procedures based on RM cells.

9.6.2 ATM Layer Performance

I.356 defines speed, accuracy, and dependability parameters for cell transfer in the ATM layer of B-ISDN. The defined parameters apply to end-to-end ATM connections and to specified portions of such connections. The parameters are defined on the basis of ATM cell-transfer reference events that may be observed at physical interfaces between ATM networks and associated customer equipment, and at physical interfaces between ATM networks. The parameters defined in I.356 may be augmented or modified based upon further study of the requirements of the services to be supported on B-ISDNs. The parameters apply to cell streams in which all cells conform to the negotiated I.371 traffic contract. The defined parameters are intended to characterize ATM connections in the available state. Availability-decision parameters and associated availability parameters and their objectives are the subject of Recommendation I.357.

I.356 recommends ATM performance values to be achieved internationally for each of the defined parameters. Some of these values depend on which QoS class the end users and network providers agree on for the connection. It defines four different QoS classes and provides guidance on the allocated performance levels that each specified portion should provide to achieve the recommended end-to-end international performance.

9.6.3   OAM Functions

To provide high quality and reliable information transmission through a B-ISDN/ATM network, operation, administration, and maintenance functions are essential. Based on the ITU-T Recommendation I.610, these functions are divided into five categories, as follows:

- *Performance monitoring:* Normal functioning of the managed entity is monitored by continuous or periodic checking of functions. As a result, maintenance-event information is produced.
- *Defect and failure detection:* Malfunctions or predicted malfunctions are detected by continuous or periodic checking. As a result, maintenance-event information or various alarms are produced.
- *System protection:* The effect of a managed entity's failure is minimized by blocking or changeover to other entities. As a result, the failed entity is excluded from operation.
- *Failure- or performance-information transmission:* Failure information is given to other management entities. As a result, alarm indications are sent to other management planes. A response to a status-report request will also be provided.
- *Fault localization:* Internal or external test systems determine which entity failed if failure information is insufficient.

OAM functions at the physical and ATM layers are performed based on five hierarchical levels, as shown in Figure 9.5. The information flow at each level is called the OAM flow, as follows:

- F1: regenerator-section level
- F2: digital-section level
- F3: transmission path level
- F4: Virtual Path (VP) level
- F5: Virtual Channel (VC) level

For the F4 and F5 flows, two types of flows are defined: the end-to-end flow defined between connection end-points and segment flow consisting of multiple links. The OAM functions at each level operate independently. The OAM functions defined here correspond to the layer management shown in the protocol reference model of I.321. Other OAM functions related to higher layers (e.g., the AAL) need further study.

9.6.4   Signaling Procedure

A B-ISDN/ATM network uses the same out-channel signaling mechanism as an N-ISDN does. This out-channel signaling system enables call control to be done independently of user-information transfer. Also, most parts of the signaling procedures are the same as in the N-

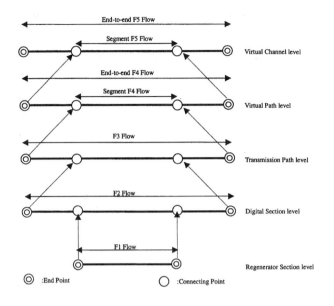

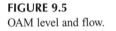

FIGURE 9.5
OAM level and flow.

ISDN. The B-ISDN signaling system uses a meta-signaling channel to control the setting up and closing of signaling channel VCC. The meta-signaling procedure is recommended in Q.2120.

The Signaling AAL (SAAL) includes the layers above the ATM and below layer 3. It uses the AAL type 5 common part and SSCOP for both the UNI and NNI. The uppermost part of the SAAL is the Service Specific Convergence Function (SSCF) that coordinates newly designed primitives provided by the SSCOP and the existing primitives required by Q.2931 and MTP-3b. Q.2130 is an SSCF for UNI, and Q.2140 is one for NNI.

A B-ISDN UNI signaling system is defined in Q.2931, and is called Digital Subscriber Signaling System No. 2 (DSSI2) after the Digital Subscriber Signaling System No. 1 (DSSI1) for the N-ISDN signaling system. The B-ISDN NNI signaling protocol is identified in Q.2761–Q.2764 as part of the No. 7 signaling system, and is called B-ISDN User Part (B-ISUP). The same set of supplementary services as the N-ISDN is defined in the Q.2951 series for the UNI and in Q.2730 for the NNI. Unidirectional point-to-multipoint connection can be established by Q.2971 signaling. This function enables the broadcast services.

9.6.5   VB5 Interfaces

The VB5 reference-point concept, based on ITU-T Recommendations G.902 and I.414, was split into two variants. The first variant, based on ATM cross-connecting with provisioned connectivity, is called the VB5.1 reference point, and is described in G.967.1. The other variant, which further enables on-demand connectivity within an Access Network (AN), is called the VB5.2 reference point and is specified in G.967.2.

The VB5.2 reference point extends the capabilities at the VB5.1 reference point to include on-demand connectivity in the AN under the control of a Service Node (SN). The major correspondence between the VB5.1 and VB5.2 reference points can be described as:

- both VB5 reference points support B-ISDN as well as narrowband and other non-B-ISDN customer-access types

- both VB5 reference points support ATM multiplexing/cross-connection in the AN at the VP and/or VC level.

It is anticipated that the Real-Time Management Coordination (RTMC) protocol for the VB5.1 reference point will be a subset of the RTMC protocol for the VB5.2 reference point. G.967.1 and G.967.2 define the physical, procedural, and protocol requirements for interfaces at the VB5.1 and VB5.2 reference points, respectively, between an AN and an SN. They also specify the flexible Virtual Path Link allocation and flexible Virtual Channel Link allocation (controlled by the Q3 interfaces) at the VB5.1 and VB5.2 reference points.

9.7   OTHER ATM NETWORK TECHNOLOGIES

9.7.1   IP Over ATM

With the rapid growth of Internet Protocol (IP)-based networks and applications in both private and public networks, it is necessary to consider arrangements to transport IP services over ATM in the public network environment. For the private network environment, the ATM Forum has specified Multi-protocol Over ATM (MPOA). The Internet Engineering Task Force (IETF) has specified Classical IP Over ATM (C-IPOA) and Multi-protocol Label Switching (MPLS). To ensure that public networks will interwork with each other supporting a set of services, and to ensure the interworking of public and private networks, it is necessary to recommend the preferred approach for transporting IP over ATM in public networks.

ITU-T's new IP-related Recommendation Y.1310 identifies generic requirements and key IP services, and determines which IP Over ATM approach is preferred for each service. Approaches taken into account include classical IPOA, MPOA, and MPLS. IP services are defined as services provided at the IP layer and do not include those at the application layer.

The mandatory generic requirements of IP Over ATM approaches are as follows:

- The recommended approach must be independent of the IP version supported.
- The recommended approach must have sufficient scalability to support large networks. Items to be taken into account regarding scalability include:
  — Use of VCI and VPI values
  — Complexity of routing calculation at layer 2 and layer 3
  — Complexity of address resolution mechanism
  — Control messaging load (e.g., frequency of setup and cleardown of ATM connections, frequency of IP-related signaling messages)
  — Complexity of the packet classification mechanism needed to support QoS. The less granularity in the QoS (e.g., from per IP flow, to per IP flow aggregation, to per service, as in Diffserv), the simpler the packet classification mechanism
- The recommended approach must include the capability for efficient and scalable solutions to support IP multicasting over ATM networks.
- The recommended approach must have sufficient robustness to support large networks. Items to be taken into account include the capability to support restoration systems.

Considering the generic requirements, Y.1310 recommends that MPLS be adopted as the single preferred approach for public networks. MPLS supports all the services identified. It is recognized that MPLS does not provide significant benefits over properly engineered classical IP Over ATM for the support of the Intserv service. However, MPLS does not offer less-than-classical IPOA for the support of Intserv, and also provides support for all other services.

9.7.2 MPEG2 Over ATM

The MPEG2 standard is usually dedicated to coding and decoding standard-definition TV signals, such as NTSC and PAL, and sound at a transmission rate of about 3.5–6 Mbit/s. As digital transmission becomes more popular than analog, ATM transmission will be increasingly useful due to its high transmission rate. However, not all ATM transmission characteristics are suitable for MPEG2-coded data streams: the cell loss and Cell Delay Variation (CDV) create problems in an MPEG2 CODEC, especially in the MPEG2 decoder. Cell loss is equivalent to a burst error for an MPEG2 decoder, because about 50 octets are lost on a single loss. Thus, the image is degraded. As well as image degradation, CDV can also cause system-clock fluctuation.

There are five types of AAL, and AAL type 1 and AAL type 5 are used for constant-bit-rate (CBR) transmission. AAL type 1 is supported in audio-visual communication systems and AAL type 5 is supported in audio-visual transmission systems (TS) such as Video on Demand (VoD) systems. AAL type 1 changes one TS packet into four ATM cells. If the interleaving method is applied, error-correcting octets are added and some overhead cells must be transmitted. On the other hand, AAL type 5 changes two TS packets into eight ATM cells. With AAL type 5, 8 octets ($8 \times 48 - 2 \times 188 = 8$) are available for cell-loss detection and transmission-error detection. AAL type 5 appears to be more effective than AAL type 1, since it uses two TS packets, the transmission jitter must be added to network one.

9.8   CONCLUDING REMARKS

This chapter has described some of the current standards and the ongoing international standardization activities concerning ATM technology by ITU-T and the ATM Forum. Technical overviews of, for example, the physical-layer, ATM-layer, and AAL standards, used for the UNI and the NNI were given. In addition to these standards, network aspects such as traffic control, performance, OAM, and signaling protocols were explained. ATM will be a key network technology for supporting multimedia services, including IP and MPEG applications, due to its traffic-control capabilities.

9.9   DEFINITIONS OF KEY TERMS

asynchronous transfer mode (ATM): A transfer mode in which the information is transferred within labeled cells; it is asynchronous in the sense that the recurrence of cells containing information from an individual user is not necessarily periodic.

broadband: Qualifying a service or system requiring transmission channels capable of supporting rates greater than the primary rate.

cell delineation: The identification of cell boundaries in a cell stream.

interface bit rate: The gross bit rate at an interface, that is, the sum of the bit rates of the interface payload and the interface overhead. Example: the bit rate at the boundary between the physical layer and the physical medium.

meta-signaling: The procedure for establishing, checking, and releasing signaling virtual channels.

multimedia service: A service in which the interchanged information consists of more than one type, such as text, graphics, sound, image, or video.

network node interface (NNI): The interface at a network node that is used to interconnect with another network node.

network parameter control (NPC): The set of actions taken by the network to monitor and control traffic at the internetwork node interface, to protect network resources from malicious as well as unintentional misbehavior by detecting violations of negotiated parameters and taking appropriate actions.

usage parameter control (UPC): The set of actions taken by the network to monitor and control traffic at the user network interface, and to protect network resources from malicious as well as unintentional misbehavior by detecting violations of negotiated parameters and taking appropriate actions.

OAM cell: An ATM cell that carries OAM information needed for specific OAM functions. OAM cells are also often called maintenance cells.

9.10   BIBLIOGRAPHY

ITU-T. 1990. Broadband aspects of ISDN, Recommendation I.121 ITU-T, Geneva, Switzerland.

ITU-T. 1995. B-ISDN asynchronous transfer mode functional characteristics, Recommendation I.150 ITU-T, Geneva, Switzerland.

ITU-T. 1995. B-ISDN ATM Layer specification, Recommendation I.361 ITU-T, Geneva, Switzerland.

ITU-T. 1996. B-ISDN ATM Adaptation Layer (AAL) specification: Type 1 AAL, Type 2 AAL, Type 3/4 AAL, Type 5 AAL, Recommendation I.363-1, -2, -3, -5 ITU-T, Geneva, Switzerland.

ITU-T. 1996. Traffic control and congestion control in B-ISDN, Recommendation I.371 ITU-T, Geneva, Switzerland.

ITU-T. 1996. B-ISDN user-network interface—Physical Layer specification, Recommendation I.432-1, -2, -3, -4, -5 ITU-T, Geneva, Switzerland.

ITU-T. 1995. Operation and maintenance principles of B-ISDN access, Recommendation I.610 ITU-T, Geneva, Switzerland.

ITU-T. 1995. Framework Recommendation on Functional Access Networks Architecture and Functions, Access Types, Management and Service Node Aspects, Recommendation G.902 ITU-T, Geneva, Switzerland.

ITU-T. 1996. Network Node Interface for the Synchronous Digital Hierarchy (SDH), Recommendation G.707 ITU-T, Geneva, Switzerland.

IU-T. 2000. Transport of IP over ATM in Public Networks, Recommendation Y.1310 ITU-T, Geneva, Switzerland.

Koichi Asatani and Yoichi Maeda, August 1998. Access Network Architectural Issues for Future Telecommunication Networks, *IEEE Communications Magazine, 36,* 8, 110–114.

Koichi Asatani et al., 1998. *Introduction to ATM Networks and B-ISDN,* New York: John Wiley & Sons.

9.11   FOR FURTHER INFORMATION

The latest information on the relevant standards is available at:

ITU-T: http://www.itu.int/ITU-T/index.html

ATM Forum: http://www.atmforum.com/

IETF: http://www.ietf.cnri.reston.va.us/home.html

CHAPTER 10

ISDN

KOICHI ASATANI
TOSHINORI TSUBOI

10.1 INTRODUCTION

10.1.1 General Features of ISDN

Integrated Services Digital Network (ISDN) is a network that provides various types of services, including telephone, data, fax, and video, through a single and universal user-network interface (UNI). Studies on ISDN were initiated in International Telegraph and Telephone Consultative Committee (CCITT) Study Group XVIII [currently Telecommunication Standardization Sector of International Telecommunications Union (ITU-T) SG13] in the 1970s, and the first set of I-series recommendations on ISDN was adopted in 1984 [1].

Before ISDN was introduced, as shown in Figure 10.1, a customer needed different and individual dedicated interfaces for telephone, packet, and video. ISDN, on the other hand, has the following features [2]:

- ISDN provides 64 Kbit/s circuit switched and nonswitched bearer capabilities, $n \times 64$ Kbit/s circuit switched and nonswitched bearer capabilities ($n = 2 - 30$), and packet switched bearer capabilities, as shown in Figure 10.2.
- Up to eight terminals are connected to a single UNI for basic access, and a customer can use various and multiple services through the interface simultaneously.
- ISDN provides end-to-end digital connectivity, enabling a customer to enjoy high-speed and high quality services.
- ISDN is associated with SS7, enabling advanced call controls such as user-to-user signaling and enhanced call handling.

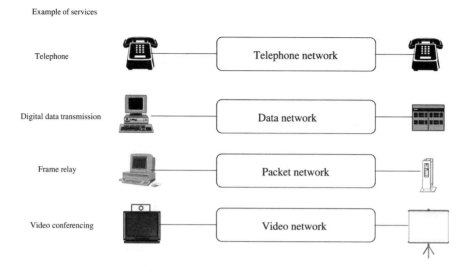

FIGURE 10.1
Access configuration before ISDN era.

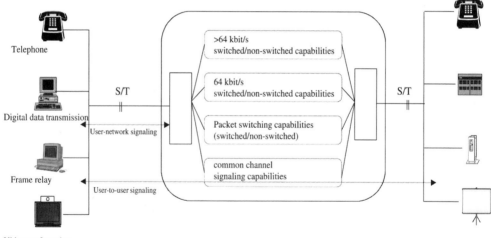

FIGURE 10.2
ISDN architectural model.

10.1.2 Service Aspects of ISDN

ISDN customers can enjoy various types of services, as listed in Table 10.1 [3]. "Speech" and "3.1 kHz audio" supports voice communication, and voice and/or modem communications over 3.1 kHz bandwidth, respectively. "Unrestricted digital information" provides end-to-end transparent digital communication.

Furthermore, ISDN can provide various supplementary services by using No. 7 common channel signaling. Examples of supplementary services are listed in Table 10.2 [4]. Supplementary services enable customers to enjoy more attractive and convenient services.

Table 10.1   Examples of Service Attributes

Attribute	Possible Values of Attribute
Information transfer mode	Circuit Packet
Information transfer rate	64 Kbit/s 384 Kbit/s 1536 Kbit/s 1920 Kbit/s
Information transfer capability	Unrestricted digital information Speech 3.1 kHz audio 7 kHz audio 15 kHz audio Video

Table 10.2   Examples of Supplementary Services

Supplementary Services	Explanation of Service
1. Number identification	
(1) Direct dial in	Enables a user to directly call another user on a PABX.
(2) Calling line identification presentation	Provides the calling party's ISDN number to the called party.
2. Call offering	
(1) Call transfer	Enables a user to transfer an established call to a third party.
(2) Call forwarding busy	Transfers all incoming calls to another ISDN number if called party is busy.
3. Call completion	
(1) Call waiting	Notifies called party that an incoming call has been received during an existing connection.
(2) Call hold	Allows a user to switch the existing connection to a second user with wait status.
4. Multiparty	
(1) Conference calling	Allows a user to communicate simultaneously with multiple parties.
5. Community of interest	
(1) Closed user group	Enables users to form groups with restrictions for access to and from users of the public network.
6. Charging	
(1) Advice of charge	Provides charge information to a user during the connection or at the end of the connection.
7. Additional information transfer	
(1) User-to-user signaling	Allows a user to send information to another ISDN user over the signaling channel.

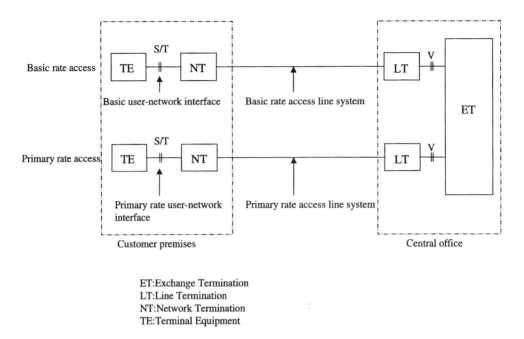

ET:Exchange Termination
LT:Line Termination
NT:Network Termination
TE:Terminal Equipment

FIGURE 10.3
Access system model of ISDN.

10.1.3   Access Features

ISDN provides an end-to-end digital connectivity by adopting digital access systems. An outline of ISDN access systems is shown in Figure 10.3. ISDN supports two types of UNIs: basic access and primary rate access. Basic access interface provides 64 and 128 Kbit/s communications for circuit mode and packet mode and 16 Kbit/s for packet mode [5]. Primary rate interface operates at 1544 and 2048 Kbit/s [6, 7]. Standard UNI at customer premises is defined at the T or S reference point. The interface between Line Termination (LT) and Exchange Termination (ET) at the central office is designated as the V reference point. The line interface for the access transmission system is defined and designated as U reference point in the United States.

10.2   ISDN USER-NETWORK INTERFACES

10.2.1   ISDN UNI Structure

The UNI specifies the interface between terminal equipment (TE) and network termination (NT). Features of an ISDN user-network interface are as follows [8]:

1. Different types of terminals and applications are supported through the same interface.
2. Multiple communications are supported simultaneously.
3. Terminal portability is supported.

Two types of interfaces are specified.

10.2.1.1   Basic Interface

The basic interface is composed of two B channels and one D channel. Each B channel is for carrying user information and its capacity is 64 Kbit/s. The D channel primarily is for carrying signaling information and its capacity is 16 Kbit/s. The interface structure is designated as 2B + D. The basic interface is mainly used for residential and small business customers.

10.2.1.2   Primary Rate Interface

The primary rate interface has the same bit rate as the primary rate of the plesiochronous hierarchy. There are two interface bit rates corresponding to the 1.5 Mbit/s system of the North American and Japanese hierarchy and the 2 Mbit/s system of the European hierarchy. An example of the 1544 Kbit/s primary rate interface structure is 23B + D with D = 64 Kbit/s. An example of the 2048 Kbit/s primary rate interface is 30B + D with D = 64 Kbit/s. The interface also supports any combination of n × 64 Kbit/s and 384 Kbit/s channels as long as the total capacity does not exceed the interface capacity. These interfaces are mainly used for large business customers and are used to support private automatic branch exchange (PABX) and videoconferencing with n × 64 Kbit/s codecs defined in H.261 [9].

10.2.2   Reference Configurations and Reference Points

Reference configurations are conceptual configuration models to identify possible access types to an ISDN. Reference points are also conceptual points dividing functional groups. Different types of actual installment are possible according to regulations and technical feasibility. Reference configurations and reference points are shown in Figure 10.4 [10]. The functional groups are defined in the following sections.

10.2.2.1   Network Termination 1 (NT1)

NT1 is a functional group that terminates layer 1 of an access transmission line and the ISDN user-network interface.

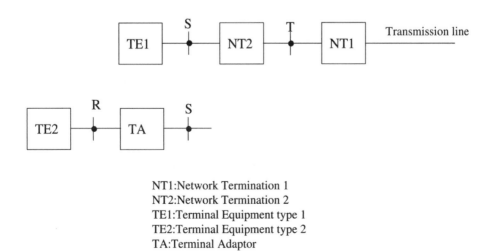

NT1:Network Termination 1
NT2:Network Termination 2
TE1:Terminal Equipment type 1
TE2:Terminal Equipment type 2
TA:Terminal Adaptor

FIGURE 10.4
Reference configurations for ISDN user-network interface.

10.2.2.2   Network Termination 2 (NT2)

NT2 is a functional group that performs processing of layers 2 and 3. An example model is a PABX that performs switching, concentration, or multiplexing.

10.2.2.3   Terminal Adapter (TA)

TA is a functional group that supports non-ISDN (e.g., V and X interface) terminal equipment in ISDN access arrangement.

10.2.2.4   Terminal Equipment 1 (TE1)

TE1 is a functional group that represents terminal devices with the ISDN UNI.

10.2.2.5   Terminal Equipment 2 (TE2)

TE2 is a function group that represents terminal devices with a non-ISDN interface.

These functional groups could be implemented in several ways, such as in one physical device.

Reference points T, S, and R are defined as shown in Figure 10.4. Reference point T is located between NT1 and TE1/NT2. Reference point S is located between NT2 and TE1/TA. Reference points T and S are identical. Reference point R is specified for the connection of non-ISDN terminals to ISDN.

Various physical configurations of UNI are allowed, including a point-to-point configuration, a point-to-multipoint star configuration, and a point-to-multipoint bus configuration, as shown in Figure 10.5. This enables ISDN to be adopted in various requirements of installments.

10.2.3   Interface Features

The main features of the ISDN user-network interface are channel types, interface structures, and access capabilities [11].

10.2.3.1   Channel Types

A *channel* is a portion that carries information. The B channel, H channel, and D channel are specified, as shown in Table 10.3. The B and H channels are used for carrying user information, while the D channel is primarily used for carrying signaling information for circuit switching and may be used for carrying packet types of user information when it is not used for signaling.

10.2.3.1.1   B Channel The B channel has 64 Kbit/s capacity. It carries a wide variety of user information streams for circuit switching, packet switching, and leased line. The B channel capacity of 64 Kbit/s is specified in harmonization with the Pulse Code Modulation (PCM) voice bit rate.

10.2.3.1.2   H Channels H channels are intended to carry high bit-rate signals, such as video and data signals. An H_0 channel and an H_1 channel are specified. The capacity of H_0 is 384 Kbit/s ($= 6 \times 64$ Kbit/s). The capacity of the H_1 channel is 1536 Kbit/s ($= 24 \times 64$ Kbit/s) for 1.5 Mbit/s hierarchy countries or 1920 Kbit/s ($= 30 \times 64$ Kbit/s) for 2 Mbit/s countries. The H_1 channel with 1536 Kbit/s is designated as H_{11}, and with 1920 Kbit/s as H_{12}.

10.2.3.1.3   D Channel A D channel primarily carries signaling information for circuit switching. The capacities of a D channel are 16 Kbit/s for a basic interface and 64 Kbit/s for a primary rate interface. User information for packet switching is also supported by the D channel.

(a) Point-to-point configuration

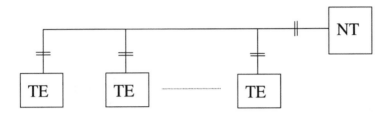

(b) Point-to-multipoint (bus) configuration

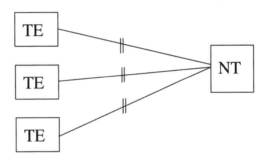

(c) Point-to-multipoint (star) configuration

Note: ‖ indicates ISDN user-network interface

FIGURE 10.5
Physical configurations of ISDN user-network interface.

Table 10.3   Channel Types

Channel Type			Channel Rate	Use
B			64 Kbit/s	• User information stream • Circuit switching/packet switching/semi-permanent connection
D			16 Kbit/s 64 Kbit/s	• Signaling information for circuit switching • Packet switched data
H	H_0		384 Kbit/s	
	H_1	H_{11}	1536 Kbit/s	• User information stream • Circuit switching/packet switching/semi permanent connection
		H_{12}	1920 Kbit/s	

Table 10.4   User-Network Interface Structure

Interface Type	Interface Rate	Interface Structure		Notes
Basic interface	192 Kbit/s	B channel interface structure	2B + D	D = 16 Kbit/s
		B channel interface structure	23 B + D (1544 Kbit/s) 30 B + D (2048 (Kbit/s)	
Primary rate interface	1544 Kbit/s or 2048 Kbit/s	H_0 channel interface structure	$4H_0$ or $3H_0+D$ (1544 Kbit/s) $5H_0$ + D (2048 Kbit/s)	D = 16 Kbit/s
		H_1 channel interface structure	H_{11} (1544 Kbit/s) H_{12} + D (2048 Kbit/s)	
		Mixture of B and H_0 channel structure	$nB + mH_0 + D$	

10.2.3.2   Interface Structure
Two types of UNI structures are specified, as shown in Table 10.4.

10.2.3.2.1   Basic Interface A basic interface supports two B channels and one D channel, designated as 2B + D where D = 16 Kbit/s. Two independent circuit mode communications are possible simultaneously by using the basic interface. Framing and maintenance bits are employed other than B and D channels, and the interface operates at the rate of 192 Kbit/s. Eight-packet mode communications are supported by using one of or any combination of B and D channels.

10.2.3.2.2   Primary Rate Interface A primary rate interface corresponds to a primary rate of the plesiochronous hierarchy, and two bit rates are specified: 1544 and 2048 Kbit/s. The following interface structures are possible.

B Channel Interface Structure For the 1544 Kbit/s interface, the interface structure of 23B + D is defined, and for the 2048 Kbit/s interface, 30B + D is specified. D channel capacity is 64 Kbit/s in both interface structures. Note that for the 1544 Kbit/s rate, the interface structure of 24B is possible, and the D channel on the different interface can support signaling for these 24B channels, such as the D channel of the different primary rate interface or the basic interface.

H_0 Channel Interface Structure For the 1544 Kbit/s interface, the interface structures of $3H_0$ + D and $4H_0$ are specified. For the 2048 Kbit/s interface, $5H_0$ + D is specified. The D channel capacity is 64 Kbit/s.

H_1 Channel Interface Structure For the 1544 Kbit/s interface, the interface structure of H_{11} is specified. For the 2048 Kbit/s interface, H_{12} + D is adopted. The D channel capacity is 64 Kbit/s.

Mixture of B Channels and H_0 Channel The interface structure of $nB + mH_0 + D$ is possible. The parameters m and n are assigned on provisioning or on demand through D channel signaling from the network.

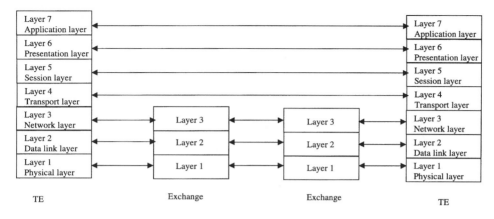

FIGURE 10.6
Layering structure of ISDN.

10.3   LAYERS 1, 2, AND 3 SPECIFICATIONS OF UNI

10.3.1   Layered Structure

The ISDN user-network interface is based on the layered model of the Open Systems Interconnection (OSI). Connection setup is made using functions of layers 1–3, as shown in Figure 10.6 [12].

- Layer 1 is a physical layer and specifies electrical and physical conditions of the interface and including interface connectors.
- Layer 2 is a data link layer and specifies assured transport of upper layer information. A link access protocol on the D channel (LAPD) is employed for ISDN.
- Layer 3 is a network layer and specifies communication protocols such as setup and release of connections using the D channel.

10.3.2   Basic Interface Layer 1

10.3.2.1   Wiring Configuration

Layer 1 of the basic interface is specified in ITU-T Recommendation I.430 [13]. Wiring configurations of point-to-point and point-to-multipoint (passive bus configuration) are defined, as shown in Figure 10.7.

The point-to-point configuration is capable of 1 km reach. Two types of bus configurations are specified: a short passive bus and an extended passive bus. The reach of a short passive bus is around 100–200 m, and eight terminals can be connected to the bus. The extended bus allows 500 m, however, terminal connection points are restricted to a grouping at the far end of the cable.

10.3.2.2   Frame Structure and Line Code

10.3.2.2.1   Frame Structure   A frame is composed of 48 bits in 250 ms, as shown in Figure 10.8, so the bit rate is 192 Kbit/s. Two B channels, a D channel, framing bits, direct current (DC) balancing bits, a bit used for activation, and D-echo channel bits used for a congestion control in the D channel are assigned in the frame.

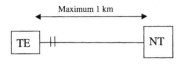

(a) Point-to-point configuration

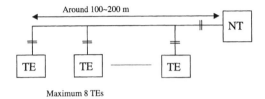

(b) Short passive bus configuration

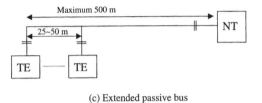

(c) Extended passive bus

FIGURE 10.7
Wiring configurations of a basic interface.

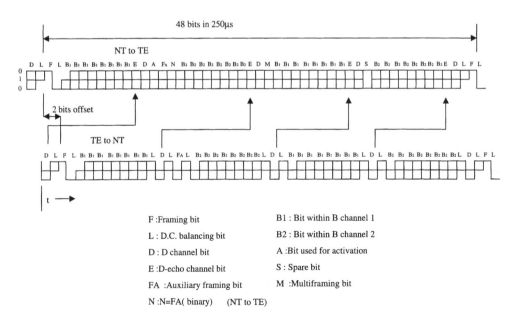

F : Framing bit   B1 : Bit within B channel 1

L : D.C. balancing bit   B2 : Bit within B channel 2

D : D channel bit   A : Bit used for activation

E : D-echo channel bit   S : Spare bit

FA : Auxiliary framing bit   M : Multiframing bit

N : N=FA(binary)   (NT to TE)

FIGURE 10.8
Frame structure of a basic interface at reference points T and S.

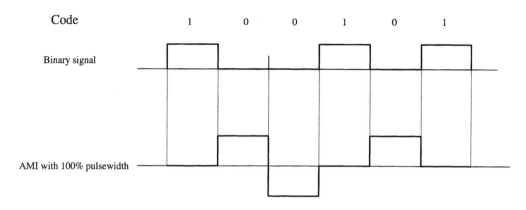

FIGURE 10.9
AMI coding rule.

TEs derive timing from the signal received from the NT and use this timing to generate a frame that is transmitted toward the NT. The first bit of the frame transmitted from a TE toward the NT is delayed by a 2-bit period with respect to the received frame in the TE.

10.3.2.2.2   Line Code   The line code is alternate mark inversion (AMI) with 100% duty, as shown in Figure 10.9. The "binary one" is represented by no line signal, whereas "binary zero" is represented by a positive or negative pulse.

The L-bit is for DC balancing. The L-bit is set binary zero if the number of binary zeros following the previous L-bit is odd. L-bit is set binary one if the number of binary zeros following the previous L-bit is even. This procedure ensures that the number of binary zeros in the frame is always even, and therefore the DC offset is minimized because the number of positive pulses and negative pulses in the line signal is equal. In the TE to NT direction, direct current (DC) balancing is required for each channel, because each TE transmits signals independently. In the NT to TE direction, DC balancing is performed for each frame.

10.3.2.2.3   Frame Alignment   The first bit of each frame is the frame bit, F, as shown in Figure 10.8. The frame bit is always set positive. Frame alignment is achieved by using a code rule violation of AMI, as shown in Figure 10.10:

- The F bit is set to have the same polarity as the preceding pulse (AMI violation).
- The L-bit adjacent to the F bit is set negative, and the first binary zero bit following the L-bit is set negative (second AMI violation).

The F_A-bit is introduced to guarantee the second violation in the frame.

For the direction from NT to TE, the N-bit is set binary opposite polarity of F_A-bit ($N = F_A$), so if there are no binary zeros after the L-bit, AMI violation occurs at the F_A-bit or the N-bit.

For the direction from TE to NT, at least the thirteenth bit following the F-bit, the F_A-bit is set binary zero, so the second AMI violation is assured. However, when multiframing is used, the F_A-bit is stolen for Q channel once in every five frames, so AMI violation is not assured in every five frames.

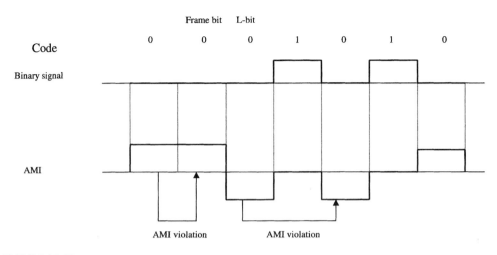

FIGURE 10.10
Frame alignment of a basic interface.

*10.3.2.2.4   **Multiframing*** Multiframing of the F_A-bit is optional to realize an auxiliary channel; 20 multiframing of the F_A-bit is specified.

For the direction from TE to NT, Qi (Q1, Q2, Q3, and Q4) is assigned to the F_A-bit once in
every five frames; thus Qi is transmitted once in every 20 frames, as shown in Table 10.5. In the

Table 10.5   Multiframe of Basic Interface

Multiframe Number	NT→TE			TE→NT
	F_A-bit	M-bit	S-bit	F_A-bit
1	Binary "1"	Binary "1"	$SC1_1$	Q1
2	"0"	"0"	$SC2_1$	Binary "0"
3	"0"	"0"	$SC3_1$	"0"
4	"0"	"0"	$SC4_1$	"0"
5	"0"	"0"	$SC5_1$	"0"
6	"1"	"0"	$SC1_2$	Q2
7	"0"	"0"	$SC2_2$	"0"
8	"0"	"0"	$SC3_2$	"0"
9	"0"	"0"	$SC4_2$	"0"
10	"0"	"0"	$SC5_2$	"0"
11	"1"	"0"	$SC1_3$	Q3
12	"0"	"0"	$SC2_3$	"0"
13	"0"	"0"	$SC3_3$	"0"
14	"0"	"0"	$SC4_3$	"0"
15	"0"	"0"	$SC5_3$	"0"
16	"1"	"0"	$SC1_4$	Q4
17	"0"	"0"	$SC2_4$	"0"
18	"0"	"0"	$SC3_4$	"0"
19	"0"	"0"	$SC4_4$	"0"
20	"0"	"0"	$SC5_4$	"0"

frame from NT to TE, M-bit is set binary one once in every 20 frames. The TE sends Q1 in the frame corresponding to the frame that received M-bit with binary one.

For the direction from NT to TE, F_A-bit is set binary one once in every five frames; the M-bit is binary one once in every 4 frames in which the F_A-bit is binary one. Then, the frame that has F_A-bit and M-bit as binary ones is the first frame of 20 multiframes. Auxiliary information is carried in five subchannels (SC) (SC1, SC2, SC3, SC4, and SC5) by utilizing the multiframe procedure, as shown in Table 10.5.

10.3.2.3   D-Channel Access Control

When a number of terminals connected to one bus sends signals over the D channel simultaneously, collisions occur. D-channel access control is required to send signals successfully.

This procedure utilizes the D channel coding rule that information on a D channel is delimited by flags consisting of binary pattern "01111110", and when consecutive 6 bits of 1's occur, a binary zero is inserted after the fifth 1. If there is no information to transmit over the D channel, terminals send binary 1's over the D channel. This procedure is called an *interframe* (Layer 2) *time fill*.

Thus, when one terminal occupies the D channel, consecutive binary ones are limited to at longest 6 bits. If consecutive binary ones of more than 7 bits are received, it is interpreted that no terminals are sending signals over the D channel.

10.3.2.3.1   D-Channel Monitoring
A terminal wanting to access the D channel monitors and counts the number of consecutive binary one bits in D-echo channel bit (E-bit) to judge if other terminals are using the D channel or not. If received E-bit is binary zero, another terminal is using the D channel; the terminal resets the counter value and restarts counting again. When the terminal detects binary one in the E-bit, the counter value increases by 1. The current counter value is designated as C. This procedure is useful for priority control and access fairness for all terminals.

10.3.2.3.2   Priority Control
Signaling information is given priority (priority class 1) over all other types of information (priority class 2). Furthermore, the terminal that has successfully completed the transmission of a layer 2 frame over the D channel is set lower priority in the same priority class in order to ensure fair access.

The threshold value X1 of the C counter for the priority class 1 (signaling information) and X2 for the priority class 2 (other types of information) are specified. Normal level and low level for each priority class are defined. The value of X1 is 8 and X2 is 10 for the normal level, and X1 is 9 and X2 is 11 for the low level. If C is equal to or exceeds the threshold value, the terminal can transmit a layer 2 frame to the D channel. When the terminal has successfully transmitted a layer 2 frame, the threshold value is changed from the normal level to the low level. If the counter value equals the low level, the threshold value is set back to the normal level.

10.3.2.3.3   Collision Detection
Even with the above procedure, a number of terminals can transmit layer 2 frames simultaneously. A collision detection mechanism and access control scheme are used to complete the transmission of information from all active terminals successfully in order.

NT sends back the received data over the D channel as an E-bit toward terminals (echo back). Each terminal compares the bit received in an E-bit with the last bit transmitted over the D channel. If both bits are the same, the terminal can continue to transmit information; otherwise the terminal stops transmitting information. At the D channel, a binary zero is coded as negative pulse, and a binary one is identified as no signal. As illustrated in Figure 10.11, when terminals

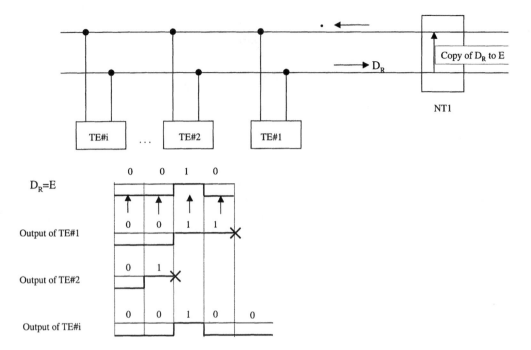

FIGURE 10.11
Illustration of D-channel congestion control.

#1 and #i send binary zeros, and terminal #2 sends a binary one, the resulting signal on the bus is perceived as binary zero. As a result, terminals #1 and #i find that the transmitted bit in the D channel and the received bit in the E-bit are the same; therefore the terminals #1 and #i continue to transmit information.

Normally transmitted bits of terminals are different; therefore finally only one terminal can continue to transmit all data over the D channel.

10.3.2.4 *Activation/Deactivation*

NT and terminals are likely to power off to reduce power consumption, when they are not used for communication. The activation procedure to power on at the time of communication start and the deactivation procedure to power off at the time of communication termination are specified.

Activation and deactivation are achieved using special signals on the UNI, as shown in Table 10.6. When a terminal requests call setup to NT, the terminal sends the INFO1 signal, as listed in Table 10.6. Because the terminal is asynchronous with the network at the time of activation, the INFO1 signal is asynchronous. An example of activation is shown in Figure 10.12.

10.3.2.5 *Power Feeding*

To support digital telephony, the network can provide DC power to NT through the access line as an option, and also to a terminal through the UNI that uses an eight-pin connector. Access

Table 10.6   Definition of Info Signals

Signals from NT to TE		Signals from TE to NT	
INFO 0	No signal	INFO 0	No signal
INFO 2	• Frame with all bits of B, D, and D-echo channels set to "0". • A-bit set to "0".	INFO 1	 Continous signal with above pattern. Nominal bit rate = 192 Kbit/s
INFO 4	• Frames with operational data on B, D, and D-echo channels. • A-bit set to "1".	INFO 3	Synchronized frames with operational data on B and D channels.

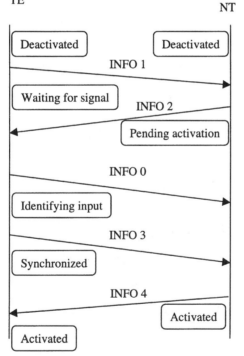

FIGURE 10.12
Example of an activation procedure (activation request from TE).

leads 3–4 and 5–6 are used for the bidirectional transmission of the digital signal. DC power can be provided using leads 3–4 and 5–6 over the digital signal, as shown in Figure 10.13. This is designated *phantom power feeding.*

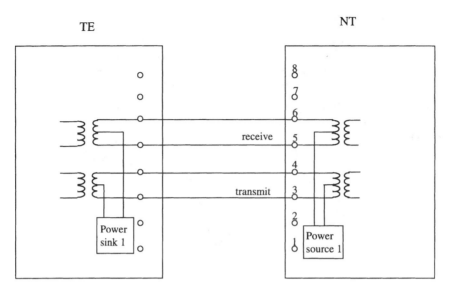

Note: Leads 3, 4, 5, and 6 are mandatory.

FIGURE 10.13
Phantom power feeding.

10.3.2.6 *Electrical Condition*
Electrical conditions specify parameters such as jitter, impedance, and output pulse mask, as shown in Table 10.7 and Figure 10.14.

10.3.2.7 *Physical Condition*
An eight-pin connector is specified for the UNI interconnection that is defined in International Organization for Standardization (ISO) 8877.

Table 10.7   Main Items of Basic Interface Electrical Characteristics

Items	Specifications
Bit rate	192 Kbit/s
Tolerance (free running mode)	±100 ppm
Pulse shape	Pulse mask specification shown in Figure 10.14
	Nominal pulse amplitude: 750 mV
Termination of line	100Ω
Line code	AMI
TE jitter characteristics	Timing extraction jitter: ±7%
NT jitter characteristics	Maximum output jitter (peak-to-peak): 5%

10.3.3   Primary Rate Interface Layer 1

A primary rate interface layer 1 is specified in ITU-T Recommendation I.431 [14]. The specification is mainly based on the primary rate interface specified for network elements in Recommendation G.704 [15] for the plesiochronous hierarchy. The main items are summarized in Table 10.8.

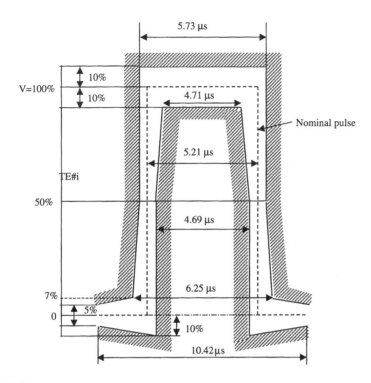

FIGURE 10.14
Transmitter ouput pulse mask.

Table 10.8   Main Layer 1 Characteristics for Primary Rate Interface

	Items	1544 Kbit/s Primary Rate Interface	2048 Kbit/s Primary Rate Interface
Physical conditions	Wiring configuration	Point-to-point 4-wire	Point-to-point 4-wire
Electrical conditions	Cable	Metallic pair cable	Metallic pair cable Coaxial cable
	Bit rate	1544 Kbit/s	2048 Kbit/s
	Clock frequency tolerance (free running mode)	± 32 ppm	± 50 ppm
	Pulse shape	Amplitude: $3V_{0\text{-}p}$ Duty: 50%	Amplitude: $3V_{0\text{-}p}$ (metallic) $2.37\,V_{0\text{-}p}$ (coaxial) Duty: 50%
	Line termination	100Ω	120Ω (metallic) 75Ω (coaxial)
	Line code	B8ZS	HDB3
	Transmission loss	0~18dB (at 772 kHz)	0~6dB (at 1.024 MHz)

The UNI interface of the primary rate specifies only point-to-point configuration, and no D-channel access control is defined. No power feeding from NT to terminal is specified, and the primary rate UNI is always active.

10.3.3.1   Wiring Configuration

The primary rate access only supports the point-to-point configuration. Twisted pair metallic cable is used for the user-network interface cabling. The UNI is four-wire and bidirectional.

10.3.3.2   Line Code

The line code of the primary rate interface uses a bit sequence independence (BSI) code, such as Bipolar with 8 Zeros Substitution (B8ZS) code or High-Density Bipolar of order 3 (HDB3) code, so that a timing signal can be extracted even if there is a run of continuous "0's". The 1544 Kbit/s interface uses B8ZS code while the 2048 Kbit/s interface uses HDB3 code. Coding rule examples are shown in Figure 10.15. When "0's" continue for more than the specified value, the sequence is replaced by a specific pattern that includes "1". This procedure guarantees timing extraction.

10.3.3.3   Frame Structure

Frame structures are shown in Figures 10.16a and b.

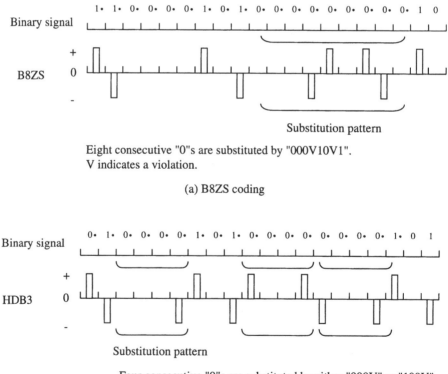

Eight consecutive "0"s are substituted by "000V10V1".
V indicates a violation.

(a) B8ZS coding

Four consecutive "0"s are substituted by either "000V" or "100V".

(b) HDB3 coding

FIGURE 10.15
Line code rule for a primary rate interface.

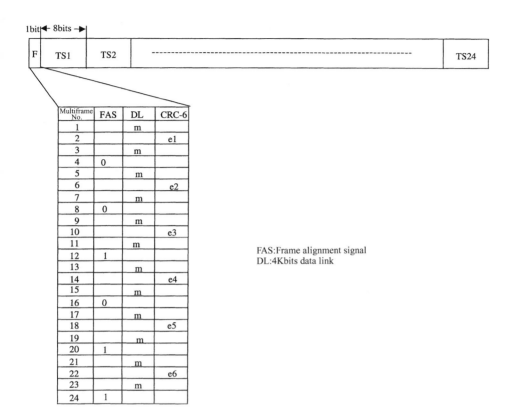

Multiframe No.	FAS	DL	CRC-6
1		m	
2			e1
3		m	
4	0		
5		m	
6			e2
7		m	
8	0		
9		m	
10			e3
11		m	
12	1		
13		m	
14			e4
15		m	
16	0		
17		m	
18			e5
19		m	
20	1		
21		m	
22			e6
23		m	
24	1		

FAS:Frame alignment signal
DL:4Kbits data link

FIGURE 10.16a
Frame structure for a 1544 Kbit/s primary rate interface.

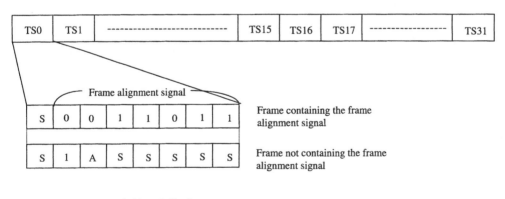

A:Alarm indication
S:Spare

FIGURE 10.16b
Frame structure for a 2048 Kbit/s primary rate interface.

10.3.3.3.1   *Frame Structure of 1544 Kbit/s Interface* A frame is composed of 193 bits in 125 μs. The first bit is a framing bit; the following 192 bits are composed of 24 time-slots numbered 1 to 24 for carrying digital signals.

The framing bit forms 24 multiframes. Six bits are for multiframe alignment, and 6 bits are for Cyclic Redundancy Check-6 (CRC-6) for error checking and protection from misframing.

If the 64 Kbit/s D channel is used, the D channel is assigned in time-slot 24. Other channels (B, H_0, H_{11}) are assigned in time-slots from 1 to 23 or 1 to 24. As for the H_{11} channel, the D channel is not assigned on the interface and, if required, the D channel is used on the associated interface such as another primary rate interface or basic interface.

10.3.3.3.2 Frame Structure of 2048 Kbit/s A frame is composed of 32 time-slots (256 bits) numbered from 0 to 31 in 125 ms. The frame alignment signal occupies time-slot 0 of every two frames, as shown in Figure 10.16b.

The D channel is assigned in time-slot 16, and other channels (B, H_0, and H_{12}) are assigned in time-slots from 1 to 15 and 17 to 31.

10.3.4   Layer 2 Specification

The layer 2 specification is common to the basic and the primary rate interfaces. The general aspects are specified by ITU-T Recommendation I.440/Q.920 [16], and the details are in Recommendation I.441/Q921 [17]. These recommendations are double numbered as series I and Q Recommendations.

The layer 2 offers a reliable container using the D channel for carrying layer 3 information between a network and terminals, such as call control signaling and packet information. Frame structure, type of commands, and information transfer procedures are specified.

The Protocol called Link Access Procedure on the D-channel (LAPD) is specified for ISDN based on the balance mode of the High-Level Data Link Control procedure (HDLC) with the enhanced applications to the bus interface.

10.3.4.1   SAPI and TEI

As described in Section 10.3.2, the basic interface adopts the bus interface and a number of terminals connected to the same bus can be active simultaneously. Therefore, there exist a number of layer 2 data links and each should be identified for each terminal, as shown in Figure 10.17. The Terminal Endpoint Identifier (TEI) identifies the data link for each terminal.

Call control information and packet information are multiplexed over the D channel. The Service Access Point Identifier (SAPI) identifies services that layer 2 provides to layer 3.

TEI and SAPI are shown in Table 10.9.

Table 10.9   TEI and SAPI

TEI/SAPI	Values	Use
TEI	0–63	Nonautomatic TEI assignment TE
	64–126	Automatic TEI assignment TE
	127	Broadcast data link
SAPI	0	Call control procedure
	16	Packet communication procedure
	63	Layer 2 managment procedure
	All others	Reserved

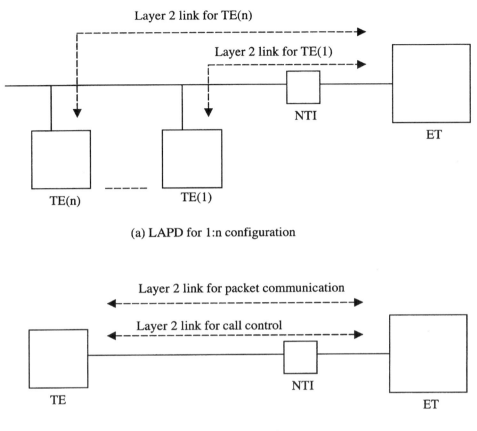

(a) LAPD for 1:n configuration

(b) LAPD for 1:1 configuration

FIGURE 10.17
A number of layer 2 links in one user-network interface.

10.3.4.2   Frame Structure

The layer 2 frame structure is shown in Figure 10.18. The frame structure is based on HDLC, and is composed of flag sequence, address field, control field, information field, and frame checking sequence (FCS) field.

10.3.4.2.1   Flag Sequence   The flag sequence indicates the start and end of the frame. The flag sequence pattern is "01111110". In other fields, when six consecutive "1's" occur, "0" is inserted after five consecutive "1's" to avoid the same pattern as the flag sequence.

10.3.4.2.2   Address Field   The address field size is 2 bytes. It is composed of TEI, SAPI, and C/R bit, which indicates the type of frame: command or response.

10.3.4.2.3   Control Field   The control field size is 1 or 2 bytes. It is used for transferring the type of command/response and sequence number. There are three control field formats: numbered information transfer (I format), supervisory functions (S format), and unnumbered information transfer and control functions (U format), as shown in Table 10.10.

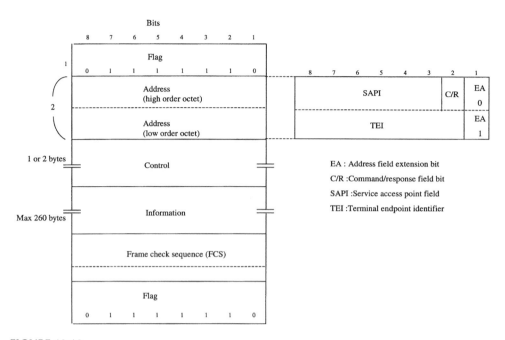

FIGURE 10.18
Layer 2 frame format.

Table 10.10   Commands and Responses

Format	Commands	Responses	Encoding 8	7	6	5	4	3	2	1
Information transfer	I (information)		N(S)							0
			N(R)							P
Supervisory	RR (receive ready)	RR (receive ready)	0	0	0	0	0	0	0	1
			N(R)							P/F
	RNR (receive not ready)	RNR (receive not ready)	0	0	0	0	0	1	0	1
			N(R)							P/F
	REJ (reject)	REJ (reject)	0	0	0	0	1	0	0	1
			N(R)							P/F
Unnumbered	SABME (set asynchronous balance mode extended)		0	1	1	P	1	1	1	1
		DM (disconnect mode)	0	0	0	F	1	1	1	1
	DISC (disconnect)		0	1	0	P	0	0	1	1
		UA (unnumbered acknowledge)	0	1	1	F	0	0	1	1
		FRMR (frame reject)	1	0	0	F	0	1	1	1

10.3.4.2.4 Frame Checking Sequence (FCS) FCS is used for error checking the frame. The generating polynomial is $x^{16} + x^{12} + x^5 + 1$.

10.3.4.3 Information Transfer Procedures

Two types of information transfer procedures are specified: unacknowledged information transfer and acknowledged information transfer procedures.

10.3.4.3.1 Unacknowledged Information Transfer Procedure This procedure does not apply to an acknowledgment for transferring information, and broadcasting is used to notify the arrival of a call to all terminals on the bus. This procedure uses an unnumbered information (UI) frame, and does not provide error control.

This is also used for TEI assignment, as described in section 10.4.4.4.

10.3.4.3.2 Acknowledged Information Transfer Procedure This procedure employs the acknowledgment of information arrival at the destination. This provides error recovery and flow control.

Establishment Procedure of Data Link   Prior to transmitting information, a point-to-point link is established between a network and a terminal with the specified TEI value. In this procedure, the Set Asynchronous Balanced Mode Extended (SABME) command with the P-bit set to "1" is first transmitted to the destination, as shown in Figure 10.19. The receiving side sends back the unnumbered acknowledgement (UA) response. These procedures establish the data link. Both sides reset states to the initial values and wait for information transferring stage.

Information Transfer Procedure   Layer 3 information is transferred using the information frame (I frame). A number of I frames can be transmitted continuously without any response from the receiving side. Retransmission of an I frame is possible to achieve error recovery and flow control. This procedure utilizes I frame sequence number and state variables of terminals and network.

Send State Variable V(S)   This value denotes the sequence number of the next in-sequence I frame to be transmitted. After the successful transmission of an I frame, this value is increased by 1.

Acknowledge State Variable V(A)   This value identifies the last frame that has been acknowledged by the remote end. This value is updated by the valid N(R) values received from the remote end. The V(A)-1 equals the N(S) as described below of the last acknowledged I frame at the remote end.

Send Sequence Number N(S)   This shows sequence number for each I frame transmitted. Only I frames contain N(S). The value of N(S) is set equal to V(S) prior to transmitting the I frame.

Receive State Variable V(R)   This value denotes the sequence number of the next-in-sequence I frame expected to be received. The value of the receive state variable is increased by 1 upon the receipt of an error-free, in-sequence I frame whose send sequence number N(S) equals V(R).

Receive Sequence Number N(R)   All I frames and supervisory frames contain N(R). Prior to sending these frames, the value of N(R) is set equal to the current value of V(R). One end acknowledges that a remote end has correctly received all I frames up to and equal to N(R)-1.

Release Procedure of Data Link   The end that requests the release of the data link should send the Disconnect (DISC) command, and the receiving end issues an Unnumbered Acknowledgement (UA) response.

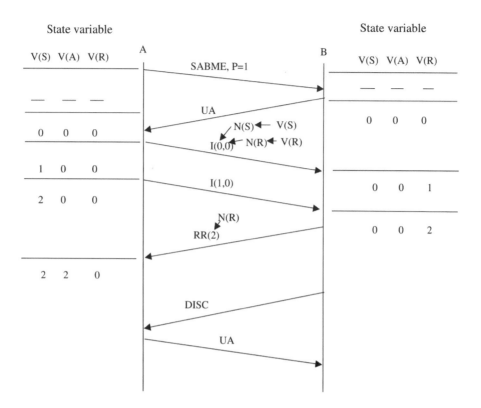

FIGURE 10.19a
Example of valid acknowledgments procedure.

An example of a valid acknowledgments procedure from link establishment to link release is shown in Figure 10.19a.

Figure 10.19b shows an example of frame loss due to transmission error. The end B identifies V(R) = 2; however, N(S) = 3 on the received I frame, then the end B identifies N(s) ≠ V(R). This means that a loss of the I frame has occurred. The end B sends REJ(2) to end A. At the end A, values of V(S) and V(A) are updated and set equal to N(R) value on REJ frame, and an I frame corresponding to V(S) is retransmitted.

10.3.4.4   Terminal Endpoint Identifier (TEI) Management Procedure
Each terminal can communicate using a point-to-point data link after being assigned a TEI. Two cases for TEI assignment procedure are described: automatic assignment and manual assignment. For *manual assignment,* the user assigns the TEI value that must be verified to ensure that the TEI is not already being used by another user's equipment. For *automatic assignment,* the TEI is assigned by the network using the TEI assignment procedure.

TEI assignment procedure, TEI check routine procedure, TEI removal procedure, and TEI identity verify procedure (optional use) are specified. These procedures are performed using the unacknowledged information transfer procedure.

10.3.4.4.1   TEI Assignment Procedure   This is a procedure that TE requests TEI assignment to network. TE sends an unnumbered information (UI) frame whose SAPI = 63, TEI = 127, and message type = Identity request (ID request). The network selects and verifies TEI value, and

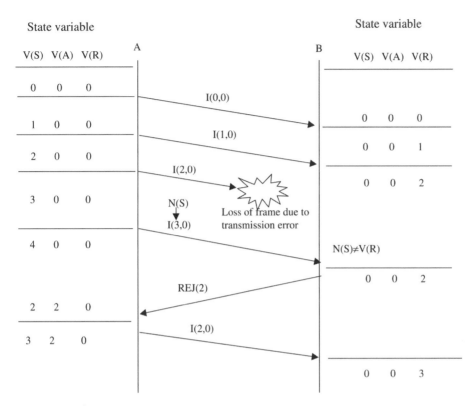

FIGURE 10.19b
Example of retransmission due to frame loss.

sends back a UI frame having the selected TEI value in action indicator (Ai) field, reference number (Ri) that was received from TE, and message type = ID assigned.

10.3.4.4.2 TEI Check Routine Procedure TE with assigned TEI may be removed from a socket or is moved to another bus. The TEI check routine procedure involves the network checking TEI considering the above situations.

When the network requests to check TEI to all TEs, the network sends UI frame with Ai = 127, and message type = ID check request. On the other hand, if the network requests to check the TEI of specific TE, the network sends a UI frame with Ai = TEI. The network can determine if the TEI is double-assigned or free to be assigned by monitoring the responses from the TEs.

10.3.4.4.3 TEI Removal Procedure When a network determines an erroneous TEI assignment such as double assignment, the network requests the TE to remove the TEI. The network sends a UI frame with the TEI to be removed in the Ai field and message type = ID remove.

10.3.4.4.4 TEI Identity Verify Procedure This procedure is optional. TE can request the network to issue a TEI check. TE sends a UI frame with TEI to be checked in Ai field, and message type = ID verify. When the network receives this request, it activates the TEI check routine procedure.

10.3.5   Layer 3 Specification

The general aspect of layer 3 is specified in Recommendation I.450/Q.930 [18], and the details are specified in Recommendation I.451/Q.931[19].

   Layer 3 specifies control information and procedures between terminal and network to establish, maintain, and terminate circuit switched connections and packet switched connections. Layer 3 is carried over D channel, as shown in Figure 10.20, and utilizes layer 2 functions such as establishing/releasing data link, data transmission, and error recovery.

   Layer 3 of ISDN is specified for various types of access and services, as in the following.

- *Service select for call by call.* Customer can select information transfer attributes on call by call bases, such as circuit switched/packet switched, bit rate, and speech/unrestricted digital/audio to use various types of services.
- *Compatibility checking.* Compatibility checking is employed to avoid misconnection between different types of services. Low-layer compatibility checking, such as user rate, and high-layer compatibility checking, such as media type, are specified.
- *Variable length format.* A variable-length format is adopted to provide unified format for all messages and future use.

10.3.5.1   *Message Format*
Message format is composed of a common part and a message-dependent part, as shown in Figure 10.21.

10.3.5.1.1   *Common Part* The common part is found in all messages.

- *Protocol discriminator.* This is used to distinguish messages for user-network call control from other messages. The code "000010000" is assigned to indicate call control in Q.931.
- *Call reference.* This is used to identify a call. The specified size is 1 byte for the basic UNI and 2 bytes for the primary rate interface. Call reference value is assigned by the side originating the call. The call reference flag (bit 8 in octet 2) is always set to "0" at the origination side and "1" at the destination side.
- *Message type.* This is used to identify functions of message, such as connect, response, and disconnect.

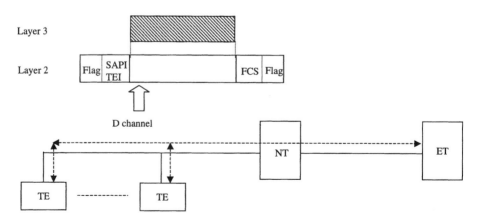

FIGURE 10.20
Layer 3 mapping.

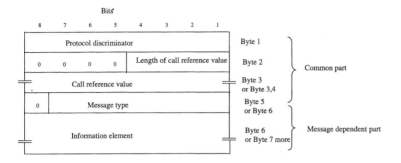

FIGURE 10.21
Message format.

10.3.5.1.2 Message-Dependent Part   The message-dependent part is composed of information elements, such as calling party number, called party number, and bearer capability. Information element format has variable length with an information element identifier, a length of contents of information element, and a contents of information element, as shown in Figure 10.22. A fixed-length format is also specified.

10.3.5.2   Examples of Call Control Procedure

10.3.5.2.1   Example of Circuit-switched Call

Basic Procedure   One example of the call control procedure for a circuit switched call is illustrated in Figure 10.23. Messages are also indicated in this figure.

User-to-User Signaling   User-to-user signaling is a supplementary service providing communications between two users using layer 3 protocol on the D channel. There are three types of user-to-user signaling associated with circuit switched calls.

Service 1. This supports exchange of information during call setup and call clearing phases, within call control messages.

Service 2. This supports exchange of information during call establishment, between alerting and connect messages, within user information messages.

Service 3. This supports exchange of information during the information transfer phase, within user information messages.

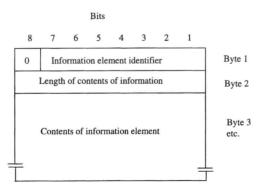

FIGURE 10.22
Format of variable length information element.

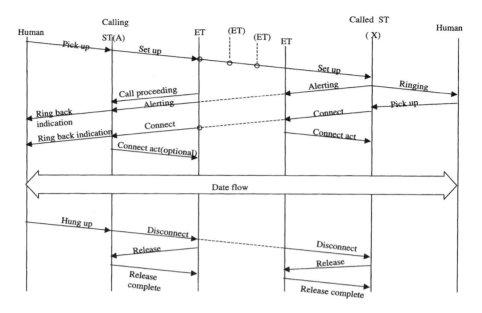

FIGURE 10.23
Example of a procedure for a circuit switched call.

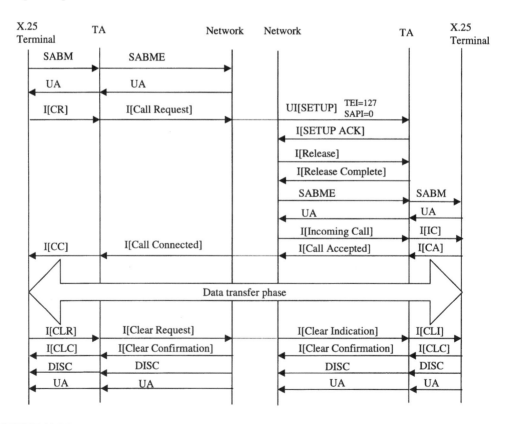

FIGURE 10.24
Example of a procedure of packet communication over the D channel.

10.3.5.2.2   Example of Packet Switched Call Packet communication is provided over the D channel and the B channel. An example of virtual calling over the D channel is shown in Figure 10.24.

10.3.5.3   ISUP (ISDN User Part)

Circuit switched call control between customer and network is performed by using the D channel. However, call control within the network is performed by using ISDN User Part (ISUP) of No. 7 Common Channel Signaling System [20], as shown in Figure 10.25. The No. 7 Common Channel Signaling System adopts the OSI 7 layer model to allow for future expansion of the functions supported. The user part corresponds to layer 3 and higher. ISUP and Telephone User Part (TSUP) are specified.

10.4   ACCESS TRANSMISSION LINE SYSTEMS

10.4.1   Outline of Transmission Line System

Layer 1 architecture for access systems between customer and central office is shown in Figure 10.26. The section between T and V reference points is designated as the access digital section.

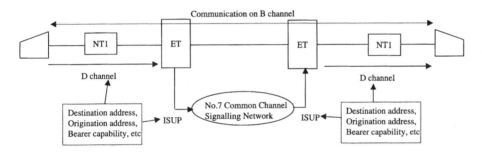

FIGURE 10.25
Transfer signaling within ISDN using ISUP.

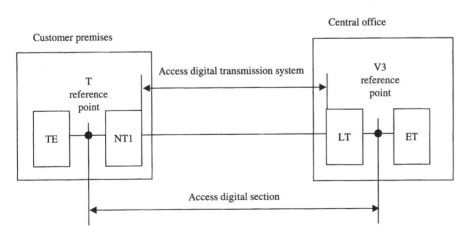

FIGURE 10.26
Layer 1 for access system.

Basic access transmission line systems utilize existing two-wire metallic cables for economical and quick penetration of ISDN. Echo cancellation (ECH) schemes and time compression multiplexing (TCM) schemes are described as examples in the Appendices to ITU-T Recommendation G.961 [21]. Two line codes are described for the ECH scheme. Common requirements for these three schemes, such as function, operation and maintenance, transmission medium, system performance, activation/deactivation, and power feeding are specified in the main body of G.961.

For primary rate access, the access digital section at 2048 Kbit/s is specified in G.962 [6], and the access digital section at 1544 Kbit/s is specified in G.963 [7]. These are specifications for T and V reference points. The existing primary rate transmission systems used for trunk lines are applied to the access transmission line.

10.4.2   Metallic Transmission Line System for Basic Access

10.4.2.1   Echo Cancellation System

10.4.2.1.1   Outline The ECH system uses a hybrid circuit and an echo cancellor circuit, as shown in Figure 10.27. Echo cancellor cancels return signals due to the imperfection in the line, including the imperfection of hybrid circuit over the used bandwidth.

The ECH system transmits signals at the same speed as the user-network interface. The transmission distance is limited by the near-end-crosstalk (NEXT). There are two line code systems. One is 2 Binary to 1 Quaternary conversion (2B1Q), and the other is Modified Monitoring State code mapping 4 bits into 3 ternary symbols with levels "+", "0", or "−" (MMS43) based on 4 Binary to 3 Ternary conversion (4B3T) code. In the United States, the 2B1Q system is specified in American National Standards Institute (ANSI) T1.601 standard [22]. In Europe both systems are recommended by the European Telecommunications Standards Institute (ETSI).

10.4.2.1.2   2B1Q Transmission System 2B1Q is four-level code for a 2-bit block, as shown in Table 10.11, so the line speed is half that of the original signal. The frame structure is shown in Figure 10.28. Twelve 2B + D ($12 \times (2 \times 8 + 2) = 216$ bits) and associated framing bits and maintenance bits compose the frame. The total bit rate is 160 Kbit/s, so the symbol rate is 80 Kbaud.

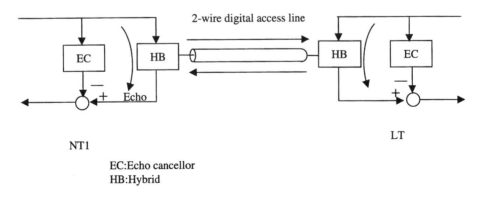

FIGURE 10.27
ECH functional diagram.

Table 10.11 2B1Q Coding

	Bits	2B1Q Quaternary Symbol
	0 0	−3
	0 1	−1
	1 0	+1
	1 1	+3

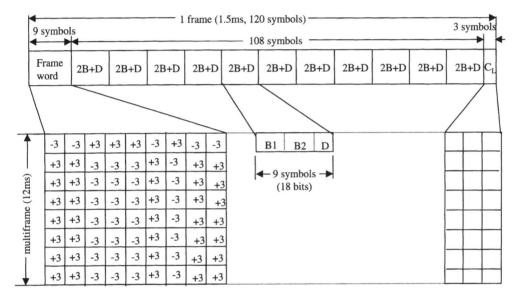

FIGURE 10.28
Frame structure of an echo cancellor transmission system (2B1Q).

The frame word is nine symbols, and is structured as eight multiframes. The C_L channel is three symbols, and is also structured as eight multiframes. The C_L channel is used for maintenance, such as error monitoring by CRC, activation/deactivation control, test control, and status indication.

10.4.2.1.3 MMS43 Transmission System MMS43 is based on 4B3T line code that converts a 4-bit block into 3 ternary symbols with positive pulse ("+"), negative pulse ("−"), or "0". The conversion rule of MMS43 is shown in Table 10.12. The number i (i = 1, 2, 3, or 4) in the columns for each S1, . . . S4 indicates the symbol number Si to be used for coding the next block of 4 bits. At first, S1 is used. For example, as shown in Figure 10.29, the first 4 bits are "1010", so the coded symbol in S1is "++−", and the number in the column is 2. Therefore, the next symbol is S2. Thus the next 4 bits are "1101", and the coded symbol is "0 + 0". This procedure is adopted to reduce DC offset.

Table 10.12   MM3 43 Coding

Bits	S1		S2		S3		S4	
			MMS43 Ternary symbol					
0001	0 – +	1	0 – +	2	0 – +	3	0 – +	4
0111	– 0 +	1	– 0 +	2	– 0 +	3	– 0 +	4
0100	– + 0	1	– + 0	2	– + 0	3	– + 0	4
0010	+ – 0	1	+ – 0	2	+ – 0	3	+ – 0	4
1011	+ 0 –	1	+ 0 –	2	+ 0 –	3	+ 0 –	4
1110	0 + –	1	0 + –	2	0 + –	3	0 + –	4
1001	+ – +	2	+ – +	3	+ – +	4	– – –	1
0011	0 0 +	2	0 0 +	3	0 0 +	4	– – 0	2
1101	0 + 0	2	0 + 0	3	0 + 0	4	– 0 –	2
1000	+ 0 0	2	+ 0 0	3	+ 0 0	4	0 – –	2
0110	– + +	2	– + +	3	– – +	2	– – +	3
1010	+ + –	2	+ + –	3	+ – –	2	+ – –	3
1111	+ + 0	3	0 0 –	1	0 0 –	2	0 0 –	3
0000	+ 0 +	3	0 – 0	1	0 – 0	2	0 – 0	3
0101	0 + +	3	– 0 0	1	– 0 0	2	– 0 0	3
1100	+ + +	4	– + –	1	– + –	2	– + –	3

Binary Code   1 0 10   1 1 0 1

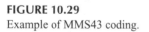

MMS43 Code

FIGURE 10.29
Example of MMS43 coding.

The total bit rate is 160 Kbit/s and the symbol rate is 120 Kbaud. Two 2B + D (36 bits then 27 symbols) forms one group. Four of these groups, C_L channel, and frame word compose the frame. There are a total of 120 symbols in one frame for 1 ms, as shown in Figure 10.30, and the frame word is different in each direction.

10.4.2.2   Time Compression Multiplexing System

TCM is a "burst mode" method. Signals are once stored in a buffer memory for each block (burst), and are read out at higher timing than two times of write timing, as shown in Figure 10.31. Bidirectional transmission over a two-wire system is possible by adopting burst multiplexing. This system is mainly used in Japan.

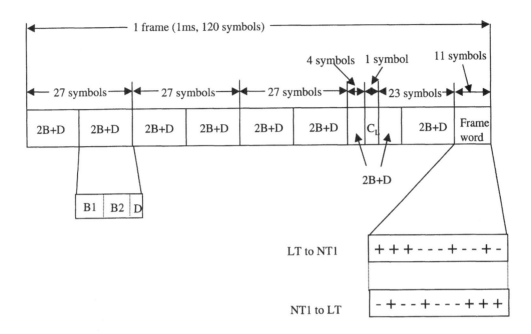

FIGURE 10.30
Frame structure of an echo cancellor transmission system (MMS43).

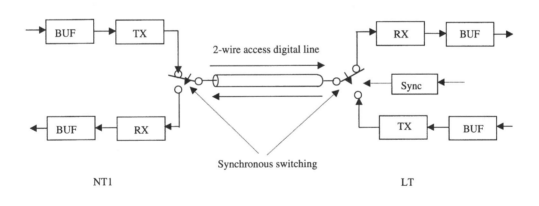

TX:Transmitter
RX:Receiver
BUF:Buffer
SYNC:Burst synchronization

FIGURE 10.31
TCM functional diagram.

One burst is 2.5 ms, as shown in Figure 10.32, and thereafter signals for each direction are switched in 2.5 ms. Twenty 2B + D, frame word, and C_L channel is 1.178 ms (377 bits) and guard time is included, and the total length is then 2.5 ms, as shown in Figure 10.33. The line rate is 320 Kbaud.

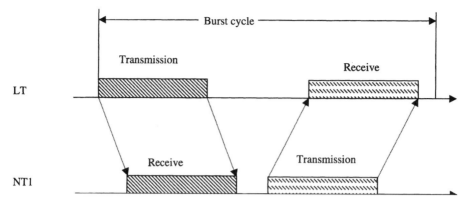

FIGURE 10.32
Time compression multiplexing.

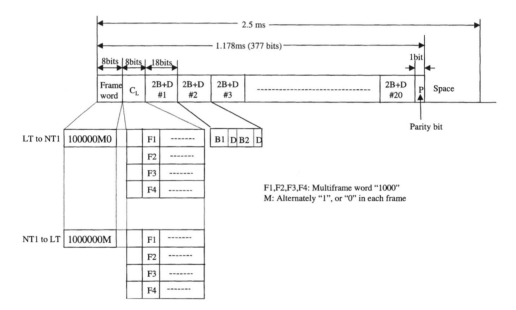

FIGURE 10.33
Frame structure of a TCM transmission system (AMI).

If burst synchronization is achieved for all signals in one cable from a central office, near-end crosstalk can be suppressed, which allows longer transmission distance.

AMI is adopted as the line code.

10.4.3   Primary Rate Transmission System

10.4.3.1   1544 Kbit/s System
In North America, the access transmission system uses a metallic cable with a bit rate of 1544 Kbit/s. The line code is AMI or B8ZS. B8ZS is the same method as used in the UNI.

There are three frame structures: T1, D4 (12 multiframe), and Extended Superframe (EFS): 24 multiframe.

Another optical transmission system that operates at a bit rate of 1544 Kbit/s is used in Japan. The line code is Coded Mark Inversion (CMI). The frame structure is 24 multiframe.

10.4.3.2   2048 Kbit/s System

The access transmission system that operates at a bit rate of 2048 Kbit/s is mainly used in Europe. Metallic pair cable and coaxial cable are used. The line code is HDB3.

The frame structure is the same as that of the trunk line system, designated as E1.

10.5   REFERENCES

[1] ITU-T I-Series Recommendations: "Integrated Services Digital Network (ISDN)," Redbook, Malaga, Spain, 1984.

[2] Recommendation I.310 : "ISDN—Network functional principle," 1993.

[3] Recommendation I.210: "Principles of telecommunication services supported by an ISDN and the means to describe them," 1993.

[4] Recommendation I.250: "Definition of supplementary services," 1988.

[5] Recommendation G.960: "Access digital section for ISDN basic rate access," 1993.

[6] Recommendation G.962: "Access digital section for ISDN primary rate at 2048 kbit/s," 1993.

[7] Recommendation G.963: "Access digital section for ISDN primary rate at 1544 kbit/s," 1993.

[8] Recommendation I.410: "General aspects and principles relating to Recommendations on ISDN user-network interfaces," 1988.

[9] Recommendation H.261: "Video codec for audiovisual at p × 64 kbit/s," 1993.

[10] Recommendation I.411: "ISDN user-network interfaces—reference configurations," 1993.

[11] Recommendation I.412: "ISDN user-network interfaces—Interface structures and access capabilities," 1988.

[12] Recommendation I.320: "ISDN protocol reference model," 1993.

[13] Recommendation I.430: "Basic user-network interface—Layer 1 specification," 1995.

[14] Recommendation I.431: "Primary rate user-network interface—Layer 1 specification," 1995.

[15] Recommendation G.704: "Synchronous frame structures used at 1544, 6312, 2048, 8488 and 44736 kbit/s hierarchical levels," 1995.

[16] Recommendation I.440/Q.920: "ISDN user-network interface data link layer—General aspects," 1993.

[17] Recommendation I.441/Q.921: "ISDN user-network interface—Data link layer specification," 1997.

[18] Recommendation I.450/Q.930: "ISDN user-network interface layer 3—General aspects," 1993.

[19] Recommendation I.451/Q.931: "ISDN user-network interface layer 3 specification for basic call control," 1998.

[20] Recommendation Q.761: "Signaling System No. 7—ISDN User Part functional description," 1997.

[21] Recommendation G.961: "Digital transmission system on metallic local lines for ISDN basic rate access," 1993.

[22] ANSI T1.601: "ISDN—Basic access interface for use on metallic loops for application on the network side of the NT (layer 1 specification)," 1992.

CHAPTER 11

Video-on-Demand Broadcasting Protocols

STEVEN W. CARTER
DARRELL D. E. LONG
JEHAN-FRANÇOIS PÂRIS

11.1 INTRODUCTION

Video-on-demand (VOD) will one day allow clients to watch the video of their choice at the time of their choice. It will be less complicated to use than a typical Web browser, and it will only require that the clients have a set-top box (STB) connected to their television set. The STB will allow the clients to navigate a VOD server's video library, and will then handle the reception and display of the video once the clients make a selection. It is possible that the VOD service will even be interactive, allowing clients to use videocassette recorder (VCR) controls such as pause, rewind, and fast forward while viewing their requests. In this case the STB will also be responsible for communicating the desired interactions to the VOD server.

Unfortunately, video is an expensive medium to distribute. Even using a compression scheme such as MPEG-2, a stream of video data will take a bandwidth of at least 4 Megabits per second (Mbit/s)—or roughly 70 times the capacity of a 56K modem. This sort of bandwidth will not necessarily be a problem for the client, but the VOD server will be limited by its bandwidth capacity, and so it must be able to service as many clients as possible in the bandwidth it has available.

Simply dedicating a unique stream of video data to each client request is not a good idea because that can quickly use up all of the available bandwidth on the VOD server. Better, more efficient strategies are needed.

One such set of strategies involves broadcasting videos to clients. Video rental patterns suggest that most client requests will be for the most popular 10–20 videos [6], and so broadcasting these videos ensures bounds on the amount of time clients must wait before they can watch their requests and on the total amount of bandwidth each video will require.

In the remainder of this chapter we will first define some common terms and concepts used by broadcasting protocols, and will describe the protocols themselves.

179

11.2 COMMON TERMS AND CONCEPTS

The broadcasting protocols described in this chapter have many concepts in common. For example, many of the protocols divide videos into segments. *Segments* are consecutive chunks of the video. By concatenating the segments together (or by playing them in order), clients will have (see) the entire video.

The *consumption rate* is the rate at which the client STB processes data in order to provide video output. For MPEG-2, this rate can be 4 Mbit/s or more. We will represent the consumption rate as b and use it as the unit of measure for VOD server bandwidth. When all video segments are the same size, we define the time it takes the STB to consume a segment as a *slot*.

Each distinct stream of data is considered to be a logical *channel* on the VOD server. These channels do not need to be of bandwidth b, and each video may have its segments distributed over several channels.

When the channel bandwidth is greater than b or when the client STB must listen to multiple channels, the STB must have some form of local storage. Since MPEG-2 requires at least 60 Megabytes per minute (Mbytes/min), the STB storage is likely to reside on disk. The storage capacity on the STB and the number of channels the STB must listen to are the two client requirements that differ among the broadcasting protocols. The fewer requirements placed on the client STB, the less it will cost to produce and the more attractive it will be to clients.

11.3 STAGGERED BROADCASTING PROTOCOLS

Staggered broadcasting [2, 4, 7] is the most straightforward of all broadcasting protocols. It simply dedicates a certain number of channels to each video and then staggers starting times for the video evenly across the channels. That is, given n channels for a video of duration D, staggered broadcasting starts an instance of the video every D/n minutes. The difference in starting times (D/n) is called the *phase offset*. All clients who request a video during the current phase offset will be grouped together to watch the next broadcast of the video, and so the phase offset is the maximum amount of time clients must wait for their request to be serviced.

Staggered broadcasting does not use the VOD server's bandwidth very efficiently. To cut the client waiting time in half, for example, the VOD provider would have to *double* the amount of bandwidth dedicated to each video. This is a much higher cost than that required by the other broadcasting protocols described in this chapter.

On the other hand, staggered broadcasting places the fewest requirements on the client STB. The STB only has to listen to a single channel to receive an entire video, and it does not require a significant amount of extra storage.

The most appealing aspect of staggered broadcasting, though, is how it can handle interactive VOD. If it were to allow interactions of arbitrary lengths—that is, allow *continuous* interactions—then each interaction would force a client out of its broadcasting group and require it to use its own unique channel. Such a strategy would quickly use up all of the available bandwidth on the VOD server and cause unacceptable delays.

By forcing clients to use multiples of the phase offset, staggered broadcasting can service interactions without creating new channels for the clients [2]. These *discontinuous* interactions can simply use the existing channels for the video. For example, each rewind interaction could cause the client to shift to an earlier channel, which would be equivalent to rewinding for an amount of time equal to the phase offset. Similarly, fast-forward interactions could be implemented by shifting the client to a later channel, and pause interactions by waiting for an earlier channel to reach the same point in the video.

11.4 PYRAMID BROADCASTING PROTOCOLS

Viswanathan and Imielinski introduced the idea of pyramid broadcasting in 1995. Their Pyramid Broadcasting protocol [19, 20] divides up each video into n segments, $S_1, \ldots, S_n$. The protocol then divides up the available bandwidth on the VOD server into n equal-bandwidth channels, $C_1, \ldots, C_n$, and broadcasts serially the ith segment of each video on channel C_i (Figure 11.1).

The video segments are not equally sized; they grow geometrically using parameter $\alpha \geq 1$. That is, given the duration of the first segment d_1, the duration of the ith segment is

$$d_i = \alpha^{i-1} d_1$$

for $1 < i \leq n$. Pyramid Broadcasting received its name because, if one were to envision the segments stacked one on top of the other, they would form the shape of a pyramid.

Pyramid Broadcasting works as follows. A client must first wait for a showing of segment S_1 on channel C_1. At that point it can begin receiving and consuming the segment. For the remaining segments, the client can begin receiving S_i, for $1 < i \leq n$, at the earliest opportunity after starting the consumption of S_{i-1}.

In order for a client to receive all data on time, it must be able to start receiving S_i before it finishes consuming S_{i-1}. In other words, Pyramid Broadcasting must have

$$\frac{md_i}{b'} \leq d_{i-1} \qquad (1)$$

where b' is the bandwidth of each channel, measured in multiples of b, and m is the number of videos to be broadcast. By substituting αd_{i-1} for d_i in Equation 1, we can solve for alpha and find that as long as

$$\alpha \leq \frac{b'}{m} \qquad (2)$$

clients will never experience a break in playback. Viswanathan and Imielinski [19, 20] always used $\alpha = b'/m$, since that resulted in the smallest d_1 and thus the shortest client waiting times. A typical value is $\alpha = 2.5$.

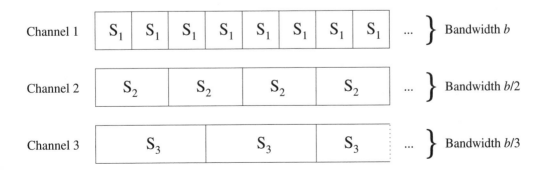

FIGURE 11.1
The segment-to-channel mapping for Pyramid Broadcasting with four videos. The superscript represents the video index.

The benefit of using Pyramid Broadcasting is that it uses VOD server bandwidth much more efficiently than staggered broadcasting protocols. This is because client waiting times only decrease linearly with bandwidth in staggered protocols but decrease exponentially with Pyramid Broadcasting. As an example, given $10b$ bandwidth per video, staggered protocols can only guarantee a maximum waiting time of 12 minutes for a two-hour video, but Pyramid Broadcasting can guarantee under two minutes.

The problem with Pyramid Broadcasting is that it makes great demands of the client equipment. Going back to Equation 2 and noting that $\alpha \geq 1$ and $m >> 1$, the bandwidth per channel will be very high, and clients will have to be able to receive up to two channels at once. So not only will clients have to handle extremely high bandwidths, but they will also have to store the large amount of data they receive early. In fact, this storage has to be large enough to hold almost the entire last two segments of the video [20], and those two segments typically account for 75–80% of the video size.

In 1997, Aggarwal et al. greatly reduced client equipment demands with their Permutation-Based Pyramid Broadcasting protocol [1]. By dividing up each channel into $p \geq 1$ subchannels for each video, and by evenly staggering the starts of segments on those subchannels, they were able to prevent clients from having to receive data from more than one subchannel at one time. This reduction in bandwidth by a factor of $2mp$ also reduced the amount of storage needed by the clients, down to about a third of that required by Pyramid Broadcasting. Unfortunately, Permutation-Based Pyramid Broadcasting requires more bandwidth than Pyramid Broadcasting to achieve the same client waiting times.

The next advance in pyramid broadcasting protocols came with Hua and Sheu's Skyscraper Broadcasting protocol [10], also in 1997. Instead of using a geometric series to determine the amount of data to place on each channel, their protocol divides up each video into n equally sized segments, $S_1, \ldots, S_n$, and then uses the series

$$\{ 1, 2, 2, 5, 5, 12, 12, 25, 25, 52, 52, \ldots \}$$

to determine the number of consecutive segments to place on each channel. Thus, channel C_1 would repeatedly broadcast segment S_1, C_2 would repeatedly broadcast S_2 and S_3, C_3 would repeatedly broadcast S_4 and S_5, and so on. This series grows much more slowly than the one employed by the previous pyramid protocols—it is analogous to using $\alpha \approx 1.5$—but each channel only requires bandwidth b, so Skyscraper Broadcasting can use many more channels. Also, the channels are constrained to a maximum number of segments (or width), so the protocol avoids the problem of having an extremely long data block on the last channel that needs to be stored on the client STB. Lastly, the client only has to receive at most two channels at once, so transfer rates are kept low. In total, Skyscraper Broadcasting manages to keep the good parts of Permutation-Based Pyramid Broadcasting—lower transfer rates and lower storage requirements—while also reducing the client waiting times found in Pyramid Broadcasting.

In 1998, Eager and Vernon improved Skyscraper Broadcasting by allowing it to dynamically schedule channels and to use a more efficient segment-to-channel series [8]. However, in both versions, since the protocols attempt to keep transfer rates low at the client end, they waste bandwidth on the VOD server. Consider, for example, a situation where a video uses five channels and 15 ($1 + 2 + 2 + 5 + 5$) segments. Segment S_{15} is broadcast every fifth time slot on channel C_5, and that is three times as often as needed.

An opposite approach to Skyscraper Broadcasting was used by Juhn and Tseng in 1997 with their Fast Broadcasting protocol [11, 14]. This protocol uses the series

$$\{ 1, 2, 4, 8, 16, 32, 64, \ldots \}$$

to determine the number of segments to broadcast on each channel, so it uses many more segments than Skyscraper Broadcasting, and its associated client waiting times are much lower. Unfortunately, clients must receive data from all channels at once, so their transfer rates are much higher and they must store up to half of the video locally. Fast Broadcasting also allows a mode where clients only receive a single channel at once and don't require local storage, but the maximum waiting time for this mode is half the duration of the video, which is likely to be too long for any video popular enough to be broadcast.

Pâris, Carter, and Long improved efficiency even further in 1999 with their Pagoda Broadcasting protocol [17, 18]. This protocol attempts to broadcast video segments as infrequently as possible while still maintaining an even transfer rate to the client. Like Skyscraper and Fast Broadcasting, Pagoda Broadcasting divides up each video into n equally sized segments, $S_1, \ldots, S_n$, but then it uses the series

$$\{\ 1, 3, 5, 15, 25, 75, 125, \ldots\ \}$$

to determine the number of segments for each channel. Unlike the previous protocols, it does not require that the segments appearing on each channel be consecutive. In fact, after the first channel, Pagoda Broadcasting uses pairs of channels when assigning segments, and so, for example, segments S_2 and S_4 are assigned to channel C_2, but S_3 is assigned to C_3 (see Figure 11.2).

The segment-to-channel mapping works as follows. The first segment S_1 is repeatedly broadcast on channel C_1. Then, given $z-1$, the highest index of a segment appearing on an earlier channel (or channels), Pagoda Broadcasting assigns $4z$ segments to the current pair of channels. The first channel contains:

- Segments S_z to $S_{3z/2-1}$, inclusive, broadcast every other time slot. Since there are $z/2$ segments, each is broadcast once every z time slots.
- Segments S_{2z} to S_{3z-1}, inclusive, also broadcast every other time slot. Each segment is broadcast once every $2z$ time slots.

Channel 1	S_1	S_1	S_1	S_1	S_1	S_1	S_1	S_1	S_1	...
Channel 2	S_2	S_4	S_2	S_5	S_2	S_4	S_2	S_5	S_2	...
Channel 3	S_3	S_6	S_7	S_3	S_8	S_9	S_3	S_6	S_7	...
...				...						
Channel i	S_z	S_{2z}	S_{z+1}	...	S_z	...	$S_{3z/2-1}$	S_{3z-1}	S_z	...
Channel i+1	$S_{3z/2}$	S_{3z}	S_{3z+1}	...	$S_{3z/2}$	...	S_{5z-2}	S_{5z-1}	$S_{3z/2}$	...
...				...						

FIGURE 11.2
The segment-to-channel mapping for Pagoda Broadcasting.

The second channel contains:

- Segments $S_{3z/2}$ to S_{2z-1}, inclusive, broadcast every third time slot. Since there are $z/2$ segments, each is broadcast once every $3z/2$ segments.
- Segments S_{3z} to S_{5z-1}, inclusive, broadcast two out of every three time slots. Thus, each segment is broadcast once every $3z$ time slots.

Figure 11.2 shows a sample of these two channels.

For a client to use Pagoda Broadcasting, it must first wait for an instance of S_1 on channel C_1. At that point it can start receiving and consuming S_1, and it can also start receiving data from every other channel dedicated to the video. Since each segment S_i is broadcast at least once every i slots of time, the client will either receive the segment ahead of time and have it in its buffer when it is needed, or it will receive the segment directly from the VOD server when it is needed.

Since Pagoda Broadcasting schedules many more segments per channel than Fast Broadcasting, the required client and VOD server bandwidths are much lower for any given maximum client waiting time. However, Pagoda Broadcasting still requires enough local storage for about half of the video.

None of the pyramid protocols described in this section have been shown to work effectively with interactive VOD.

11.5 HARMONIC BROADCASTING PROTOCOLS

The first harmonic protocol was Juhn and Tseng's Harmonic Broadcasting protocol [12], introduced in 1997. This protocol divides each video into n equally sized segments, $S_1, \ldots, S_n$, and continuously broadcasts those segments on their own data channels. In particular, segment S_i is broadcast on channel C_i using bandwidth b/i (Figure 11.3). The sum of the channel bandwidths is

$$\sum_{i=1}^{n} \frac{b}{i} = b \sum_{i=1}^{n} \frac{b}{i} = bH(n)$$

where $H(n)$ is the harmonic number of n. It is this series that gives the protocol its name.

For a client to use Harmonic Broadcasting, it must first wait for the start of an instance of S_1 on C_1. Then it can begin receiving and consuming S_1 while simultaneously receiving and storing data from every other channel. Once the client has a complete segment S_i, it can stop listening to channel C_i, but it still must support a bandwidth of $bH(n)$, since that is how much it will receive during the first slot of time.

The appeal of Harmonic Broadcasting is that the harmonic series grows very slowly. As with other protocols, the maximum client waiting time is the duration of a segment, but Harmonic Broadcasting can use hundreds of segments and still not require much bandwidth. For example, with a bandwidth of only $5b$, it can support $n = 82$ segments and have a maximum client waiting time of under a minute and a half for a two-hour video.

On the down side of Harmonic Broadcasting is the fact that it sends data to clients (well) before it is needed, and so the clients must have some sort of local storage. In fact, Harmonic Broadcasting requires storage for about 37% of the video as long as $n \geq 20$ [12].

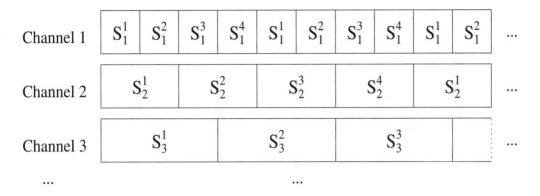

FIGURE 11.3
The segment-to-channel mapping for Harmonic Broadcasting.

One other problem with Harmonic Broadcasting is that it does not quite work as originally presented. Consider again Figure 11.3 and suppose a client arrives in time to start receiving the second instance of S_1. During that time it will receive and store the second half of S_2, but when it comes time to consume S_2 it will only be receiving the first half at half the consumption rate, and so much of the data will arrive after it is needed.

A straightforward way to fix Harmonic Broadcasting is simply to force the client to delay for a slot of time after receiving the first segment of the video [15]. Then the client will have each segment completely in its local storage before starting to consume it, and no data will arrive late. The only drawback to this Delayed Harmonic Broadcasting protocol is that its maximum client waiting time is twice that of the original harmonic protocol.

In 1998, Pâris, Carter, and Long presented three harmonic protocols that fixed the correctness problem of Harmonic Broadcasting but used bandwidth more efficiently than Delayed Harmonic Broadcasting. With the Cautious Harmonic Broadcasting protocol [15], C_1 stays the same but C_2 alternately broadcasts S_2 and S_3, and C_i, for $3 \le i < n$, broadcasts S_{i+1} at bandwidth b/i. This scheme guarantees that a client will either receive a segment at full bandwidth when it is needed or have the segment in its local storage before it is needed. Since C_2 uses bandwidth b instead of $b/2$, Cautious Harmonic Broadcasting requires about $b/2$ more bandwidth than Delayed Harmonic Broadcasting, but its maximum client waiting time is only one slot instead of two.

The second protocol, Quasi-Harmonic Broadcasting [15], divides each of the n segments into a number of subsegments (or *fragments*), and, while each segment is broadcast on its own distinct channel, the subsegments are not broadcast in order. By rearranging the subsegments and by broadcasting the first subsegment twice as often as the other subsegments, clients can start consuming a segment while still receiving parts of it, and still receive all of the data on time. As with Cautious Harmonic Broadcasting, the maximum client waiting time is one slot, but the bandwidth converges to $bH(n)$ as the number of subsegments increases.

Polyharmonic Broadcasting [16] uses an integral parameter $m \ge 1$ and forces each client to wait m slots of time before consuming data. While waiting, the client can receive data, and so each segment can be broadcast with a lower bandwidth than that used in the other harmonic protocols. In particular, each channel C_i, for $1 \le i \le n$, can use bandwidth $b/(m + i - 1)$ to continuously broadcast segment S_i. Since a great deal of the total value of $H(n)$ comes from its first few terms, and since Polyharmonic Broadcasting bypasses some of those terms when $m > 1$, even though clients must wait m slots of time, Polyharmonic Broadcasting can use m times as many segments, and so the real waiting time is still the same as Cautious and Quasi-Harmonic

Broadcasting. For any fixed delay d and video duration D, the bandwidth used by Polyharmonic Broadcasting converges to $\log(D/d + 1)$ as m increases, and, for realistic parameters, uses less bandwidth for a given maximum client waiting time than Quasi-Harmonic Broadcasting.

Juhn and Tseng also presented a new harmonic protocol in 1998: the Enhanced Harmonic Broadcasting protocol [13]. This protocol uses a strategy similar to Polyharmonic Broadcasting, but it suffers from the same correctness problem as their original harmonic protocol.

None of the harmonic protocols described in this section have been shown to work effectively with interactive VOD.

11.6   SUMMARY

The work of recent years has led to the the development of many efficient broadcasting protocols for VOD. These protocols use a fixed amount of VOD server bandwidth to guarantee a maximum waiting time before clients can begin viewing their requests.

Although the protocols appear to be different, most share a fundamental strategy: that if clients frequently request the same video, and if the clients have some form of local storage, then data occurring later in the video does not need to be broadcast as often as data occurring earlier in the video. By taking advantage of this strategy, the broadcasting protocols can save bandwidth on the VOD server and either allow more videos to be broadcast or allow the VOD server to be manufactured with less expense.

Client STB requirements for the broadcasting protocols are summarized in Table 11.1, and the VOD server bandwidth requirements for the protocols are shown in Figure 11.4. We assumed that Skyscraper Broadcasting had a maximum width of 52 and that Polyharmonic Broadcasting used $m = 4$. These are representative values, and they should not greatly affect the results. Also, for Figure 11.4 the waiting times are for a two-hour video.

Between Table 11.1 and Figure 11.4, there is no clear "best" protocol. Polyharmonic Broadcasting poses the lowest bandwidth requirements on the VOD server, but its numerous data channels per video may be too much for a client STB to stage into its disk drive. Pagoda Broadcasting has a low VOD server bandwidth as well, but the amount of bandwidth it requires the client STB to receive may be too much for the client's network connection or disk drive to handle. Staggered broadcasting has by far the worst VOD server bandwidth requirements, but it

Table 11.1   Client STB Requirements for the Broadcasting Protocols

Broadcasting Protocol	Storage Requirement (Percent of Video)	Bandwidth Requirement (Multiples of b)
Staggered	0	1
Pyramid	75	4–5[a]
Permutation-Based Pyramid	20	2–3
Skyscraper	10	2
Fast	50	6–8
Pagoda	45	5–7
Harmonic	40	5–6

[a]Per video broadcast.

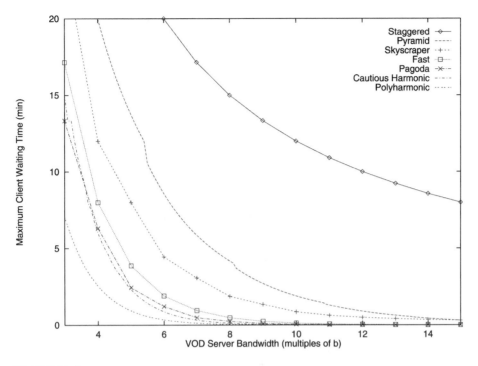

FIGURE 11.4
VOD server bandwidth requirements for the broadcasting protocols.

puts no extra load on the client STB, and it is the only broadcasting protocol that as yet allows for interactive VOD.

There are several open issues relating to broadcasting protocols. First, none of the most efficient protocols have been shown to work with interactive VOD. It is not clear whether such extensions are not possible, are not efficient, or just have not yet been explored. Second, the broadcasting protocols assume (implicitly, if not explicitly) that the video signal has a fixed bit rate. This is likely not to be the case. The Moving Pictures Experts Group (MPEG) compression schemes all use variable bit rates, for example. One solution is to reserve enough bandwidth in a channel for the maximum bit rate, but that will waste a lot of bandwidth. Are there better solutions? Finally, broadcasting protocols do not handle well fluctuations in video popularity. If horror videos, for example, are more popular at night than in the day, is there a graceful way for the broadcasting protocol to change the amount of bandwidth the videos use? Staggered broadcasting has the easiest solution to this problem—it can simply change the phase offset (although even this might be a problem with interactive VOD)—but what of the other protocols?

11.7   DEFINITIONS OF KEY TERMS

Channel: A stream of data on the VOD server.
Consumption rate: The rate at which the client STB processes data in order to provide video output.
Interactive video-on-demand: A VOD service that also allows clients to use VCR controls such as pause, rewind, and fast forward.
Phase offset: In staggered broadcasting, the amount of time between starts of a particular video.
Segment: A consecutive chunk of a video.

Set-top box (STB): The piece of equipment linking a client to a VOD server.

Slot: The amount of time it takes for an STB to consume a segment of data.

Video-on-demand (VOD): A service allowing clients to watch the video of their choice at the time of their choice.

Width: In Skyscraper Broadcasting, the maximum number of segments per channel.

11.8 REFERENCES

[1] Charu C. Aggarwal, Joel L. Wolf, and Philip S. Yu. A permutation-based pyramid broadcasting scheme for video-on-demand systems. In *Proceedings of the International Conference on Multimedia Computing and Systems,* pages 118–126, Hiroshima, Japan, June 1996. IEEE Computer Society Press.

[2] Kevin C. Almeroth and Mostafa H. Ammar. The use of multicast delivery to provide a scalable and interactive video-on-demand service. *IEEE Journal on Selected Areas in Communications,* 14(5):1110–1122, August 1996.

[3] Steven W. Carter and Darrell D. E. Long. Improving bandwidth efficiency on video-on-demand servers. *Computer Networks and ISDN Systems,* 30(1–2):99–111, January 1999.

[4] Asit Dan, Perwez Shahabuddin, Dinkar Sitaram, and Don Towsley. Channel allocation under batching and VCR control in video-on-demand systems. *Journal of Parallel and Distributed Computing,* 30(2):168–179, November 1995.

[5] Asit Dan and Dinkar Sitaram. Buffer management policy for an on-demand video server. Technical Report RC 19347, IBM Research Division, T. J. Watson Research Center, January 1993.

[6] Asit Dan, Dinkar Sitaram, and Perwez Shahabuddin. Scheduling policies for an on-demand video server with batching. In *Proceedings of the 1994 ACM Multimedia Conference,* pages 15–23, San Francisco, CA, October 1994. Association for Computing Machinery.

[7] Asit Dan, Dinkar Sitaram, and Perwez Shahabuddin. Dynamic batching policies for an on-demand video server. *Multimedia Systems,* 4(3):112–121, June 1996.

[8] Derek L. Eager and Mary K. Vernon. Dynamic skyscraper broadcasts for video-on-demand. In *Proceedings of the 4th International Workshop on Advances in Multimedia Information Systems,* pages 18–32, Berlin, Germany, September 1998, Springer-Verlag.

[9] Leana Golubchik, John C. S. Lui, and Richard R. Muntz. Adaptive piggybacking: A novel technique for data sharing in video-on-demand storage servers. *Multimedia Systems,* 4(30):140–155, June 1996.

[10] Kien A. Hua and Simon Sheu. Skyscraper Broadcasting: A new broadcasting scheme for metropolitan video-on-demand systems. In *Proceedings of SIGCOMM 97,* pages 89–100, Cannes, France, September 1997. Association for Computing Machinery.

[11] Li-Shen Juhn and Li-Meng Tseng. Fast broadcasting for hot video access. In *Proceedings of the 4th International Workshop on Real-Time Computing Systems and Applications,* pages 237–243, Taipei, Taiwan, October 1997. IEEE Computer Society Press.

[12] Li-Shen Juhn and Li-Ming Tseng. Harmonic broadcasting for video-on-demand service. *IEEE Transactions on Broadcasting,* 43(3):268–271, September 1997.

[13] Li-Shen Juhn and Li-Ming Tseng. Enhanced harmonic data broadcasting and receiving scheme for popular video service. *IEEE Transactions on Consumer Electronics,* 44(2):343–346, May 1998.

[14] Li-Shen Juhn and Li-Ming Tseng. Fast data broadcasting and receiving scheme for popular video service. *IEEE Transactions on Consumer Electronics,* 44(1):100–105, March 1998.

[15] Jehan-François Pâris, Steven W. Carter, and Darrell D. E. Long. Efficient broadcasting protocols for video-on-demand. In *Proceedings of the International Symposium on Modelling, Analysis, and Simulation of Computing and Telecom Systems,* pages 127–132, Montreal, Canada, July 1998. IEEE Computer Society Press.

[16] Jehan-François Pâris, Steven W. Carter, and Darrell D. E. Long. A low bandwidth broadcasting protocol for video on demand. In *Proceedings of the 7th International Conference on Computer Communication and Networks,* pages 609–617, Lafayette, LA, October 1998. IEEE Computer Society Press.

[17] Jehan-François Pâris, Steven W. Carter, and Darrell D. E. Long. A hybrid broadcasting protocol for video on demand. In *Proceedings of the 1999 Multimedia Computing and Networking Conference,* pages 317–326, San Jose, CA, January 1999.

[18] Jehan-François Pâris, Steven W. Carter, and Darrell D. E. Long. A simple low-bandwidth broadcasting protocol for video-on-demand. In *Proceedings of the 8th International Conference on Computer Communication and Networks,* pages 118–123, Boston-Natick, MA, October 1999, IEEE Press.

[19] S. Viswanathan and T. Imielinski. Pyramid broadcasting for video-on-demand service. *Proceedings of SPIE—the International Society for Optical Engineering,* 2417:66–77, 1995.

[20] S. Viswanathan and T. Imielinski. Metropolitan area video-on-demand service using pyramid broadcasting. *Multimedia Systems,* 4(4):197–208, August 1996.

11.9   FOR FURTHER INFORMATION

Broadcasting protocols are proactive in that bandwidth is reserved in advance. There are also several reactive protocols that reserve bandwidth on the fly. Some of these protocols include *adaptive piggybacking* [9], *interval caching* [5], and *stream tapping* [3]. Since their bandwidth requirements depend on the access patterns of the videos, reactive protocols tend to be less efficient than broadcasting protocols for popular videos, but can be very efficient for "lukewarm" or cold videos.

Current research on broadcasting protocols can be found in a variety of places. Relevant journals include *IEEE Transactions on Broadcasting, Computer Networks and ISDN Systems,* and *Multimedia Systems.* Conferences include *ACM Multimedia; Infocom; the Symposium on Modeling, Analysis and Simulation of Computer and Telecommunication Systems (MASCOTS); the International Conference on Computer Communication and Networks (ICCCN);* and the *Multimedia Computing and Networking Conference (MMCN).*

CHAPTER 12

Internet Telephony Technology and Standards Overview

BERNARD S. KU

12.1 INTRODUCTION

In recent years, Internet Telephony (IPTel) has evolved from the "hobbyist" level to a technology that sooner or later may replace the Public Switched Telephone Network (PSTN) to become the foundation for building a new era of businesses and networks. According to recent marketing studies, IPTel has been estimated to constitute a $560 million market by the end of 1999. According to a 1998 International Data Corporation report, 28% of international voice traffic and 8% of domestic voice traffic will migrate to IP by the year 2002. Figure 12.1 illustrates the trend of the future traffic growth.

Traditional telephone companies (telcos) and Internet Service Providers (ISPs) are seeking to incorporate the Voice Over IP (VoIP) services into their existing applications. New operators such as Qwest and Level 3 are dropping the traditional circuit switched telephony network and deploying large-scale Internet Protocol (IP)-based networks with rapid speed.

On the other hand, VoIP is just the core service of IPTel. In general, IPTel refers to the transport of real-time media—such as voice, data, and video conferencing—over the Internet to provide interactive communications among Internet users. For example, a customer connects to the Internet and uses the Web browser to search for a special product such as a Christmas gift. After locating the product, she would like to know details of the product, so she makes a VoIP call to reach a sales representative. The sales representative can show the potential buyer a real-time promotion video to show how easy to operate the product. An instant price quote chart can also be displayed to the customer regarding the final price including selected options and the extended warranty contract. The customer agrees and makes the purchase over the Internet. The electronic commerce web site confirms the transaction and emails the customer the receipt and also the tracking number for the express delivery. More and more business transactions of such a kind can easily make the traditional telecommunications network an extinct dinosaur in just a

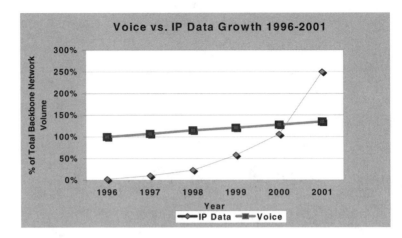

FIGURE 12.1
Voice versus IP data growth.

few years. In addition, there are a myriad of value-added services such as Fax Over IP, Internet Call Waiting, Unified Messaging, and others enhancing the versatility and profitability of the telecommunications industry.

Section 12.2 provides an overview of the various components that make up the integrated Internet Telephony network architecture. Section 12.3 surveys the IPTel standardization bodies and their areas of concentration. Section 12.4 summarizes the major IPTel signaling/connection control protocols, namely, H.323, Session Initiation Protocol (SIP), Media Gateway Control Protocol (MGCP), and also the upcoming IETF Media Gateway Control (MEGACO)/H.GCP standards. Section 12.5 explains how VoIP works by exploring three different gateway configurations. Section 12.6 highlights some of the open issues of IPTel, such as billing, quality of service, and directory services. Section 12.7 discusses one critical deployment issue, Intelligent Network (IN)/IP integration, to allow IN services to be deployed seamlessly across the PSTN and the IPTel network. Section 12.8 focuses on the integration of Signaling System 7 (SS7) and the IPTel network to allow service interoperability. Finally, Section 12.9 concludes by highlighting the future issues that need to be addressed to make IPTel a reality. The emphasis of this chapter will be on discussing relevant standards and on highlighting the key technology advances that will enhance IPTel to be successfully deployed in the public network in the current millennium.

12.2 INTERNET TELEPHONY ARCHITECTURE OVERVIEW

Given the billions of dollars of investment in the legacy circuit switched telephony and also IN services such as Emergency 911, we can expect a long period of time during which IPTel will coexist with PSTN, the IN, and the Integrated Digital Service Network (ISDN). The real challenge of IPTel to the service provider is to lower and leverage the cost of switching and service development and the integration of heterogeneous network technologies. Supporting IPTel requires a lot of interworking protocols in various areas, from transport, to call setup, to billing. The overall architecture of IPTel is subject to changes as work on new and refined protocols progresses at the Internet Engineering Task Force (IETF) and the International Telecommunications Union—Technical (ITU-T).

As illustrated in Figure 12.2, elements in the integrated IPTel architecture using traditional and Internet devices are the following:

- Plain telephones or Time Division Multiplexing (TDM) phones
- PSTN switch
- Media gateways
- Media gateway controllers, or call agents
- Service Control Points (SCPs) for PSTN and IN functions
- IP client and IP phone
- IP telephony server
- Application protocols
- Media protocols
- Signaling protocols
- Interworking protocols

There are an increasing number of products in the industry for these elements. Of all these elements, gateways play a vital role in bridging the gap of PSTN to IPTel. They often represent the major IPTel equipment sold today. The bearer portion of the gateway is composed of one or more media gateways to translate voice signal of the TDM format from the PSTN to Real-Time Protocol (RTP) packets over the IP. This gateway is capable of processing audio, video, and T.120 alone or in any combination., It is also capable of full-duplex media translations, playing audio/video messages and performing other Interactive Voice Response (IVR) functions, or even

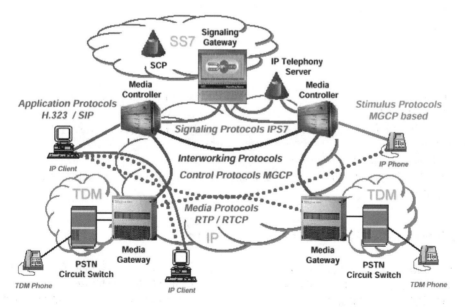

FIGURE 12.2
Internet telephony architecture—convergence of IP, PSTN, and IN.

media conferencing. The media gateway controller controls the parts of the call state that pertain to connection control for media channels in a media gateway.

On the other hand, the signaling portion of the gateway is composed of one or more signaling gateways to exchange the SS7 of the PSTN to the IP signaling based on either IETF Session Initiation Protocol (SIP) or ITU-T H.323. Signaling gateways also control the media gateways for interworking simple to complex services over the two networks.

Gateways can further be classified as residential or trunk depending on where they are located. Residential Gateway (RGW) is responsible of interfacing with an IPTel subscriber on all related call events and sending the signals of these events to a call agent. Trunk Gateway (TGW) is responsible for bridging between the PSTN and the IP network. Both RGW and TGW take instructions from the call agent on how to react or proceed with the call events. The intelligence resides in the call agent. MGCP is the first but not the only candidate interface between the gateway and the call agent. As of July 1999, the IETF Media Gateway Control (MEGACO) working group has been collaborating with ITU-T Study Group (SG) 16 to develop a combined protocol specification based on inputs from H.GCP (H.248). Similar to the telephone switches, call agents use the SS7 ISDN User Part (ISUP) for setting up the calls. The call agents also support legacy IN services such as 800 calls, credit calls, and local number portability by interfacing with the SCP through a signaling gateway. ISUP may have to be extended to support more complex IPTel services. More detailed descriptions of the protocols are found in Section 12.4.

Examples of new IPTel services can be the following:

- Voice over IP
- Click-to-dial
- Fax over IP
- Internet call waiting
- Unified messaging
- Network-based call distribution
- Multimedia conferencing
- Internet call centers
- Voice enhancements
- Never-busy voice

The architecture is incorporating many protocols, for example, Transmission Control Protocol (TCP), User Datagram Protocol (UDP), and the protocols at the Open Systems Interconnection (OSI) data link and physical layers to send/receive data to or from various locations on the Internet.

In this chapter, we limit our discussions to only the higher-layer signaling and connection control protocols that are specific to IPTel services. Table 12.1 shows the various areas and the supporting protocols.

12.3   RELATED INTERNET TELEPHONY STANDARDS

Before we explore the IPTel-related protocols, it is important to know the various domestic and international standard bodies and their activities on IPTel technology. Resolution of the technical issues of IPTel is moving at a very fast pace, first in IETF, then spreading to the national standards T1 committee, and to international standards such as ITU-T and ETSI TIPHON.

Table 12.1 Internet Telephony Protocols

Areas	Protocols
Signaling/Connection Control	H.323 Real-Time Streamlining Protocol (RTSP) Session Initiation Protocol (SIP) Media Gateway Control Protocol (MGCP) Media Gateway Device Protocol (MGDP) MEGACO/H.GCP (H.248)
Media Transport	Real-Time Transport Protocol (RTP)
Quality of Service (QoS)	Resource Reservation Protocol (RSVP) RTP Control Protocol (RTCP)
Session Content Description	Session Description Protocol (SDP)
Gateway Discovery	Gateway Location Protocol
Directory Service	Lightweight Directory Access Protocol (LDAP)
Authentication, Authorization, and Accounting (AAA)	RADIUS DIAMETER Open Settlement Protocol (OSP) Common Open Policy Protocol (COPS)

12.3.1   IETF

SIGnaling TRANsport (SIGTRAN) group: To address the transport of PSTN signaling, for example, SS7, ISDN over IP networks.

MEdia GAteway COntrol (MEGACO) group: To develop standards detailing the architecture and requirements for controlling Media Gateways from external control elements such as a Media Gateway Controller. A *Media Gateway* is a network element that provides conversion between the information carried on telephone circuits (PSTN) and data packets carried over the Internet or over other IP networks.

IP TELephony (IPTEL) group: To develop "peripheral" protocols, that is, call processing syntax and gateway attribute distribution protocol, to enable call setup between IP clients as well as between IP and PSTN networks.

PSTN/Internet Interface (PINT) group: To study architecture and protocols to support services where Internet users initiate a telephone call to a PSTN terminal. Specifically, to develop standards specifying a Service Support Transfer Protocol (SSTP) between Internet applications and servers and PSTN IN nodes.

12.3.2   ETSI Telecommunications and Internet Protocol Harmonization Over Networks (TIPHON)

The TIPHON project is chartered to identify requirements and develop global standards for various aspects of communications between an IP network-based user and PSTN-based user. This includes a scenario where either a PSTN or an IP network is the origination, termination, or transit network for the call. The aspects to be studied are:

- requirements for service interoperability
- reference configurations and functional models, including position of gateway
- functions between IP-based networks and PSTN and interfaces at these gateways
- call control procedures, information flows, and protocols
- address translation between E.164 and IP addresses
- technical aspects of charging/billing
- technical aspects of security
- end-to-end quality-of-service aspects

12.3.3   ITU-T

Anticipating the rapid growth of IP networks, ITU-T has launched an IP Project to encompass all the ITU-T IP-related work. The project will be regularly updated as the work progresses and as the various ITU-T SGs expand their activities in support of the IP-related work.

The following 12 work areas have been identified as being of current major concern to the ITU-T:

- Area 1: Integrated architecture
- Area 2: Impact to telecommunications access for IP applications
- Area 3: Interworking between IP-based network and switched circuit networks, including wireless-based networks
- Area 4: Multimedia applications over IP
- Area 5: Numbering and addressing
- Area 6: Transport for IP-structured signals
- Area 7: Signaling support, IN, and routing for services on IP-based networks
- Area 8: Performance
- Area 9: Integrated management of telecom and IP-based networks
- Area 10: Security aspects
- Area 11: Network capabilities including resource management
- Area 12: Operations and Maintenance (OAM) for IP

For each of these we will describe the scope and focus of the work area, the issues, the current work in the ITU-T SGs, and related work in the IETF. Particularly in relation to the IETF work, an analysis of the IETF Working Group Charters is currently being undertaken to identify areas of potential overlap and areas where the ITU-T could collaborate with the IETF. Figure 12.3 shows a schematic representation of the scope of the ITU-T IP Project.

12.3.3.1   SG2/SG3
There is a Project 11 called "Voice Over IP" (VoIP) that examines how IPTel users can be reached from regular PSTN phones; the major issue it addresses is the decision on whether E.164 Country Codes can be reserved for voice over IP applications.

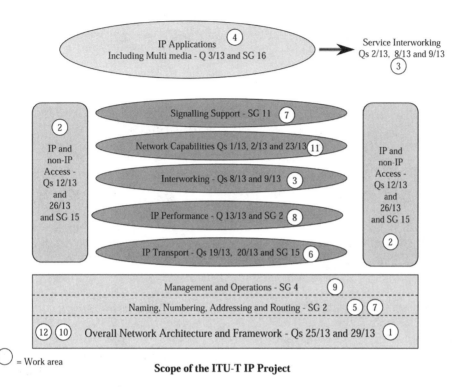

Scope of the ITU-T IP Project

FIGURE 12.3
Scope of the ITU-T IP Project.

12.3.3.2   SG4

SG4 has launched a working group working on integrated service and network management of PSTN- and IP-based networks. It is in a close working relationship with T1M1.

12.3.3.3   SG11

SG11 has created a "Signaling Support of the Internet" (SoI) group to identify all signaling issues, that is, ISDN, SS7, and Asynchronous Transfer Mode (ATM), that require resolution to enable the interworking of Internet- and PSTN-based services. In addition, all modifications/ enhancements to IN and Broadband ISDN (BISDN) model and protocol to enable Internet interworking will also be addressed by SG11.

12.3.3.4   SG13

ITU will not duplicate what IETF is doing. But it will pick up the projects that IETF is not handling (for instance, interworking) and that ITU has the expertise on. ITU designated SG-13 to do an inventory on what has been done within and outside ITU on IP networks and to identify the gaps that need to be resolved. Therefore, SG13 is the lead SG for Internet standards. There are liaisons from many other SGs, including SG2, SG4, SG7, SG11, SG15, and SG16. At the last ITU Chair meeting in January 1999, it was determined that the Internet is one of the two "hottest items" and will be handled aggressively in the forthcoming ITU meetings.

12.3.3.5   SG16

SG16 is collaborating with the IETF MEGACO Working Group to provide a joint contribution of gateway protocol specifications based on its H.GCP (H.248).

12.3.4   T1S1

T1S1 is involved in developing U.S. positions on signaling issues related to PSTN and Internet interworking. The focus so far has been on SS7 over IP, and the creation or modification of the IN and BISDN model to enable interworking.

12.4   CURRENT AND DEVELOPING INTERNET TELEPHONY PROTOCOLS

In this section we will cover in detail some of the current and developing IPTel protocols, such as H.323, H.248 (MEGACO/H.GCP under development in SG16), SIP, and MGCP.

12.4.1   H.323

H.323 has been developed by ITU-T as a suite of protocols for multimedia communication between two or more parties over the IP network. It includes H.245 for control, H.255.0/Q.931 for call setup, H.332 for large conferences, H.450.1, H.450.2, and H.450.3 for supplementary services, H.235 for security, and H.246 for interoperability with circuit switched service. H.323 is by far the most popular protocol for IPTel and videoconferencing applications. Figure 12.4 shows the entire family of H.323 protocols. A first version was released in 1996, followed by a second version in 1998, and a third version is due for release in late 1999.

The following is a selected list of well-defined H.323 family protocols.

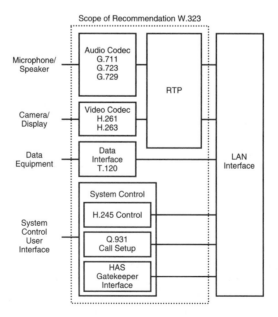

FIGURE 12.4
H.323 protocols.

H.323 Entity: Any H.323 component, including terminals, Gateways (GWs), Gatekeepers (GKs), Media Gateway Controllers (MGCs), and Media Gateways (MGs).

Call: Point-to-point multimedia communication between two H.323 endpoints. The call begins with the call setup procedure and ends with the call termination procedure. The call consists of the collection of reliable and unreliable channels between the endpoints. A call may be directly between two endpoints, or may include other H.323 entities such as a Gatekeeper or MGC. In case of interworking with some Switch Circuit Network (SCN) endpoints via a Gateway, all the channels terminate at the GW where they are converted to the appropriate representation for the SCN end system. Typically, a call takes place between two users for the purpose of communication, but there may be signaling-only calls. An endpoint may be capable of supporting multiple simultaneous calls.

Call Signaling Channel: Reliable channel used to convey the call setup and teardown messages (following Recommendation H.225.0) between two H.323 entities. H.323 signaling is very similar to ISDN signaling as specified in the Q.931 recommendation. H.323 offers two basic call processing mechanisms: one is gatekeeper-routed call signaling and the other is direct endpoint call signaling. Reservation, admission, and status channels are carried on a User Datagram Protocol (UDP) protocol stack, whereas call signaling channels are carried on the more reliable Transport Control Protocol (TCP) protocol stack.

Multipoint Control Units (MCUs): Control device used to assist in creating multiparty communications.

Gatekeeper: The Gatekeeper (GK) is an H.323 entity on the network that provides address translation and controls access to the network for H.323 terminals, GWs, and MCUs. The Gatekeeper may also provide other services to the terminals, GWs, and MCUs such as bandwidth management and locating GWs.

Gateway: An H.323 GW is an endpoint on the network that provides for real-time, two-way communications between H.323 terminals on the packet-based network and other ITU terminals on a switched circuit network, or to another H.323 GW. Other ITU terminals include those complying with Recommendations H.310 (H.320 on B-ISDN), H.320 (ISDN), H.321 (ATM), H.322 GQOS-LAN, H.324 General Switched Telephone Network (GSTN), H.324M (Mobile), and V.70 Digital Simultaneous Voice and Data (DSVD).

RAS (Reservation, Admission, and Status): The RAS signaling function uses H.225.0 messages to perform registration, admissions, bandwidth changes, status, and disengage procedures between endpoints and GKs.

RAS Channel: Unreliable channel used to convey the registration, admissions, bandwidth change, and status messages (following Recommendation H.225.0) between two H.323 entities.

Terminal: An H.323 terminal is an endpoint on the network that provides for real-time, two-way communications with another H.323 terminal, GW, or Multipoint Control Unit (MCU). This communication consists of control, indications, audio, and moving color video pictures, or data between the two terminals. A terminal may provide speech only, speech and data, speech and video, or speech, data, and video.

Zone: A *zone* (Figure 12.5) is the collection of all terminals (Tx), GWs, and MCUs managed by a single GK. A zone includes at least one terminal, and may or may not include GWs or MCUs. A zone has one and only one GK. A zone may be independent of network topology and may be composed of multiple network segments that are connected using routes (R) or other devices.

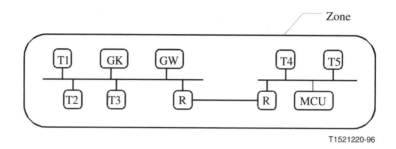

T1521220-96

FIGURE 12.5
H.323 zone.

The following is a selected list of IPTel protocols being developed in ITU-T SG16.

H.323 Service Control Protocol: This is work under study in SG16, intended to enhance service-related information transfer to and from the GK, which is currently limited to RAS. This work is expected to be strongly influenced by the IN-IP interworking model and the joint work of SG16/SG11 in general.

Composite Gateway: A *Composite Gateway* is a logical entity composed of a single MGC and one or more MGs that may reside on different machines. Together, they preserve the behavior of a gateway as defined in H.323 and H.246.

H.323 Call Routing Models: The location of the different functional entities in physical entities depends on the routing model used. In H.323 two models exist, the so-called Gatekeeper-Routed Call (GRC) model and the so-called Direct-Routed Call (DRC) model.

DRC: The terminals or gateways exchange call control signaling (H.225, H.245) directly with each other. Interaction between terminal/gateway and the GK is only via RAS signaling.

GRC: In addition to RAS, the terminals or gateways exchange call control signaling via the Gatekeeper, which acts as a signaling proxy. The Gatekeeper may alter the signaling information.

12.4.2   Session Initiation Protocol (SIP)

A *Session Initiation Protocol* (SIP) is a simple signaling protocol for creating, modifying, and terminating phone calls and multimedia sessions. It is a signaling protocol alternative to ITU-T H.323, developed within the IETF Multiparty Multimedia Session Control (MMUSIC) working group. It is a lightweight client server protocol, reusing the Internet addressing and the HyperText Transfer Protocol (HTTP). SIP requests and responses are all text-based and contain a header field to convey call properties and address information. SIP relies on a host of Internet protocols, including the Real-Time Transport Protocol (RTP) for data transport.

There are two kinds of components in a SIP system, user agents and network servers. The user agent is an end system that acts on behalf of a user and can exist in either client or server parts. The User Agent Client (UAC) initiates a SIP request, whereas the User Agent Server (UAS) receives and responds to the requests. There are also two kinds of network servers, proxy and redirect servers. A SIP proxy server forwards requests to the next server. A SIP redirect server does not forward requests to the next server. Instead it sends a redirect response back to the client containing the address of the next server to contact.

Like HTTP, the SIP client invokes several methods on the server:

- INVITE: initiates a session or a phone call
- BYE: terminates a connection between two users
- OPTIONS: solicits information on user capabilities
- STATUS: informs another server about the progress of signaling
- CANCEL: terminates a search for a user
- ACK: acknowledges a reliable message transaction for invitations
- REGISTER: provides user location information to a SIP server

SIP uses Uniform Resource Locators (URLs) as addressing with the general form as user@host. The host part is either a domain name or a numeric network address, for example, sip:john@doe.com or sip:johndoe@166.7.6.1.

Unlike the H.323, which has been designed for multimedia communications over the Local Area Network (LAN), SIP works over the PSTN as well. The ITU-T H.323 sets offer a rigid advanced service architecture that adopts a "service-by-service" specific deployment approach. SIP is in general more generic and flexible and relies on an end-to-end paradigm. There is still a lot of improvement to be made in SIP to support the creation and management of advanced communication services for IPTel.

Figure 12.6 illustrates the chronological steps of a call setup based on SIP signaling.

The following is an explanation of the ISDN User Part (ISUP) and SIP message sequence for the call setup.

1. User initiates a call by dialing a Called Party address digits.
2. PSTN formulates an ISUP Initial Address Message (IAM) and sends it to the Signaling Gateway.
3. Signaling Gateway converts the IAM to a SIP Invite message and sends to the SIP Proxy Server corresponding to the Called Party.

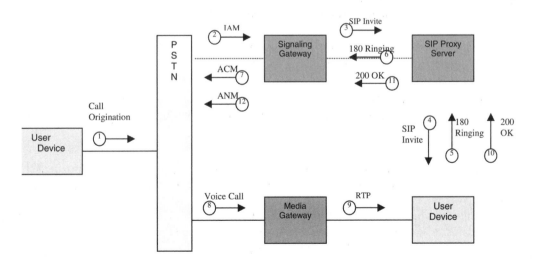

FIGURE 12.6
A call setup based on SIP signaling.

4. SIP Proxy Server forwards the SIP Invite message to the Called Party User Device.

5. User Device starts ringing and responds to the SIP Proxy Server with SIP 180 Ringing message.

6. SIP Proxy Server forwards the SIP 180 Ringing message to the Signaling Gateway.

7. Signaling Gateway converts the SIP 180 Ringing message to an ISUP Address Complete Message (ACM) and sends it to the PSTN.

8. PSTN then sets up the Time Division Multiplexing (TDM) Voice path to the Media Gateway.

9. Media Gateway converts the TDM Voice path to the RTP stream in IP and sets up the RTP path to the User Device.

10. When Called Party answers the call, the Called Party User Device sends a SIP 200 OK message to the SIP Proxy Server.

11. SIP Proxy Server forwards the SIP 200 OK message to the Signaling Gateway.

12. Signaling Gateway converts the SIP 200 OK message to an ISUP Answer Message (ANM) and sends it to the PSTN.

At the end of the Answer Message (ANM), the Calling and Called parties start their conversation.

12.4.3   Media Gateway Control Protocol (MGCP)

MGCP is a result of combining two previous protocols, Simple Gateway Control Protocol (SGCP) and Internet Protocol Device Control (IPDC). It is based on the IN concept of separating the call control from the switching and routing. In MGCP, the call control intelligence is resident in the external call agents, leaving the MG to be responsible for connection control only. MGCP communicates between the MGC or call agents and the media gateway in a master/slave manner. The slave media gateways execute the commands sent by the master call agents. As a result of gateway decomposition, IPTel services can be developed more cost effectively, more rapidly, and in a more scalable fashion.

MGCP exercises its control over the following types of MGs:

• Trunk gateway: operates between the IPTel network and the PSTN
• Residential gateway: operates between the IPTel network and the PSTN end user via the analog RJ11 interface
• ATM gateway: operates between the IPTel network and the ATM network
• Access gateway: operates between the IPTel network and the digital interface of a Private Branch Exchange (PBX)

Strictly speaking, MGCP is a connection control protocol. The call agents assume the signaling function to convey call state information from one entity to another and to exchange/negotiate the connection status. MGCP does not compete with the other two well-known signaling protocols in IPTel, H.323, and SIP. It complements them by allowing either H.323 or SIP to set up a call between the call agents and the IPTel client. The interworking of MGCP with PSTN, H.323, and SIP can be summarized in the following.

• *Interworking with PSTN:* The MGCP is used by the call agents to control the end user through the residential gateway, to connect the IP network to PSTN through the trunk gate-

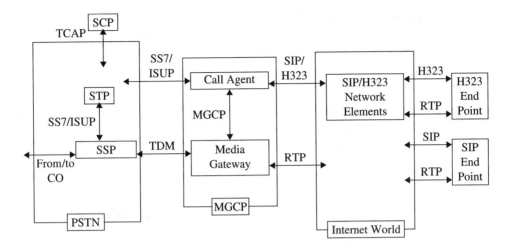

FIGURE 12.7
MGCP functional architecture: interworking with IP/PSTN. (SSP = Service Switching Point; STP = Signal Transfer Point; TCAP = Transactional Capabilities Application Part.)

way, and to exchange the ISUP message over the SS7 signaling gateway to terminate a call in the PSTN.

- *Interworking with H.323:* The MGCP call agent functions as a H.323 gatekeeper. Current work focuses on MGCP/H.323 interworking to investigate any performance issues for wide-area IPTel deployment as H.323 now runs over Transmission Control Protocol/ Internet Protocol (TCP/IP).

- *Interworking with SIP:* The MGCP has additional functionality to control media gateways and allocate resources, and both MGCP and SIP are text-based and share the use of SDP, so MGCP is compatible with SIP. However, SIP will need to be optimized to support call agent communications.

12.4.4   MEGACO/H.248 (H.GCP)

To put all interworking functionality in one single box, that is, the gateway, soon proved not to be a scalable and efficient approach. Different standards bodies such as IETF, ETSI TIPHON, and ITU-T are looking at ways to decompose the gateway into a set of standard interfaces for dedicated functions. At a higher level, the gateway can be decomposed into the following entities:

Media Gateway (MG): The MG converts media provided in one type of network to the format required in another type of network. For example, an MG could terminate bearer channels from an SCN and media streams from a packet network (e.g., RTP streams in an IP network). This gateway may be capable of processing audio, video, and T.120 alone or in any combination, and will be capable of full-duplex media translations. The MG may also play audio/video messages and perform other Interactive Voice Response (IVR) functions, or may perform media conferencing.

Media Gateway Controller (MGC): Controls the parts of the call state that pertain to connection control for media channels in an MG. It is also referred as the call agent.

Signaling Gateway (SG): Constitutes the interface to the SCN's out-of-band signaling network.

In general, the MEGACO/H.GCP describes a control model and protocol for an MGC to control an MG. An MGC-MG association resembles the behavior of an H.323 gateway. H.248 is currently being developed in SG16, in cooperation with IETF MEGACO, to provide a single, international standard for MGC. The combined MEGACO/H.GCP identifies a group of interfaces and functions to be used to decompose gateways.

Decomposed gateways using MEGACO/H.GCP may employ a wide variety of interfaces, which leads to considerable variation in network models. A commonly envisioned gateway would interface the SCN network to a packet or cell data network. In Figure 12.8 the packet/circuit media component terminates SCN media channels and converts the media streams to packet-based media on the packet network interface. The protocol for interface A is used to create, modify, and delete gateway media connections. The control logic component accomplishes signaling interworking between the SCN and packet sides of the gateway.

Interfaces B1 and B2 form the packet signaling transport/interworking interface between the gateway entities on the IP network and the decomposed gateway controller function. Interface C describes the ISDN-type call control function between the Facility Associated Signaling (FAS)-based SCN services and the gateway control logic. Non-Facility-Associated Signaling (NFAS)-based SCN is conveyed between the SCN signaling transport termination and the SCN signaling termination on interface D. The B1 and B2 interfaces represent the H.245 and H.225 and similar signaling interfaces, respectively.

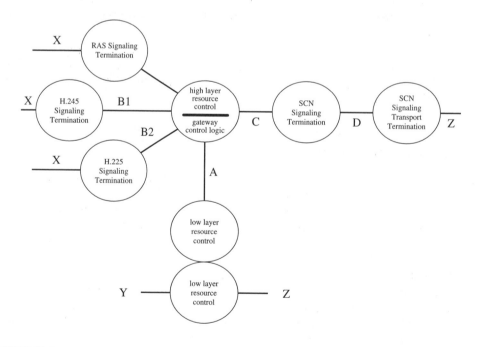

FIGURE 12.8
Functional architecture of MEGACO/H.GCP.

The Resource control elements differentiate between a high-level understanding of resources in the gateway controller and a lower-level understanding of resources in a gateway device.

The SCN interfaces are described as a low-level interface that transports signaling and a high-level SCN signaling termination that interfaces with the controller of this gateway. This can be FAS signaling such as ISDN Primary Rate Interface (PRI) or NFAS signaling such as SS7.

Figure 12.8 does not represent a physical decomposition at this point. The challenge for gateway vendors is to group these components into physical devices and implement the associated interfaces to produce highly scalable, multivendor gateways. MEGACO/H.GCP will define these interfaces to facilitate implementation of a wide range of decomposed gateways. The X indicates the external network gateway or terminal interface, the Y the external packet interface, and the Z the external SCN interface. Figure 12.8 illustrates the functional architecture of MEGACO/H.GCP.

12.5   HOW VOICE OVER INTERNET PROTOCOL (VoIP) WORKS

Voice Over Internet Protocol (VoIP) can be defined simply as the transmission of voice over IP networks. Originating and terminating devices can be traditional telephones, fax machines, multimedia personal computers (PCs), or a new class of "Internet-aware" telephones with Ethernet interfaces. Even though the above definition holds true in general, the technology can be categorized into three classes:

- PSTN gateways that provide carrier-based dial-tone solutions
- VoIP gateways that provide enterprise-based dial-tone solutions
- IP telephony gateways that provide enterprise-value additions and route the data over the Internet

12.5.1   PSTN Gateways

PSTN gateways provide an IP interface to PSTN voice services. The main requirement of these gateways is that they are virtually indistinguishable from the PSTN services in terms of voice quality. Consumers may be willing to tolerate minor inconveniences such as two-stage dialing if they perceive that they are obtaining significant price reductions. These gateways are configured with the called number to destination IP address and route the call based on the routing table entries.

From a system perspective, PSTN gateways need to be deployed in large scale. They have termination points close to the majority of the call destinations and can attract enough traffic to gain economies of scale sufficient to achieve low cost. They should offer high density—hundreds of ports per system—at the lowest possible per-port cost. They also require real-time monitoring tools and a sophisticated billing system. Figure 12.9 shows a PSTN gateway.

All the voice and fax calls are routed to the gateway via the trunk interface to the Central Office (CO). The gateway performs some or all of the following functions:

- *Call signaling and routing:* Progressing the incoming and outgoing calls to the remote gateway that is closest to the destination phone/fax number with the standard signaling procedures like H.323 and routing the calls to multiple IP networks with different routing algorithms

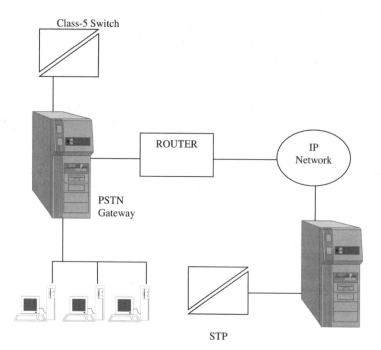

FIGURE 12.9
PSTN gateway.

- *Voice/fax compression:* Using any of the standard coding algorithms, such as G.711, G.723, ADPCM, or fax relay
- *Packetization:* Formatting the compressed data into IP packets that contain routing and sequence information
- *Quality management:* A variety of techniques, including buffering, interleaving, and bandwidth reservation through RSVP that compensate for delay, packet loss, and congestion that may occur in IP-based networks
- *Management interface:* Signaling Network Management Protocol (SNMP) or Common Management Information Protocol (CMIP) interface for routing and resource configurations and to support value-added applications

12.5.2 VoIP Gateways

VoIP gateways seek to save toll charges by routing long-distance calls over dedicated data lines between company offices.

Even external calls may be routed over the company network to the network location nearest the call destination, thereby saving the toll charges. In terms of quality and user experience, the requirements of these solutions are similar to those of carrier-based systems. Additionally, these systems must support existing PBX-based functions, such as call forwarding and conference calling.

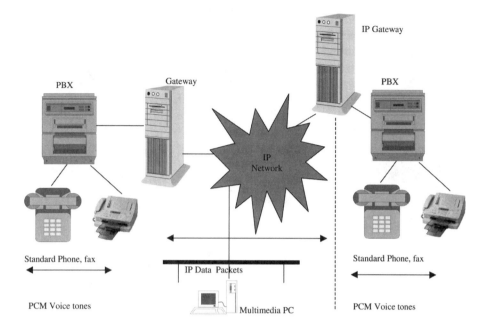

FIGURE 12.10
VoIP gateway.

These gateways interface with the office PBX and the IP routers through a trunk (T1) interface. The IP network may be an intranet maintained by the company or may use an ISP's network. The functions of the VoIP gateways are almost the same as those in PSTN gateways except for the additional signaling needed for the class 5 switches. Each voice sample creates a string of 0s and 1s containing information on the strength and frequency of the sound. Silent and redundant information is compressed. Figure 12.10 shows a VoIP gateway.

12.5.3   IPTel Gateways

These gateways route the voice, fax, or data calls over the Internet. It could be an interface to PSTN to route the carrier calls on the Internet or it could be a corporate Private Branch Exchange (PBX) or LAN interface to send the data over the Internet through ISPs with or without a Virtual Private Network (VPN). These gateways provide some value-added functions related to the connection such as messaging and Internet call waiting.

IPTel systems are generally smaller in scale than PSTN gateways These systems should support corporate billing and accounting functions that feed the appropriate service usage data into the carrier's mainline billing systems. The call signaling and routing functions are the same as in the other gateways. Figure 12.11 shows an IPTel Gateway with calls from a PBX.

The following are the four essential functions of an IPTel Gateway.

- *Lookup and connection functions:* When an IPTel gateway receives an incoming call, it gets the dialed number from the signaling message and converts it into the IP address of the destination GW.
- *Digitizing and demodulation functions:* Analog telephony signals coming into a trunk on the GW are digitized by the gateway into a format useful to the GW, usually 64 Kbit/s

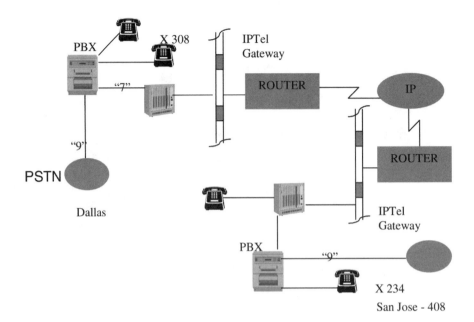

PSTN

FIGURE 12.11
IPTel gateway with calls from PBX.

PCM. This requires the GW to interface with a variety of analog telephony signaling conventions.

- *Compression function:* When the signal is determined to be voice, it is usually compressed by a DSP from 64 Kbit/s PCM to one of several compression/decompression formats and put into IP packets. Good voice quality and low digitizing delay are the primary goals. The ITU mandates the use of G.711 for voice if H.323 is used for signaling. The voice packet is constructed as an RTP/UDP/IP packet, with provisions for sequencing and in-order delivery.

- *Decompression/remodulation functions:* A gateway may support multiple compression/ decompression algorithms. At the same time it performs the above functions, the gateway is also receiving packets from other IP gateways and decompressing voice information back to the format, to be connected to the appropriate analog telephony interface. Or it can be remodulating a digital fax signal back to the incoming format, and then into the appropriate telephony interface.

The IPTel Gateway Call Flow can be summarized in the following, assuming that H.323 is used for signaling purposes:

- Incoming IAM from the CO is received by SS7 stack. If the Carrier Identification Code (CIC) is available, SS7 sends a SETUP message to call control with called party number.
- Call Control (CC) receives the SETUP message and performs incoming resource hunt, translation, and routing, and picks up an RTP port number and forwards this information to the H.323 stack.

- The H.323 stack receives the SETUP message and checks if it has a route to the destination IP address given by CC.
- If a route is found, it originates a SETUP message to the router to send it to the destination.
- If one of the above sequences fails, CC will generate a RELEASE message to the CO.
- When H.323 receives a CONNECT from the destination, it informs the TA to set up the UDP/RTP/IP channel to the destination. Two RTP connections are established per call.
- When the call is released in either direction, the SS7 stack, CC, and H.323 stack release all the resources pertaining to that call through the TA.

12.6 OPEN ISSUES IN INTERNET TELEPHONY

The following open issues in deployment of IPTel are currently being debated.

Economics, Billing, and Pricing: In theory, packet telephony offers substantial economic and cost advantages over traditional TDM voice networks because of

- the reduction of operational and management costs with reduction in network elements
- the potential to send voice packets "in between" data bursts for bandwidth efficiency, leveraging the latest in Digital Speech Processing (DSP) technology

However, only a few service providers have recently had an opportunity to begin deploying real packet telephony networks and test some of these assumptions. While there is little doubt that packet telephony will continue to be driven by application opportunities, the true costs of packet telephony with quality of service might be so high as to negate theoretical cost savings. Open Settlement Protocol (OSP) is used for setting interdomain billing issues such as authentication, authorization, and accounting standards. The major difficulties of measuring the economics of deploying IPTel are twofold: (1) how to evaluate the long-term cost of software that has never been priced, supported, or tested; (2) how to build a new service business case if you can't measure the costs of delivering the service. The real economics of IPTel may not be known until the 21st century.

Home Networking and Residential Access: As we plan to deliver IPTel to the home, a number of fundamental questions need to be addressed for home networking and residential access.

Many service providers like BellSouth, GTE, and Bell Atlantic reported the successful deployment of IPTel over Asymmetric Digital Subscriber Line (ADSL) architecture. The key attributes of ADSL can be summarized in the following:

- Leverages existing copper pairs to provide broadband access
- Incrementally augments existing copper-based subscriber network and backbone ATM data network assets
- Utilizes point-to-point architecture to permit dedicated customer bandwidth
- Provides ATM-based QoS to achieve performance objectives
- Capable of utilizing existing PSTN switch-based features
- Provides graceful evolution to ATM-based fiber-to-the-customer alternatives

Quality of Service (QoS): Delivering a PSTN level of QoS is one of the substantial challenges facing IPTel. QoS means availability, reliability, integrity, and performance in large-scale deployment. Two of the tools in the arsenal for QoS are integrated services and differentiated services. The Integrated Services (Int-serv) model is characterized by network elements that accept and honor commitments to deliver specific levels of service to individual traffic flows. In contrast, Differentiated Services (Diff-serv) are characterized by network elements that classify and mark packets to receive a particular per-hop forwarding behavior on nodes along their path. Sophisticated classification, marking, policing, and shaping operations have to be implemented at network boundaries or hosts. The fundamental problems of IPTel QoS can be summarized in the following:

- The Internet provides no guarantees on the loss or delay packets will experience
- Real-time audio and video applications run adequately under low load, but inadequately under congestion (i.e., normally)
- To introduce QoS "guarantees" into IP without damaging best effort

At IETF, there are comparisons of Int-serv and Diff-serv models on their use and applicability and how the two QoS models can be deployed together. At the present time, a combination of RSVP, Int-serv, and Diff-serv promises scalability, solid guarantees, and admission control, and may set the future research and development (R&D) directions in IPTel QoS.

Directory Services: For IPTel to succeed on the scale of today's traditional circuit switched network, a number of directory services issues must be solved. This includes the capability to uniquely identify an IPTel user whom you desire to reach. Traditional telephony uses a phone number, but this may not be appropriate in the Internet environment. What if you're trying to reach the IPTel user from a traditional phone? Once you've identified the user, how do you find the user's current location? What if the client's IP address changes every time he or she logs in? How can users control when, where, and how they receive calls? Last but not least, how will directory service providers make money in the new world of IPTel? What are the opportunities and viable business models allowed by this new technology? The concept of Value-Added Directory Service has to be addressed with the following features:

- Identify subscriber and the calling party
- Communicate in any media (voice, text, video)
- Follow-me service (real-time, message, etc.)
- Preserve privacy of contact

Table 12.2 summarizes the major highlights of the current open issues in IPTel.

12.7 IN/IP INTEGRATION

IN is highly regarded as a network architecture for rapid service creation and delivery and it aims to support both service and network independence. Examples of IN services include freephone, calling card, Virtual Private Network (VPN), number portability, and mobility services. Future IPTel service creation will be based on the IN Service Creation Environment (SCE) and the Enterprise Java Bean technologies.

Table 12.2   Open Issues of IPTel and Findings

Topic	Highlights
Economics, Billing, and Pricing	• No true comparison of the cost of 1 min of IPTel calls and 1 min of circuit switched calls because of different architectural options • IPTel networks will be built anyway, as data is the driver • The real economics might not be known until the 21st century
Home Networking and Residential Access	• Asymmetric Digital Subscriber Line (ADSL) as the preferred last-mile delivery mechanism for broadband ISDN access to the home • ATM as the chosen Layer 2 technology • ADSL architecture is evolvable for convergence of IPTel and multimedia broadband services with assured quality
Quality of Service	• RSVP + Int-serv + Diff-serv promise a combination of scalability, solid guarantees, and admission control
Directory Services	• No real solution existing today • Value-Added Directory Service advocated as a Personal Communications Portal in the future

ITU-T SG11 and IETF PINT groups are working together to consider the use of IN as a means to provide advanced services (e.g., local number portability, multiparty conferencing the same as in PSTN) in IPTel networks. This implies that ultimately IN services can be provisioned uniformly across the SCNs and the IPTel networks.

12.7.1   New Elements/Functions Required

To enhance IN/IN interworking, the following key elements and functions have been introduced:

PINT Server: A PSTNI Internet Interface (PINT) server accepts PINT requests from PINT clients. It processes the requests and returns responses to the clients. A PINT server may perform these functions as a proxy server or a redirect server. A proxy server makes requests to another PINT server on behalf of its clients; a redirect server returns to its clients addresses of other PINT servers to which requests can be redirected. Additionally, this function transfers data (e.g., fax data) between IP networks and the IN, and associates IP-network entities with the related entities in gateway function.

Service Control Gateway Function (SCGF): The Service Control Gateway Function allows the interworking between the service control layer in Intelligent Networks (INS) and IP networks. For IN CS4 on the service control level, the relations between the IN and the following entities in the IP-network are supported:

• PINT Server
• H.323 GateKeeper Function

Functions Related to PINT Server: The gateway function receives service requests from the PINT Server in the IP-network domain and delivers them to the SCF. It provides the SCF with the necessary information to control service requests, identify users and authenticate data, and protect the IN from misuse or attacks from the IP network. Furthermore, it hides the Shared Channel Feedback/Specialized Resource Function (SCF/SRF) from entities in the IP-network domain and acts as a mediation device between the IP network and the IN. It also relays requests from an SCF to the IP-network domain to perform services (e.g., user notification).

H.323 GateKeeper Function: An H.323 Gatekeeper function could be seen as a logical switch (CCF). Call control signaling (H.225, Q.931-like) and connection control signaling (H.245) for VoIP is transferred via the Gatekeeper, which makes network routing decisions. A GK can require SCF assistance for these routing decisions, for example, for 1-800 numbers, number portability, user profile consultation, and VPN support.

General Functions: The following functions need to be supported by this gateway function:

- Data filtering, parsing, and mapping
- Security/Authentication
- Real-time data collection (billing and parsing) triggering of services (in the IN domain or in the IP-network domain)
- Configuration and dimensioning
- Flow control
- Feature Interaction Management

Call/Bearer Control Gateway Function (C/B GF): This GW is equivalent to a composite function combining both the MG and the MG controller. This GW supports the following functions:

- Access to a packet network through the PSTN, for example, Internet dial-up access via a modem connection
- Interworking of VoIP calls with PSTN calls

Management Gateway Function (MGF): The GW function is required for management purposes.

12.7.2   Special Extensions Required

There will also be necessary extensions to the following existing IN functional entities:

Specialized Resource Function (SRF): This function has to be extended by capabilities to exchange data with GW functions to IP networks. Additionally, for some of the services it needs to support specialized resources with media transformation functions such as "text to fax" and "text to speech."

Service Data Function (SDF): For some services there may be a need for the SCF to access a database type of entity with service-related information to be shared between the IN and the IP network (as for Internet dial-up access, Internet call waiting, such as the association between a PSTN number and an IP address, or the state of C/B GF resource, etc.).

Service Control Function (SCF): To be explored.

Service Switching Function (SSF): To be explored.

Call Control Function (CCF): To be explored.

Management Functions [Single Mode Fiber (SMF), Service Management Access Function (SMAF), Service Creation Environment Function (SCEF)]: To be explored.

12.7.3   New IN/IP Interworking Interfaces

In Figure 12.12, the following interfaces are currently being considered:

IF1: PINT Server-to-Service Control Gateway interface: This interface will relay requests either from the IN or the IP network. Since this interface is purely modeling the information relay, it is not expected to require standardization. The IETF PINT working group is developing a protocol set based on a modification of the Session Initiation and Session Description Protocols [SIP and Session Description Protocol (SDP)]. The architectural configuration envisaged is that end users will make service requests. These requests will be marshalled and converted into SIP/ SDP messages by a dedicated PINT client that will be sent to a PINT Server. The PINT Server will further relay the service requests to the Service Control Gateway Function (SCGF). From the perspective of the IP-network requesting user, this Service Control Gateway Function is responsible for

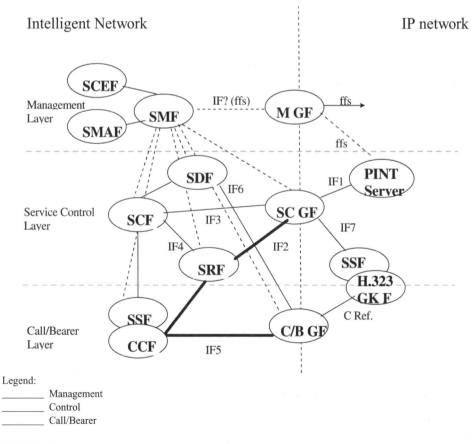

FIGURE 12.12
Enhanced functional architecture for IN support of IP networks.

processing and executing their service feature request; any entities (such as the IN entities) are "hidden" behind this SC Gateway Function, and their operation is transparent to the IP-network users.

IF2: Service Control Gateway-to-SRF interface: May not require standardization, as it will be a data stream to, for example, the SRF text conversion function.

IF3: Service Control Gateway-to-SCF interface: May require ITU-T standardization cooperatively between the IETF and the ITU-T, to reflect the requirements pertinent to the IF1 reference point.

IF4: SCF-SRF interface: Will require enhancements to the existing ITU-T standard for this reference point.

IF5: CCF-C/B GF interface: May require standardization but is not expected to be IN specific. Work is progressing on this in ETSI TIPHON, IETF, and SG16.

IF6 : SDF–C/B GF interface: May require standardization. This interface is required to control Internet access (availability control, etc.) for Internet dial-up access.

IF7: (H.323 GKF) SSF–SC GF interface: May require standardization. This interface is required to trigger and control value-added services from an H.323 GK function in the IP network, for example, for multimedia access from the Internet dial-up access.

Management IF: From the SMF to the management gateway may require standardization based on the SG4 × reference point requirements.

12.7.4 Information Flow for Click-to-Dial (CTD) Service

The information flow for Click-to-Dial (CTD) (phone-to-phone) service is shown in Figure 12.13. The detailed information between SC GF and SCF (IF3) can be deduced from the mapping of IETF defined in the SIP extended protocol for PINT. This flowchart applies to CTFB as well. This example represents stable requirements based on IN CS-2, though interworking is not defined for IN CS-3 operations.

A brief description of the information flow sequence is as follows:

1. Personal computer (PC) user requests CTD service

2. Server sends CTD request

3. SCGF sends CTD service request to SCF

4. SCF initiates call attempt to DN1 and requests DN1_answered event report

5, 6. Connection established between SSF/CCF and phone A using existing ISDN signaling

7. SSF/CCF reports to SCF phone A answered

8. SCF instructs SSF/CCF to connect phone A and SRF

9. SCF instructs SRF to play announcement

10. SCF initiates call attempt to DN2 and requests DN2_answered event report

11, 12. Connection established between SSF/CCF and phone B using existing ISDN signaling

13. SSF/CCF reports to SCF phone B answered

14. SCF instructs SSF/CCF to disconnect phone A and SRF and to merge phone A and phone B legs

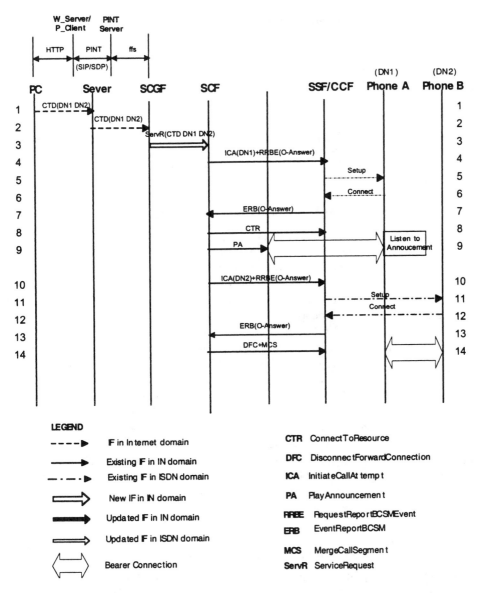

FIGURE 12.13
Information flow for Click-to-Dial (phone-to-phone service).

12.8   SS7/IP INTEGRATION

SS7 is an out-of-band signaling scheme designed with signaling being on a separate channel from the bearer connection. Coupled with the higher signaling speed, this can provide the capability to offer enhanced services by carrying important call-related information on the signaling channel during or after the call setup phase. A related requirement for the PSTN has been that the signaling transport should be very reliable and robust with very little message loss and erroneous message delivery. These factors have been built into the transport layer, that is, the Message Transfer Part of SS7 protocol. The applications layers residing over Message Transfer

Part provides functionality for call setup and transaction-oriented services, for example, 800, calling card, voice mail, and caller ID.

Delivery quality of service is one of the biggest challenges facing IPTel deployment. Two models, the integrated and differentiated service models, will be presented with a focus on their use and applicability and how they can be used together. Finally, the signaling mechanism of a telephone network determines the efficiency of the network plus all the value-added features that are supported. The PSTN has used SS7 for many years. As IP-based telephony systems are being developed, signaling is becoming a key component of all proposed solutions. Current debates are on the protocols, services, placement, and development of IP-signaling mechanisms such as H.323, SIP, MGCP, and MEGACO/H.GCP, and how they will integrate with SS7.

12.8.1   Transport of SS7 Over IP-Related Protocols

In today's PSTN, the network interconnections are typically based on SS7, since most local and IntereXchange Carriers have ubiquitously implemented SS7 in their networks. However, Internet Service Providers (ISPs) are mostly IP-based networks. As the demand for IPTel grows, interconnections between PSTN and ISPs will increase. The ISPs, many small entities, would like to avoid the expense of SS7 infrastructure. Therefore, it is only logical that some way of transporting SS7 application protocols, for example, ISDN User Part (ISUP) and Transaction Control Application Protocol (TCAP), over IP protocols will have to be provided. This will enable both PSTN and ISPs to communicate at the application level via a Signaling Gateway without worrying about the transport. It will also enable two PSTN networks to use an IP network as a transit network for interconnection as a cheaper alternative.

ITU-T SG11 and IETF SIGTRAN (Signal Transmission) working groups are investigating the issue of the transport of SS7 over IP protocol. ITU-T is looking into replacing SS7 MTP transport by introducing an adaptation layer over either UDP or TCP to interface with SS7 application protocol. IETF is evaluating a generic mechanism to carry major existing SS7 as well as ISDN protocols over IP. Solution(s) from both groups are expected by year-end 2000. A major consideration for both groups will be to guarantee end-to-end reliability and robustness of the existing PSTN SS7 network when interworking with IP networks that traditionally have provided best-effort grade of service.

12.8.2   Interworking of SS7 with IP-Related Protocols

In addition to transporting SS7 over IP, there is also an issue of how the SS7 application protocols that were designed for PSTN will interwork with a growing number of IP-related protocols under consideration in various IETF and ITU working groups. ITU-T has published the H.323 set of recommendations, for example, H.245 and H.225 for multimedia communication over IP networks. IETF is working on SIP for IP-based call setup and MGCP for communication between MGC and MG.

For interworking between PSTN- and IP-based networks, these protocols and SS7-based protocols, for example, ISUP and Intelligent Network Application Protocol (INAP), will be required not only for call setup but also for access to IN based services in PSTN. Efforts are under way in these standard bodies and other forums, for example, the IN Forum, to describe architectures and call scenarios for ordinary call setup and access to the number of services already handled by PSTN, for example, toll-free number portability. There will be pressure on SS7 application protocols to accommodate interworking with these and other IP-based protocols yet to emerge.

12.8.3   Future of IP/SS7

In the context of PSTN and Internet integration, there is ongoing debate whether SS7 has a future or not. Some believe that the future of SS7 is limited since it is complex and expensive, and was designed for typical PSTN services where the service nodes are limited in number in a network. People holding this view also believe that SS7 does not provide call and connection separation today that will be required for the multimedia services of tomorrow. Others believe that IETF protocols, for example, SIP, can be more easily extended to accommodate the distributed environment of the Internet service where a typical client may require services from a number of servers.

There is also talk of extended SIP, dubbed SIP+, desized to encapsulate SS7 protocols such as ISUP to be carried transparently over the Internet for the interconnection of end-PSTN nodes. However, it must be remembered that many carriers in the world today have existing PSTN that will have to be supported for many years to come.

SS7 today provides the glue for keeping together the PSTN, but is also getting increased acceptance in wireless networks and their interconnection with the PSTN. Standards bodies are also at work to enhance SS7 ISUP to make it bearer independent (call and connection separation). The initial application of the enhanced ISUP (ISUP+) will be for ATM trunking for narrowband applications, but in theory it can also be used to set up IP calls.

One must remember that the Internet is still in its infancy, and has yet to prove itself as a reliable and scalable alternative for the real-time communications that PSTN provides today. The technology debate will probably go on about what is the best protocol as the PSTN and Internet converge. But it is only logical to assume today that SS7 will exist for a number of years to come for the same reason that PSTN as we know it today will also exist for a while.

12.9   CONCLUDING REMARKS

The preceding sections describe some of the current standards, technologies, and ongoing efforts in IPTel. The current research effort is targeted toward achieving a PSTN level of QoS and also the seamless transition to an all IP-based network and integration of existing PSTN/IN services. New and evolved networks are already being introduced to support new voice, data, and multimedia services. In the future, IPTel network architecture will be positioned for uncertain protocol evolution and development and uncertain service innovation. The winners in the telecommunications industry in the current millennium will be those companies that can integrate PSTN, IN, wireless, access, mobility, carrier-class networking, and IP successfully in their products and services.

12.10   GLOSSARY

ATM: Asynchronous Transfer Mode
CTD: Click to Dial
DRC: Direct-Routed Call
DTMF: Dual Tone Multi-Frequency
ETSI: European Telecommunications Standards Institute
GK: Gatekeeper
GRC: Gatekeeper-Routed Call
GW: Gateway
HTTP: HyperText Transfer Protocol

IETF: Internet Engineering Task Force
IN: Intelligent Networks
IP: Internet Protocol
IPDC: IP Device Control
IPTel: Internet Telephony
ISDN: Integrated Services Digital Network
ISP: Internet Service Provider
ISUP: ISDN User Part
ITU: International Telecommunications Union
ITU-T: International Telecommunications Union—Technical
IVR: Interactive Voice Response
LAN: Local Area Network
MCU: Multipoint Control Unit
MEGACO: Media Gateway Control
MGCP: Media Gateway Control Protocol
MTP: Message Transfer Protocol
OSP: Open Settlement Protocol (ETSI/TIPHON)
PBX: Private Branch Exchange
PINT: PSTN/Internet Interface
POTS: Plain Old Telephone System
PSTN: Public Switched Telephone Network
QoS: Quality of Service
RGW: Residential Gateway
RSVP: Resource Reservation Protocol
RTCP: RTP Control Protocol
RTP: Real-Time Transport Protocol
RTSP: Real-Time Streaming Protocol
SCN: Switch Circuit Network
SCP: Service Control Point
SDP: Session Description Protocol
SG: Study Group
SGCP: Simple Gateway Control Protocol
SIGTRAN: Signaling Transport
SIP: Session Initiation Protocol
SS7: Signaling System Number 7
TCAP: Transaction Control Application Protocol
TCP: Transport Control Protocol
TDM: Time Division Multiplexing
TGW: Trunk Gateway
TIPHON: Telecommunications and Internet Protocol Harmonization Over Networks
UAC: User Agent Client
UAS: User Agent Server
UDP: User Datagram Protocol
URL: Uniform Resource Locator
VoIP: Voice Over IP
VPN: Virtual Private Network
VRU: Voice Response Unit

12.11 DEFINITIONS OF KEY TERMS

ADPCM: Adaptive Differential PCM (codec)
G.711: 64 kbps PCM half-duplex codec (high quality, high bandwidth, minimum processor load)

G.723: 6.4/5.3 kbps MP-MLQ codec (low quality, low bandwidth, high processor load due to the compression)

Gatekeeper: An H.323 entity on the LAN which provides address translation and controls access to the LAN for H.323 terminals, gateways and MCUs

Gateway: An endpoint on the LAN that provides for RT 2-way communications between H.323 Terminal on the LAN and other ITU terminals (ISDN, GSTN, ATM, . . .) on WAN or to another H.323 gateway

H.225: Protocols (RAS, RTP/RTCP, Q.931 call signaling) and message formats of the H.323 are covered in this standard

H.245: Protocol for capability negotiation, messages for opening and closing channels for media streams, etc. (i.e. media signaling)

H.248: Also called H.GCP (Gateway Control Protocol—to be merged with IETF MEGACO into a single protocol standards based on further decomposition of gateway)

H.323: An umbrella standard for audio/video conferencing over unreliable networks; architecture and procedures are covered by this standard; H.323 relies on H.225 and H.245

MC: The Multipoint Controller provides the capability negotiation with all terminals taking part in a multipoint conference

MCU: The Multipoint Control Unit is an endpoint on the LAN which provides the capability for three or more terminals and gateways to participate in a multipoint conference. The MCU consists of a mandatory MC and optional MPs

MP: The Multipoint Processor provides for centralized processing (mixing, switching,) of audio, video, or data streams in a multipoint conference

PCM: Pulse Code Modulation (codec)

Q.931: ISDN call signaling protocol (in an H.323 scenario this protocol is encapsulated in TCP and sent to the well-known port 1720)

RAS: Registration, Admission, Status—management protocol between terminals and gatekeepers

T.120: Data conference protocol

12.12 ACKNOWLEDGMENTS

The author wishes to thank Yatendra Pathak and David Devanathan of MCI WorldCom for their technical comments in this chapter.

12.13 BIBLIOGRAPHY

1. *Alcatel Telecommunications Review,* Special Issues on "Reliability" and on "Real-time and Multimedia IP," Second Quarter, 1999.
2. *Bell Labs Technical Journal,* Special Issues on Packet Networking, Vol. 3, No. 4, October–December 1998.
3. Black, U., *Voice Over IP*, Prentice Hall, 7/99, ISBN 0-13-022463-4.
4. ETSI TIPHON web site http://www.etsi.org/tiphon/
5. *IEEE Network Magazine,* Special Issue on "Internet Telephony," May/June 1999, Vol.13, No. 3.
6. Institute of Electrical and Electronics Engineers IEEE First International Workshop on Internet Telephony, June 1999.
7. IETF web site http://www.ietf.org
8. ITU-T web site http://www.itu.int
9. MEGACO/H.GCP web site: http://www.ietf.org/html.charters/megaco-charter.html
10. SIP Home Page http://www.cs.columbia.edu/~hgs/sip/

CHAPTER 13

Wideband Wireless Packet Data Access

JUSTIN CHUANG
LEONARD J. CIMINI, JR.
NELSON SOLLENBERGER

13.1 INTRODUCTION

Wideband Wireless Packet Data Access is expected to become a major factor in communications. The rapid growth of wireless voice subscribers, the growth of the Internet, and the increasing use of portable computing devices suggest that wireless Internet access will grow to be a large market. Rapid progress in digital and radiofrequency (RF) technology is making possible highly compact and integrated terminal devices, and the introduction of sophisticated wireless data software is providing more value and making wireless Internet access more user friendly. Data transmission rates are growing rapidly in fixed networks with the use of Wavelength Division Multiplexing in backbone fiber networks and the introduction of cable modems and High-Speed Digital Subscriber Line technology in the fixed access networks. In parallel with the expanding availability of high-speed transmission capabilities, increasingly demanding Internet applications and user expectations have emerged. Experience with laptop computers and Personal Digital Assistants has shown that many end users desire their portable equipment to provide essentially the same environment and applications that they enjoy at their desks. Experience with wireless access has also demonstrated the singular importance of widespread coverage and anywhere/anytime access. In this section, we first discuss some of the motivation for higher-bit-rate wireless services, and then present a summary of the current and emerging state of wireless data systems.

13.1.1 The Wireless Data Opportunity

Current cellular radio and paging services have been very successful in providing untethered communications, and wireless access to the Public Switched Telephone Network has been growing at rates of 30–50% per year [1]. The introduction of the Global System for Mobile Communications (GSM) and other digital cellular technologies has accelerated the growth of wireless communications throughout the world. With the advent of new personal communications services, wireless access is expected to become even more popular, with more than 400 million subscribers expected worldwide in the year 2000 [2]. Personal computers (PCs) and Internet services have experienced even more rapid growth than wireless services due to low-cost, high-performance computer technologies and attractive network applications. While interest in wireless Internet access was initially generated by the opportunities for e-mail and messaging, the explosion in popularity of the World Wide Web suggests broader long-term opportunities [3].

When you combine the success of cellular services with the increased presence of laptop computers and the rapid growth in the number of Internet users, wireless data should have a very bright future. Nevertheless, today, the number of wireless data subscribers is small, with the most formidable obstacle to user acceptance being the performance limitations of existing services and products [4], including link performance (data rate, latency, and quality); network performance (access, coverage, spectrum efficiency, and quality of service); and price. Therefore, it is timely to consider a synergistic combination of wireless, computer, and Internet technologies to provide a wide-area packet-data service with large improvements in performance beyond existing approaches.

13.1.2   Current Wireless Data Systems

Current wireless data systems fall into two categories: (1) Wide-area and microcellular services (designed for metropolitan area coverage) offering limited bit rates, on the order of 10–100 Kbit/s. These include RAM Mobile Data, Advanced Radio Data Information System (ARDIS), Cellular Digital Packet Data (CDPD), and digital cellular data services. (2) Products that provide much higher rates (1–10 Mbit/s) but have small coverage areas, usually limited to within a building. These include WaveLAN and RangeLAN2. In the next sections, we briefly review these wireless data options.

13.1.2.1   Wide-Area Services

The ARDIS mobile data network [5], providing coverage in over 400 metropolitan areas in the United States, uses 25 kHz channels in the 800 MHz band and provides 4.8 Kbit/s transmission rates in most areas. A protocol for 19.2 Kbit/s has been defined and has been deployed in some areas. The RAM Mobile Data Network, based on the MOBITEX protocol [6, 7], uses 12.5 kHz channels in the 800 and 900 MHz bands and provides 8 Kbit/s transmission rates. The ARDIS and RAM networks are designed for messaging; their limited bandwidth and large packet transmission delays make interactive sessions infeasible. CDPD [8, 9] is an overlay of AMPS cellular networks using paired 30 kHz channels to provide a transmission rate of 19.2 Kbit/s, with the user experiencing a maximum throughput of about half that rate. CDPD provides an Internet capability coupled with a strong synergy with cellular networks and terminals.

GSM, the Time Division Multiple Access (TDMA)-based European Digital Cellular Standard, supports a variety of circuit switched data modes using internetworking functions. The current transmission rates range from 2.4–9.6 Kbit/s, with planned upgrades to 14.4 Kbit/s. Several higher-bit rate services, with peak data rates of more than 100 Kbit/s, are emerging, including a packet-switched data service, General Packet Radio Service (GPRS) [10]. IS-136, the TDMA-based North American Digital Cellular Standard, supports 9.6 Kbit/s circuit data and fax access. By aggregating time slots and by introducing higher-level modulation, such as 8-PSK, data rates of 40–60 Kbit/s can be achieved (IS-136+) [11]. Personal Digital Cellular

(PDC), the TDMA-based Japanese Digital Cellular Standard, is very similar to IS-136 and has deployed a packet data service [private communication from M. Taketuga, NEC Corp., May 1997]. For IS-95, the CDMA-based North American Digital Cellular Standard, developers and operators are providing variable bit-rate access with peak rates of 9.6 and 14.4 Kbit/s [12]. CDPD emulation is also possible with the CDMA air interface.

13.1.2.2 *Microcellular Services*

Metricom has deployed a microcellular packet data system (called Ricochet) in the San Francisco Bay area and other areas [13]. It operates in the 902-928 MHz ISM band with a 160-kHz channel spacing and uses frequency hopping to provide peak transmission rates of about 100 Kbit/s, with Internet connectivity. Other wireless data access systems that may provide some Internet capability include digital cordless or microcellular systems, such as the Personal Handyphone System (PHS) in Japan [14], the Digital European Cordless Telecommunications System (DECT) [15], and the Personal Access Communications System (PACS) [16]. These systems were designed primarily for voice service for pedestrians within buildings or over dense areas, but, with the larger bandwidths available in these systems, data services in the 32–500 Kbit/s range could be supported.

13.1.2.3 *Local Area Network Products*

Wireless Local Area Networks (WLANs) have been available for several years. Most operate in the unlicensed Industrial Scientific Medical (ISM) bands (902–928 MHz, 2.4–2.4785 GHz, and 5.725–5.850 GHz); provide a limited transmission range, generally less than 100 m in an office environment; and use spread spectrum modulation. Products include Lucent's WaveLAN [17], which supports speeds up to 2 Mbit/s for standard products and up to about 10 Mbit/s with enhancements, and Proxim's RangeLAN2 [18]. A standard has also been completed by IEEE802.11 [19]. While WLANs generally support Internet connectivity, so far, they have been most successful in niche markets, such as inventory and LAN adjuncts in buildings where added wiring is particularly difficult.

13.1.3 Emerging and Future Wireless Data Options

The wireless data market is currently small compared to cellular voice service. To realize the enormous potential that has been predicted, the performance of current systems must be significantly improved. In the rest of this section, we summarize some of the wide array of proposals for achieving higher bit rates or covering wider areas.

13.1.3.1 *Local Area Networks*

There have been several standards activities addressing the need to push WLANs to a higher level of performance. The HIgh PErformance Radio Local Area Network (HIPERLAN) standard defined by the European Telecommunication Standards Institute (ETSI) in 1996 operates in the 5 GHz band with transmission rates of over 20 Mbit/s [20, 21]. To spur such activities in the United States, the Federal Communications Commission (FCC) allocated 300 MHz (5.15–5.35 GHz for indoor use and 5.725–5.825 GHz for both indoor and outdoor use) for unlicensed operation, in what is called the National Information Infrastructure (NII) band. IEEE802.11 recently proposed a multicarrier system, Orthogonal Frequency Division Multiplexing (OFDM), as a standard for WLANs operating in the NII band, providing up to about 30 Mbps [22]. Both ETSI, in the form of the HIPERLAN/2 standard [23], and the Association of Radio Industry and Businesses (ARIB) in Japan have also selected OFDM technologies for their high-bit-rate WLANs.

13.1.3.2   Third-Generation Systems

The most extensive efforts to provide higher bit rates can be found among the so-called Third-Generation (3G) projects around the world (e.g., see [24]). The International Telecommunications Union (ITU) program called IMT-2000 (International Mobile Telecommunications in Year 2000), formerly called Future Public Land Mobile Telecommunications System (FPLMTS), was initiated in 1986 with a goal of producing a global standard for third-generation wireless access, using the frequency bands 1885–2025 and 2110–2200 MHz.

The main design objective for 3G systems is to extend the services provided by current second-generation systems with high-rate data capabilities. Specifically, the stated goals [25] are to provide 144 Kbits (preferably 384 Kbit/s) for high-mobility users with wide-area coverage and 2 Mbits for low-mobility users with local coverage. The main application will be wireless packet transfer, in particular, wireless Internet access. However, supported services include circuit voice, circuit video, e-mail, short messages, multimedia, simultaneous voice and data, and broadband Integrated Services Digital Network (ISDN) access. In the search for the most appropriate multiple access technology for 3G wireless systems, a wide range of new multiple access schemes has been proposed to the ITU (see [26] for more details). Here, we will very briefly describe the efforts in Europe, Japan, and the United States.

Third-Generation activities in Europe, called the Universal Mobile Telecommunications Systems (UMTS) program [27, 28], started in 1992 with initial research efforts concentrating on two approaches: (1) Advanced TDMA (ATDMA) [29], an adaptive modulation scheme that is similar to GSM for large cells but achieves higher data rates using linear modulation for small cells; and (2) COde DIvision Testbed (CODIT) [30], a direct-sequence spread-spectrum scheme with bandwidths of 1, 5, and 20 MHz and transmission rates up to 2 Mbit/s. These projects were followed by a second phase concentrating on wideband CDMA (WCDMA), called FRAMES Multiple Access 2 (FMA2) [31–33]. The current proposal for UTRA [Universal Mobile Telecommunications System (UMTS) Terrestrial Radio Access] includes both a frequency-division duplex (FDD) mode and time-division duplex (TDD). The FDD mode is based on pure WCDMA, while the TDD mode includes an additional TDMA component (TD/CDMA).

In parallel with the European efforts, Japan produced a number of 3G proposals including both TDMA and CDMA schemes [34, 35]. This technology push from Japan accelerated standardization in Europe and in the United States. The Japanese research resulted in a WCDMA concept developed and submitted to ITU by ARIB. During 1997, joint parameters for Japanese and European WCDMA proposals were agreed upon, and harmonization work has continued in 1997 and 1998.

In the United States, 3G activities have included studies of both TDMA and CDMA approaches. The Telecommunications Industry Associations (TIA) TR45.3 Committee, responsible for IS-136 standardization, adopted a TDMA-based 3G proposal, UWC-136, based on the recommendation from the Universal Wireless Communications Consortium (UWCC) in February 1998. UWC-136 proposes adaptive modulation using wider-band TDMA carriers with bandwidths of 200 kHz and 1.6 MHz. The 200 kHz carrier, (vehicular/outdoor) has the same parameters as the GSM Enhanced Data Rates for GSM Evolution (EDGE) air interface and combines that with GPRS networks to provide packet data service at speeds up to 384 Kbit/s. On the CDMA front, TIA TR45.5, responsible for IS-95 standardization, adopted a framework for WCDMA backward-compatible with IS-95, CDMA2000, in March 1998. A similar WCDMA concept has also been developed and submitted to the International Telecommunications Union (ITU) by the North American T1P1 standards body. Recently, there has been an agreement to harmonize CDMA2000 and WCDMA.

13.1.4 Summary and Outline of the Chapter

In summary, strong wireless voice growth, strong portable computing growth, plus strong fixed Internet growth lead to a strong wireless data opportunity. Existing technologies that provide about 10 Kbit/s wireless data access in cellular systems will evolve to provide about 100 Kbit/s over the next few years. In the remainder of this chapter, we will describe some of the proposals for achieving higher bit rates for future wireless packet data services. In Sections 13.2 and 13.3, respectively, the 3G technologies, WCDMA and EDGE, with target bit rates of about 384 Kbit/s with wide-area coverage, are briefly described.

Even higher bit rates, say up to 5 Mbit/s, will be required in the near future to provide user experiences for existing services, such as Web browsing with a portable computer, that are similar to local area network (LAN) connectivity and will allow new mobile-specific applications, such as digital maps and driving directions. In Section 13.4, we present a more detailed discussion of wideband OFDM. This latter approach has the potential to provide even higher bit rates, on the order of 2–5 Mbit/s.

13.2 PACKET DATA ACCESS USING WCDMA

WCDMA is an emerging wireless technology that has been the focus of intensive research and development. (We focus on the WCDMA system proposed by ETSI and ARIB; other CDMA systems have similar characteristics.) WCDMA is based on 5-MHz channels using direct-sequence spread spectrum with a chip rate of 4.096 Mcps operating with paired uplink/downlink channels in an FDD mode. (Recently, the chip rate has been reduced to 3.84 Mcps, but the system characteristics remain the same.) The use of CDMA provides flexibility in mixing of services with different bit rates and Quality of Service (QoS) requirements. However, this also results in several significant obstacles to supporting high-speed packet data services: (1) A spreading factor of 85 (13 Kbit/s voice) or 128 (8 Kbit/s voice) is used with second-generation IS-95 CDMA systems to provide about 20 dB of processing gain. At much higher bit rates, CDMA systems must either reduce the processing gain or expand the bandwidth. The nominal maximum rate for a typical macrocell with 5 MHz total spectrum is 384 Kbit/s, although higher rates are considered in other environments. (2) CDMA requires precise power control so that the signals from all terminals arrive at a base station at about the same power level. The power control problem is more easily managed for circuit modes of operation by combining a channel probing procedure, which slowly raises the transmission power for initial access, with rapid feedback power control once a channel is established. With packet transmission, channel probing is undesirable because it will cause delay; it is also not efficient for short packets. Power control is also important for downlinks to achieve high efficiency with CDMA, and timely information on optimal power levels for packet transmission needs to be provided. In the following, we highlight the issues and possible solutions for WCDMA-based packet data services. Technical issues that are common to CDMA systems and packet techniques for CDMA proposals can be found in [32, 33, 36].

13.2.1 Variable-Rate Packet Data

To support a wide range of services and bit rates, WCDMA is designed to provide nominal bit rates for individual users that range from 1 Kbit/s up to 384 Kbit/s. This is accomplished using different spreading rates, different coding rates, multiple spreading codes, and multiple time slots to achieve the required bit rate and reliability. The use of a multifinger RAKE receiver and a bandwidth of 5 MHz provides a strong frequency diversity effect in many environments, particularly large cell environments. WCDMA uses a frame size of 10 ms with 16 time-slots per frame. Orthogonal Walsh functions are used to separate channels within one transmitted signal and long pseudorandom codes are used to separate signals from different sources. Since multiple signals from one source experience a common delay upon arriving at a receiver, they can be efficiently separated using functions that are orthogonal over a given period for synchronous signals (although this is compromised in a dispersive multipath environment). But, signals from different sources will generally experience different delays, and so random sequences are preferred to provide suppression of interfering signals via receiver processing gain.

For packet data transmission, variable-rate orthogonal Walsh functions can be used to simultaneously provide a wide range of bit rates. Orthogonal Variable Spreading Factor (OVSF) codes preserve orthogonality between downlink channels of different rates and spreading factors [37]. The OVSF codes can be defined using the code tree of Figure 13.1.

Each level in the code tree defines channelization codes of length Spreading Factor (SF), corresponding to a spreading factor of SF. All codes within the code tree cannot be used simultaneously within one cell. A code can be used in a cell if and only if no other code on the path from the specific code to the root of the tree or in the subtree below the specific code is used in the same cell. This means that the number of available channelization codes is not fixed but depends on the rate and SF of each physical channel. For example, a two-chip Walsh function can be defined that, with 64 repetitions, is orthogonal to each of sixty-four 128-chip sequences, but the two-chip function supports 64 times the bit rate of each 128-chip function. Variable-rate Walsh function encoding to support packet data transmission can also provide an advantage relative to multicode transmission by resulting in a smaller peak-to-average power ratio that strongly impacts RF amplifier requirements for terminals. Terminals may use a single or very few Walsh codes of variable length for transmission, but base stations may use many simultaneous Walsh codes for transmission to communicate with many terminals.

Direct-sequence CDMA systems require tight transmitter power control to achieve good performance. The best performance for power control is obtained using iterative feedback from a receiver to a transmitter through a return link to adapt a transmit power level to a good operating point with just enough received power to maintain a target error rate. With packet operation, iterative feedback power control is not efficient for short bursts because of the time required to achieve convergence. A hybrid scheme has been proposed that operates without feedback for

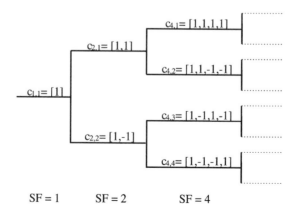

$c_{4,1} = [1,1,1,1]$

$c_{2,1} = [1,1]$

$c_{4,2} = [1,1,-1,-1]$

$c_{1,1} = [1]$

$c_{4,3} = [1,-1,1,-1]$

$c_{2,2} = [1,-1]$

$c_{4,4} = [1,-1,-1,1]$

SF = 1   SF = 2   SF = 4

FIGURE 13.1

Code tree for generation of Orthogonal Variable Spreading Factor (OVSF) codes. (SF-spreading factor.)

short bursts on a random access channel, but operates with feedback power control on a traffic channel for long data bursts to achieve good efficiency.

Two different packet data transmission modes are considered. Short data packets can be appended directly to a random access burst. This method, called common channel packet transmission, is used for short infrequent packets, where the link maintenance needed for a dedicated channel would lead to an unacceptable overhead. Also, the delay associated with a transfer to a dedicated channel is avoided. Note that for common channel packet transmission only open-loop power control is in operation. Figure 13.2 illustrates packet transmission on a common channel for short packets. Larger or more frequent packets are transmitted on a dedicated channel. A large single packet is transmitted using a scheme in which a dedicated channel is released immediately after the packet has finished transmitting. In a multipacket scheme, a dedicated channel is maintained by transmitting power control and synchronization information between subsequent packets.

In WCDMA, the widely varying levels of interference in cellular systems from different sources are reduced significantly after despreading in WCDMA. As a result, each signal (within certain limits) experiences an average level of interference impairment. WCDMA takes advantage of this interference-averaging characteristic to efficiently support multiple access. One of the characteristics of packet data networks is high peak data rates associated with individual users. This means that mobiles that are in areas of near equal signal strength from two or more cells will be required to decrease peak transmission rates significantly to avoid creating or receiving excessive interference. This problem can be mitigated by introducing some form of dynamic resource management between neighboring cells. It can also be mitigated by introducing interference suppression or multiuser detection that can manage the effects of dominant interference in a CDMA system, albeit with increased complexity. While WCDMA offers the flexibility to provide a wide range of services, simultaneous operation of different types of services will require careful management of resources and power control to maintain good quality of service.

Handoff has a significant impact on CDMA, especially when a single terminal communicates with multiple base stations in the soft handoff mode. In WCDMA, which does not require base station synchronization, asynchronous base station operation must be considered when designing soft handoff algorithms. For CDMA2000, which is based on the evolution of IS-95 and uses

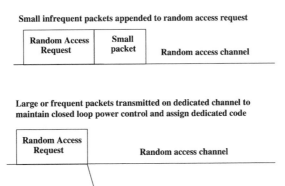

FIGURE 13.2
Dual-mode packet transmission for WCDMA.

multicarrier operation for backward compatibility, interfrequency handoffs with BS synchronicity must be considered. The design must also consider the utilization of hierarchical cell structures; macro-, micro-, and indoor cells. Several carriers and interfrequency handoffs may also be used for meeting high-capacity needs in hot spots and interfrequency handoffs will be needed to handoff to or from second-generation systems, like GSM or IS-95.

Similar to TDMA systems, an efficient method is needed for making measurements on other frequencies while still having an active channel operating on a current frequency. Two methods are considered for interfrequency measurements in WCDMA: (1) dual receiver and (2) slotted mode. The dual receiver approach is considered suitable if the mobile terminal employs antenna diversity. While performing interfrequency measurements, one receiver branch is switched to alternative frequencies for measurements, while the other receiver continues on the current operating frequency. The loss of diversity gain during measurements needs to be compensated with higher downlink transmission power. The advantage of the dual-receiver approach is that there is no break in the current frequency connection. Fast closed-loop power control is running all the time. The slotted mode approach is similar to a TDMA system where an idle period is used to measure other frequencies. This is attractive for mobile stations without antenna diversity.

In summary, WCDMA offers flexibility in supporting a wide range of bit rates from 1–384 Kbit/s in macrocellular environments. The wide bandwidth provides frequency diversity benefits to mitigate fading effects. Some of the challenges for packet data operation include peak data rates, power control for bursty traffic, and soft handoffs.

13.3 PACKET DATA ACCESS USING EDGE

The GSM system is the most popular second-generation wireless system. It employs TDMA technology to support mobile users in different environments. This system, initially standardized and deployed in Europe, has been deployed worldwide. Each carrier is 200 kHz wide, which can support eight simultaneous full-rate circuit voice users using eight TDMA bearer slots. It also supports a circuit-data capability at 9.6 Kbit/s. To support higher data rates, ETSI has generated the GPRS [38] standard, which employs variable-rate coding schemes and multislot operation to increase peak rates. Using packet access further enhances system throughput and spectrum efficiency. However, the peak rate for GPRS is limited to about 144 Kbit/s. ETSI is also developing the EDGE standard to enhance GSM packet data technology with adaptive modulation to support packet data access at peak rates up to about 384 Kbit/s. The UWCC has also adopted EDGE [39–41]. GPRS networking is proposed to support EDGE-based wireless packet data access. This system will adapt between Gaussian Minimum Shift Keying (GMSK) and 8-PSK modulation with up to eight different channel coding rates [43] to support packet data communications at high speeds on low-interference or low-noise channels and at lower speeds on channels with heavy interference or noise. (An additional rate has since been added to make a total of nine different channel coding rates.)

EDGE will use linearized GMSK pulse shaping and a data rate equal to GSM of 270.833 kbaud. The linearized GMSK pulse shaping means that each 8-PSK constellation point at the transmitter is associated with a complex Gaussian pulse at baseband, and it is specified to result in a spectral occupancy very similar to GSM to maximize commonality with GSM. It results in a low peak-to-average power ratio of the transmitted signal of only 2.5 dB, which is particularly advantageous for terminal RF power amplifiers. However, this pulse shaping does result in nonzero intersymbol interference (ISI). So, equalization at a receiver is required both for purposes of mitigating multipath dispersive fading and ISI introduced by filtering. Receiver equalization is a challenge for EDGE, because maximum-likelihood-sequence estimation, which is typically used for GSM reception, becomes very complex with eight states per baud instead of only two. An equalizer that spans five symbol periods (almost 20 μs) would require 8^4 states (4096 states), instead of the 16 states required for GSM (2^4). So, reduced complexity equalizer structures or less stringent performance requirements than GSM are likely to be used.

EDGE can be deployed with conventional 4/12 frequency reuse, that is, the same set of frequencies is reused every 4 base stations or 12 sectors, as is used for GSM. However, 1/3 reuse is also proposed to permit initial deployment with only 2×1 MHz of spectrum (which provides two guard channels as well as three operating channels). A combination of adaptive modulation/coding (referred to as link adaptation or LA), partial loading, and efficient Automatic Retransmission ReQuest (ARQ) permits operation with very low reuse factors. Incremental redundancy (IR) or type II hybrid ARQ has been proposed to reduce the sensitivity of adaptive modulation to errors in estimating channel conditions and to improve throughput [44].

An additional challenge for 1/3 reuse is the operation of control channels that require robust performance. Control channels include broadcast information that is transmitted without the benefit of a reliable link protocol. A reliable control channel can be provided by synchronizing the frame structures of base station transceivers throughout a network and using time reuse to achieve adequate protection for control channels. For example, by allocating one time-slot in a frame for control channel operation, and operating each individual base station on 1/4 of the slots allocated for the control channel, an effective reuse of 4/12 can be achieved to provide adequate robustness with a bit rate of 1/4 that of a conventional GSM control channel. By allocating two time-slots for control channel operation, a 1/2-rate control channel can be provided while maintaining an effective reuse of 4/12. One of the complications of these arrangements is the requirement that a terminal must achieve initial base station selection and synchronization on downlink channels that are not continuous.

EDGE radio links will tend to be dominant-interference-limited, so adaptive antenna array processing techniques and smart antennas can provide large gains with EDGE. Switched-beam techniques are attractive for base station transmission, and two- and four-branch adaptive antenna array techniques are attractive for base station reception. Since EDGE uses a relatively narrowband channel of 200 kHz, it is subject to narrowband Rayleigh fading with some limited frequency diversity possible through sequence estimation equalizers. Larger gains are possible with frequency hopping for the lower channel coding rates. Terminal diversity may be attractive to achieve high performance for stationary operation of terminals, and to provide robust performance with low-frequency reuse through two-branch interference rejection combining.

13.3.1 Link Adaptation and Incremental Redundancy

Radio link control selects among the modulation/coding options, in response to the Signal-to-Interference Ratio (SIR) and other radio link conditions. LA explicitly changes the modulation and coding based on link quality estimates. Incremental redundancy (IR) transmits increments of redundant bits after errors are observed. IR adaptively changes an effective data rate based on the results of actual transmissions. On the other hand, LA relies on channel estimation to determine a suitable rate, which may not be the proper choice when transmission occurs because of measurement errors or latency. Therefore, IR can achieve better aggregate throughput, possibly at the expense of extra delay and higher memory requirements in implementation [44].

One criterion for selecting a particular data rate for LA is to maximize throughput defined as $S = R_c (1 - BLER_c)$ where R_c and $BLER_c$ are the data rate and block error rate, respectively, for the transmission mode chosen. Figure 13.3 shows an example of the throughput as a function of SIR for different modes. It is found that this threshold criterion is generally effective in achieving a high *aggregate* system throughput, but the QoS for *individual users* can be significantly

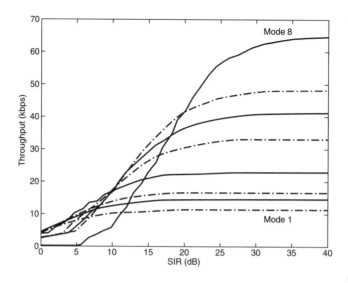

FIGURE 13.3
Throughput as a function of SIR for different transmission modes.

degraded as the system load increases [45]. Furthermore, link adaptation requires the receiver to continuously perform quality measurements and provide timely feedback to the transmitter. The GSM system has the basic capabilities to support these functions, but optimization of link adaptation is a function of parameters such as user speed, frequency hopping, and multipath delay spread profiles, which in general are highly dynamic. One useful technique for packet data transmission is to introduce a "nontransmission" mode or "mode-0" when the SIR is below a threshold to provide an admission policy on a per-packet basis. Significant performance gain is obtained, particularly for 1/3 frequency reuse. The performance sensitivity to measurement errors can also be reduced [46].

The proposed radio link scheme for EDGE is based on combining LA and IR. The initial code rate is selected based on link quality measurements as determined by the throughput, S, or a similar function. IR operation is enabled by puncturing different parts of a convolutional channel code for retransmissions. Different switching points for selecting the initial code rate may be used in the LA and IR mode.

Table 13.1 Adaptive Modulation/Coding Modes for EDGE

Scheme	Modulation	Maximum Rate (Kbit/s)	Code Rate	Family
MCS-8	8-PSK	59.2	1.0	A
MCS-7		44.8	0.76	B
MCS-6		29.6	0.49	A
MCS-5		22.4	0.37	B
MCS-4	GMSK	17.6	1.0	C
MCS-3		14.8	0.85	A
MCS-2		11.2	0.66	B
MCS-1		8.8	0.53	C

The different Modulation and Coding Schemes (MCSs) are depicted in Table 13.1 and Figure 13.4. The MCSs are divided into different families: A, B and C, with different basic units of payload: 37, 28, and 22 octets, respectively. Different code rates within a family are achieved by transmitting an appropriate number of payload units within a 20-ms block. For families A and B, 1, 2, or 4 payload units are transmitted; for family C only 1 or 2 units are transmitted. For example, a payload rate of 22.4 Kbit/s per time-slot is achieved in mode MCS-5 by transmitting 2 payload units of 28 octets each from family B with a channel coding rate of 0.37 using 8-PSK. For MCS-1 through MCS-6, each 20-ms frame contains one channel codeword per time-slot, but for MCS-7 and MCS-8, each 20-ms frame contains two channel codewords.

For initial transmissions, any MCS can be selected based on the current link quality. For retransmissions in the LA mode that requires increased robustness, two alternatives exist: If MCS-7 or MCS-8 was initially used, the block can be retransmitted at half the original code rate using one MCS-5 or MCS-6 block, respectively. If MCS-4, MCS-5, or MCS-6 was initially used, the block can be retransmitted using two MCS-1, MCS-2, or MCS-3 blocks, respectively. In the latter case, 2 bits in the header will indicate the order of the parts and that the Radio Link Control block has been split. For example, by applying both these steps, a block that was initially transmitted using uncoded 8-PSK can be retransmitted using the GMSK-based MCS-3. Using a different puncturing pattern in the retransmitted packets allows for possible soft combing and decoding for the IR mode.

In summary, a major feature of EDGE is the feasibility of initial packet data deployment with peak data rates of about 384 Kbit/s with 1/3 reuse requiring only 2 × 1 MHz of spectrum. Another major feature of EDGE is that it builds on very mature GSM technology. Frame synchronization of base stations and time reuse provides for robust control channels, and partial loading with hybrid ARQ or incremental redundancy and link adaptation are used for traffic channels. Some of the challenges with EDGE include equalization with 8-PSK modulation, accuracy in link adaptation, and robust performance with slow Rayleigh fading.

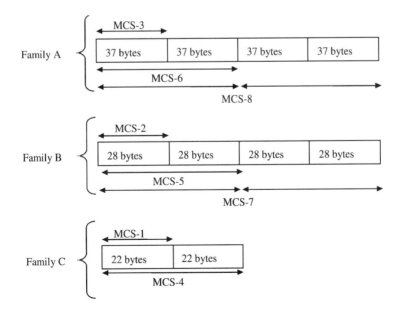

FIGURE 13.4
Proposed modulation and coding schemes for EDGE data source organization.

13.4 PACKET DATA ACCESS USING WIDEBAND OFDM

This section considers the possibility of a future high-speed wireless packet data system with a wideband 5-MHz channel using OFDM modulation to mitigate multipath fading while providing peak data rates up to about 5 Mbit/s [47]. A frequency reuse of one is considered with dynamic resource management to achieve high spectral efficiency for packet data access.

13.4.1 Physical-Layer Techniques

The radio link performance in a mobile environment is primarily limited by propagation loss, fading, delay spread, and interference. Here, we concentrate on the link-budget and dispersive-fading limitations.

Because of the large path loss encountered in serving wide areas, the link budget is challenging for multi-Mbit/s data rates. For example, consider a target data rate of 4 Mbaud, which is about 100 times that of a TDMA cellular channel. Since the SNR is inversely proportional to the baud, this corresponds to a 20 dB increase in the required transmitted power to achieve the same bit error performance and cover the same area as a typical cellular voice circuit. Clearly, the coverage and performance of such systems will be severely limited without the introduction of new techniques. This is especially true for the uplink where a mobile terminal cannot overcome the link-budget limitations and still maintain a reasonable level of complexity and power consumption.

In addition to the link-budget limitations, the bit rate is also limited by the multipath nature of the radio environment. Physically, the received signal can be considered as the summation of a large number of signals that have traveled different paths to the receiver and, therefore, have different time delays, amplitudes, and phases. Depending on the extent of the channel impulse response, called the delay spread, and the resulting intersymbol interference (ISI), the maximum data rate can be severely limited. This delay spread, in a macrocellular environment, could be as large as 20–40 μs, limiting the data rate to about 5–10 kbaud if no measures are taken to counteract the resulting ISI. In addition, the channel is time-varying, with Doppler rates as high as 200 Hz, if operating in the 2 GHz PCS band for vehicles traveling at highway speeds.

13.4.2 Physical-Layer Solutions

13.4.2.1 Asymmetric Service

Since portable terminals must be powered by batteries and their transmit power is limited to about 1 W, achieving transmission rates beyond 50–384 Kbit/s in large-cell environments is impractical. On the other hand, base stations usually are powered from commercial main power systems and can transmit with higher power and, subsequently, bit rates of 2–5 Mbit/s may be possible. Therefore, we consider an asymmetric service: a high-speed downlink with 2–5 Mbit/s peak data rates for macrocellular environments and up to 10 Mbit/s for microcellular environments, and a lower speed, 50–384 Kbit/s uplink. This alleviates the severe power problem at the mobile terminal and should be suitable for the most attractive new applications that would be supported by wideband packet OFDM.

In addition, one possible configuration is to combine 3G wireless packet data technologies such as EDGE or WCDMA with a wideband packet OFDM downlink. This approach might also permit more flexible use of spectrum than a symmetric access solution since only wideband channels would be required for the downlink. Web browsing and information access, which have caused the recent explosion in Internet usage, are highly asymmetric in transmission requirements. Only the transmission path to the subscriber needs to be high speed. Also, thin client-server computing architectures may be attractive for mobile applications in which primarily display information is transmitted to mobile terminals and applications reside within a fixed corporate or public network. This results in highly asymmetric transmission speed requirements that match the limitations of practical terminals. Many other services provided over the Internet can be provided with low to moderate bit rates transmitted from subscribers. Only video telephony and large file transfers in the direction from a terminal toward the network are envisioned to require high-speed transmission from terminals. Furthermore, advances in client-server applications where only file differences need to be transferred over wireless uplinks may minimize the need for high-speed uplinks where applications reside on terminals.

13.4.2.2   Multicarrier Modulation
One possibility for overcoming the delay spread limitations on the downlink is to use a single-carrier system modulated at about 4 Mbaud with equalization and coding (for example, see [48]). This equalizer could require 80–100 taps and must be updated at the highest Doppler rate. In addition, the extensive period required to train the equalizer could be a major source of inefficiency in a packet-based system.

An alternative approach, and the one taken here, is to use a multicarrier system. The basic concept is to divide the total bandwidth into many narrowband subchannels that are transmitted in parallel. The subchannels are chosen narrow enough so that the effects of multipath delay spread are minimized. The particular multicarrier technique used here is called Orthogonal Frequency Division Multiplexing (OFDM) (e.g., see [49, 50]) and is the standard for Digital Audio Broadcasting (DAB) in Europe [51] and for Asymmetric Digital Subscriber Line (ADSL) in the United States [53] and has been proposed for several other standards, including IEEE802.11 and HIPERLAN type 2.

To ensure a flat frequency response and achieve the desired bit rate, 400–800 subchannels are required, each modulated at 5–10 kbaud. With 5–10 kbaud subchannels and guard periods of 20–40 µs, a delay spread as large as 40 µs can be accommodated with little or no ISI. Since no equalization is required, OFDM alleviates the need for a long training period for purposes of equalization, but channel estimation is needed for coherent detection, which requires some overhead.

13.4.2.3   Diversity and Coding
To reduce the link-budget shortfall in the downlink, techniques for reducing the required SNR must be incorporated. To eliminate some of the 20-dB link-budget shortfall, switched-beam base-station transmit antennas can be used, with each 120° sector containing four beams, and two-branch antenna diversity is used at the mobile. With a transmit bandwidth of 5 MHz, the fading is significantly randomized across the OFDM bandwidth. To realize gain from the frequency diversity potential of the wideband signal, Reed-Solomon or convolutional coding is used across subchannels, using a combination of erasure correction, based on signal strength, and error correction for Reed-Solomon coding or sequence estimation for convolutional coding. With two receive antennas at the mobile combined with the frequency diversity of a wideband OFDM signal realized with the help of channel coding, the required SNR at the receiver can be reduced by about 10 dB compared to an analog cellular radio link. The use of four-beam antennas at a base station can provide up to 6 dB gain on a downlink, and the total transmitted power can be increased to about that of four ordinary analog cellular transmitters to realize an additional 12 dB gain to offset the link-budget shortfall. The use of switched beam antennas will also about double spectral efficiency by reducing interference.

13.4.3   Frequency Reuse and Spectral Efficiency

Very high spectrum efficiency will be required for wideband OFDM, particularly for macrocellular operation. First-generation cellular systems used fixed-channel assignment. Second-generation cellular systems generally use fixed-channel assignment or interference averaging with a spread spectrum. WCDMA will also use interference averaging. Interference avoidance or dynamic channel assignment (DCA) has been used in some systems, generally as a means of automatic channel assignment or local capacity enhancement but not as a means of large systemwide capacity enhancement. Some of the reasons for not fully exploiting the large potential capacity gain of DCA are the difficulties introduced by rapid channel reassignment and intensive receiver measurements required by a high-performance dynamic channel assignment or interference avoidance algorithm. OFDM is promising in overcoming these challenging implementation issues. It was shown in [53] that interference averaging techniques can perform better than fixed-channel assignment techniques, whereas interference avoidance techniques can outperform interference averaging techniques by a factor of 2–3 in spectrum efficiency. Figure 13.5 (see [53]) compares the performance of interference-averaging CDMA with several interference-avoidance DCA algorithms.

For existing second-generation systems, the achieved spectrum efficiency measured in bits per second per Hertz per sector (assuming three sectors per cell) is much lower than what is shown in Figure 13.5, which was obtained under idealized conditions. IS-136 TDMA today provides a spectrum efficiency of about 4% (3×8 Kbit/s/30 kHz $\times$ 1/21 reuse). GSM also provides a spectrum efficiency of about 4% (8×13 Kbit/s/200 kHz $\times$ 1/12 reuse). IS-95 CDMA provides a spectrum efficiency of 4–7% (12 to 20×8 Kbit/s/1250 kHz $\times$ 1 reuse $\times$ 1/2 voice activity). Dynamic channel assignment combined with circuit-based technology (which has generally been the approach taken to date) can provide some benefits. However, it cannot provide large-capacity gains, because of the dynamics of interference in a mobile system as well as the difficulty in implementing rapid channel reassignment. In the circuit-based systems, channel variations, especially those caused by the change of shadow fading, are frequently faster than what can be adapted by the slow assignment cycle possible in the circuit services. As a result, the DCA gain is limited to somewhat better traffic resource utilization, which may be achieved at the cost of nonoptimal interference management.

To achieve the potential of DCA gain, channel reassignments must take place at high speed to avoid rapidly changing interference. Dynamic Packet Assignment (DPA) based on properties

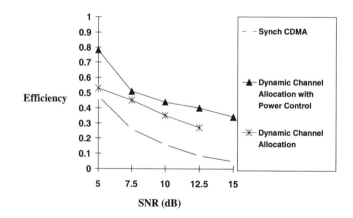

FIGURE 13.5

Spectrum efficiency in bits/s/Hz/sector for interference avoidance and interference averaging.

of an OFDM physical layer is proposed, which reassigns transmission resources on a packet by packet basis using high-speed receiver measurements to overcome these problems [54]. Since it has orthogonal subchannels well defined in time-frequency grids, OFDM has a key advantage here with the ability to rapidly measure interference or path loss parameters in parallel on all candidate channels, either directly or based on pilot tones. One of the benefits of DPA based on interference avoidance is that it is relatively insensitive to errors in power control, and it provides good performance even without power control. Figure 13.5 shows that dynamic channel assignment without power control decreases capacity. However, even without power control, interference avoidance can outperform interference averaging with power control. This is particularly advantageous for packet transmission where effective power control is problematic due to the rapid arrival and departure of interfering packets.

13.4.4   Dynamic Packet Assignment Protocol

The protocol for dynamic packet assignment for a downlink comprises four basic steps: (1) a packet page from a base station to a terminal; (2) rapid measurements of resource usage by a terminal using the parallelism of an OFDM receiver; (3) a short report from the terminal to the base station of the potential transmission quality associated with each resource (a unit of bandwidth that is separately assignable); and (4) selection of resources by the base and transmission of the data. This protocol could be modified to move some of the over-the-air functions into fixed-network transmission functions to reduce the wireless transmission overhead at the cost of more demanding fixed-network transmission requirements. In that case, the protocol would consist of the following steps: (1) mobile stations would report path loss measurements for all nearby base stations to a serving base station on a regular basis at a rate of about once per second; (2) serving base stations would forward those measurements to all neighboring base stations to allow each base station to build a data base of all nearby mobile stations and their path losses to both serving and nearby base stations; (3) base stations would maintain a second data base of occupancy of each resource in nearby base stations; (4) base stations would allocate resources for packet transmission to mobile stations using both data bases to minimize interference created for other transmissions and to minimize received interference for the desired transmission; and (5) base stations would forward information about the occupied resources to all neighboring base stations.

The frame structures of adjacent base stations are staggered in time, that is, neighboring base stations sequentially perform the different DPA functions outlined above with a predetermined rotational schedule. This avoids collisions of channel assignments, that is, the possibility for adjacent base stations to independently select the same channel, thus causing interference when transmissions occur. In addition to achieving much of the potential gain of a rapid interference avoidance protocol, this protocol provides a good basis for admission control and mode (bit-rate) adaptation based on measured signal quality.

The remainder of this section focuses on dynamic packet assignment based on over-the-air measurements. Performance with a network-centric algorithm may be better, since a network-based algorithm could more easily consider the potential impact of new interference on existing transmissions at resource assignment time than an algorithm based on over-the-air measurements.

13.4.5   Dynamic Packet Assignment Performance

Figure 13.6 shows the performance of this algorithm with several modulation/coding schemes and with either two-branch maximal-ratio-combining or with two-branch receiver interference

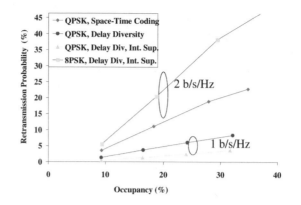

FIGURE 13.6

Performance as a function of occupancy for different modulation and diversity schemes.

suppression using packet traffic models based on Internet statistics [55]. Results with interference suppression for space-time coding are not included because each transmitted signal appears as multiple signals, which significantly limits the suppression of interference. These results are based on an OFDM radio link with a bandwidth of about 800 kHz and the bit rates in the following discussion are scaled up for an occupied bandwidth of 4 MHz (5 MHz with guard space). A system is considered with three sectors per base station, each having a transceiver. All base stations share one wideband OFDM RF channel by using DPA to avoid co-channel interference. DPA enables frequency reuse in the time domain among all radio transceivers. Occupancy is defined to be the fraction of slots in use. As traffic intensity increases, occupancy increases, which results in higher interference and more retransmissions. Power control was not used to obtain these results.

These results show that good performance is obtained with 1 bit/s/Hz coding even at an average occupancy per base station of 100% (33% per sector). With two-branch interference suppression and 1 bit/s per Hz coding, the average retransmission probability is only about 3% throughout the system with the average delivered bit rate of about 2.5 Mbit/s per base station. Using ARQ at the radio link layer will permit Internet service at this retransmission probability with good QoS. Higher retransmission probability may be acceptable at the expense of longer packet delay. Peak rates up to 5 Mbit/s are possible with lower occupancies using 2 bits per second per Hertz coding.

13.4.6 Radio Link Resource Organization

Considerations for organization of resources are the resolution of the resource size, the overhead required to allocate individual resources, and the expected size of objects to be transmitted over a resource. Minimization of overhead can be achieved by organizing the available bandwidth into large resources, but, if many objects are relatively small in size or if higher-layer protocols generate small objects that the lower layers must carry, there will exist a need to allocate small resources to achieve good efficiency. Also, streaming data may require resources that are small locally in time to avoid the need for buffering source bits before transmission, which causes delay. A 2-Mbit/s system with 20–25 resources would support about 80–100 Kbit/s rates locally in time. This rate would be suitable for high-bit-rate data services. If supporting about 10 Kbit/s locally in time were desirable (e.g., voice or audio services of 8 Kbit/s with additional coding for error correction in wireless channels), that would be equivalent to about 200 resources. The next section considers large resource assignment and then small-resource assignment is considered using only one of the eight slots in a 20-ms frame. Frequency hopping over different slots is employed to gain frequency diversity for large resources. To achieve this frequency diversity for small resources, a slot is divided into mini-slots, at the cost of reduced efficiency due to TDMA overhead.

13.4.6.1 High-Peak-Rate Data Services: Large Radio Resources

Consider as an example, 528 subchannels (4.224 MHz) that are organized into 22 clusters of 24 subchannels (192 kHz) each in frequency and eight time-slots of 13 OFDM blocks each within a 20-ms frame of 128 blocks. Figure 13.7 shows this resource allocation scheme. The control channel functions are defined in [56]. This allows flexibility in channel assignment while providing 24 blocks of control overheard to perform the DPA procedures.

This arrangement of tone-clusters is similar to the arrangements in the Band Division Multiple Access (BDMA) proposal by Sony [57]. Figure 13.8 depicts this operation. Each tone-cluster would contain 22 individual modulation tones plus 2 guard tones, and an OFDM block would have a time duration of 156.25 μs with 31.25 μs for guard time and ramp time to minimize the effects of delay spread up to about a 20-μs span. Of the 13 OFDM blocks in each traffic slot, two blocks are used as overhead, which includes a leading block for synchronization (phase/frequency/timing acquisition and channel estimation) and a trailing block as guard time for separating consecutive time slots. A single radio resource is associated with a frequency-hopping pattern, by which the packets are transmitted using eight different tone-clusters in each of the eight traffic slots. Coding across eight traffic slots for user data, as shown in Figure 13.8, exploits frequency diversity, which gives sufficient coding gain for performance enhancement in the fading channel. This arrangement supports 22 resources in frequency that can be assigned by DPA. Taking into account overhead for OFDM block guard time, synchronization, slot separation, and DPA control, a peak data rate of 2.1296 ($3.3792 \times 22/24 \times 11/13 \times 104/128$) Mbit/s is available for packet date services using all 22 radio resources, each of 96.8 Kbit/s.

For the base station, where uplink transmission for all radio resources is asynchronous, a receiver may separate 192-kHz clusters with filters followed by independently synchronized

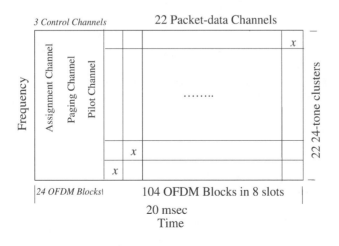

FIGURE 13.7

Division of radio resources in time and frequency domains to allow DPA for high-peak-rate data services.

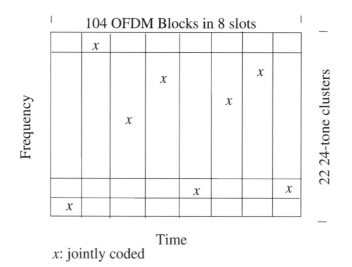

FIGURE 13.8
Coding of large radio resource with clustered OFDM and frequency hopping in a frame.

demodulators. For the mobile terminal, where downlink transmission for base-station radio resources is typically synchronous, a receiver may use a single demodulator with receiver windowing to result in strong attenuation of undesired clusters. However, adjacent clusters may be asynchronous, if transmitted by different base stations.

13.4.6.2 Low-Delay Services: Small Radio Resources

Similar to subsection 13.4.6.1, 528 subchannels (4.224 MHz) are organized into 22 clusters of 24 subchannels (192 kHz) each in frequency and eight time-slots of 13 blocks each within a 128-block (20-ms) frame. A difference is that these time-slots are further divided into four mini-slots for frequency hopping. A group of eight small radio resources can be obtained by allocating one large radio resource. A small radio resource uses a mini-slot from every other time-slot in the large radio resource frame for a total of four mini-slots. Therefore, the frame structure is the same as that for the large-resource case (shown in Figure 13.8, in subsection 13.4.6.1), except that there are 176 (8 × 22) small resources and each resource bit rate is reduced by the additional TDMA overhead needed at the beginning and end of a mini-slot. The same control channel can be used to assign both large and small resources using staggered-frame DPA.

Figure 13.9 depicts the coding scheme for the small-resource case. Each tone-cluster would contain 22 individual modulation tones plus 2 guard tones. Of the 13 OFDM blocks in each traffic slot, three blocks are used as overhead. This includes a total of two leading blocks (duration of half of a block for each of the four mini-slots) for synchronization and a trailing block (a quarter of the block for each of the four mini-slots) as guard time for separating consecutive mini-slots. A single radio resource is associated with a frequency-hopping pattern, by which the packets are transmitted using four different tone-clusters in each of the four mini-slots. Coding across four mini-slots for user data, as shown in Figure 13.9, exploits frequency diversity and interleaving in time. Taking into account the overhead for OFDM block guard time, synchronization, slot separation, and DPA control, a peak data rate of 1.936 ($3.3792 \times 22/24 \times 10/13 \times 104/128$) Mbit/s is available using all 176 radio resources, each of 11 Kbit/s.

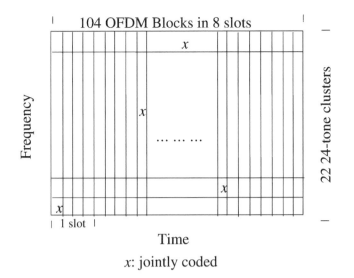

x: jointly coded

FIGURE 13.9
Coding of small radio resource with clustered OFDM and frequency hopping in a slot.

13.4.7 Frame Structure for Dynamic Packet Assignment

The frame structure described in this section supports both control information, which is needed to perform the DPA procedure, as well as the bearer traffic. A frame is 20 ms. The control part uses a staggered schedule, in which only one base station at a time, from a group of four adjacent base stations, transmits information for DPA. The bearer traffic, on the other hand, is transmitted on the assigned radio resources ("channels") without a staggered schedule. To implement the staggered schedule, four frames (80 ms) are grouped as a "superframe." Effectively, this achieves a reuse factor of 4 for the control information while allowing a reuse factor of 1 for bearer traffic by using DPA, that is, all traffic channels can be used everywhere. A reuse factor of 4 and three sectorized antennas in each base station provide extra error protection for the control channels, whereas interference avoidance based on DPA with admission control provides good quality for the traffic channels.

The downlink structure is shown in Figure 13.10. The uplink structure is similar but the control functions are slightly different. At the beginning of each frame, the control channels for both the uplink and downlink jointly perform the four DPA procedures, described in subsection 13.4.4, sequentially with a predetermined staggered schedule among adjacent base stations. This achieves a spectrum reuse of 4 in the time domain. Some control-channel overhead is included to allow three sectors to perform DPA at different time periods, thus obtaining additional SIR enhancement for the control information. This frame structure permits SIR estimation on all unused traffic slots. The desired signal is estimated by the received signal strength from the two OFDM blocks used for paging, while the interference can be estimated by measuring three blocks of received pilot signals. The pilot channels are generated by mapping all the radio resources currently in use onto corresponding pilot subchannels, thus providing an "interference map" without monitoring the actual traffic subchannels. The OFDM scheme can process many subchannels in parallel, which provides a mechanism for very fast SIR estimation. In addition, since a total of 528 subchannels are available to map 22 large resources and 176 small resources over 3 OFDM blocks, significant diversity effects are achieved to reduce measurement errors.

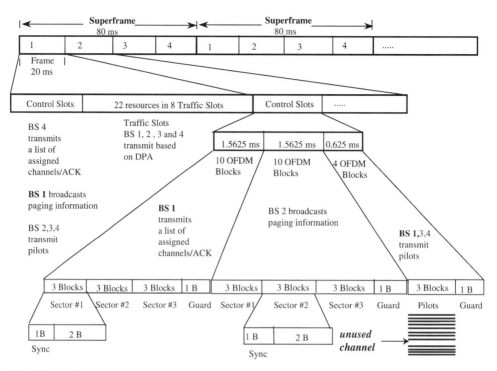

FIGURE 13.10
Staggered frame structure for DPA downlink performance for high-peak-rate data services.

The estimated SIR is compared against an admission threshold (e.g., 10 dB in our example), so channel occupancy can be controlled to achieve good QoS for the admitted users.

In the following, downlink performance is studied by large-scale computer simulations. Only the downlink simulation results are shown here, since downlink transmission requires a higher RF bandwidth and its information bandwidth demand in popular applications (such as Web browsing) is also higher. Although the uplink efficiency could be reduced by collisions, the downlink spectrum efficiency is the crucial factor in system deployment.

13.4.8 Simulation Model

To characterize the DPA performance, a system of 36 base stations arranged in a hexagonal pattern is assumed, each having three sectors using idealized antennas with 120° beamwidths and a 20-dB front-to-back ratio. The mobile antennas are assumed to be omnidirectional. In each sector, one radio provides 22 resources to deliver downlink traffic packets. The same channel can be used in different sectors of the same base station as long as the SIR at the DPA admission process exceeds 10 dB. Based on the downlink frame structure shown in Figure 13.10, four base stations in each reuse area take turns performing the DPA procedure and the assignment cycle is reused in a fixed pattern. The co-channel interference limited case is considered, that is, noise is ignored in the simulation. In the propagation model, the average received power decreases with distance d as d^{-4} and the large-scale shadow-fading distribution is log-normal with a standard deviation of 10 dB. Rayleigh fading is ignored in the channel assignment, which approximates the case where antenna diversity is employed and sufficient averaging in both the time and frequency domains is achieved in signal and interference estimations.

Uniformly distributed mobile stations (MSs) receive packets, which are generated from the network and arrive at different base stations (BSs). A data-service traffic model, described in [55], based on wide-area network traffic statistics, which exhibit a "self-similar" property when aggregating multiple sources, was used to generate packets. A radio source ("channel") is statistically multiplexed to deliver packets for different MSs. MSs are fairly allocated to as many unused radio channels as possible, provided that SIR exceeds 10 dB for resources. When the number of pending packets exceeds the number of channels assigned, they are queued for later delivery. The assigned channels are reserved for the same MS until all packets are delivered or the DPA algorithm reassigns radio channels in the next superframe. ARQ is employed, assuming perfect feedback, to request base stations for retransmission when a packet ("word") is received in error, which is simulated based on a word error rate (WER) curve versus SIR. If a packet cannot be successfully delivered in 3 s, which may be a result of traffic overload or excessive interference, it is dropped from the queue. The control messages are assumed to be error-free in the designated control slots.

We consider two radio-link enhancement techniques to study the DPA performance: (1) beam-forming and (2) interference suppression. Both beamforming and interference suppression employ two receive antennas with Minimum Mean Square Error (MMSE) combining to improve SIR. Downlink beamforming is performed at the BS using four transmit antennas to form four narrow beams. By using narrow beams to deliver packets for MSs, SIR is enhanced. For beamforming, each 120° sector is simply divided into four 30° beams (with a 20-dB front-to-back ratio and idealized antenna pattern) and the assumption is made that a packet is delivered using the beam that covers the desired MS.

13.4.9 Simulation Performance Results

Figure 13.11 shows the overall average probability of packet retransmission as a function of occupancy. With a 3–6 % target retransmission probability, about 15–50 % occupancy per radio in each sector is possible with this DPA scheme. This result is significantly superior to the efficiency provided by current cellular systems. The corresponding average packet-dropping probability is shown in Figure 13.12. Notice that both interference suppression and downlink beamforming are effective in improving retransmission probability. However, the improvement in packet-dropping probability for interference suppression is somewhat limited because interference suppression is not employed in SIR estimation, which is used for admission control. Specifically, some of the packets are delayed if the SIR estimated during resource assignment does not exceed 10 dB, although SIR may be acceptable with interference suppression that is performed in the demodulation process after admission is granted. Based on the results of Figure 13.12, it appears that the reasonable operating region of occupancy is about 20–25% and 30–35% occupancy per radio for cases with and without beam-forming, respectively. Under this condition, interference suppression or beam-forming can achieve acceptable retransmission probability, providing good QoS exists. If neither enhancement is employed, the traffic capacity must be lowered to ensure good performance. When both techniques are employed, three radios in three sectors can utilize 100% of radio resources in every base station. Finally, Figure 13.13 shows that 2–3 Mbit/s can be successfully delivered by each base station with an average delay on the order of 60–120 ms. This indicates that OFDM and DPA combined enable a spectrally efficient air interface for broadband services, even for macrocell environments, providing complementary high-bit-rate data services beyond what 3G systems can offer.

Based on the performance shown here and the coding/modulation alternatives discussed in Section 13.3, it is reasonable to expect that an 8-PSK-based modulation can deliver 5 Mbit/s in peak-rate packet data access. The wideband OFDM technology discussed here can provide high peak rates with robust performance that is not achievable in second- or third-generation technologies. However, it is a less mature technology that requires more research and development effort. A summary of attributes for wideband OFDM and two leading technologies is listed in Figure 13.14. Combining a high-speed OFDM downlink with 3G wireless systems such as EDGE or WCDMA may be attractive, providing an asymmetric access arrangement and requiring a new spectrum only for downlinks.

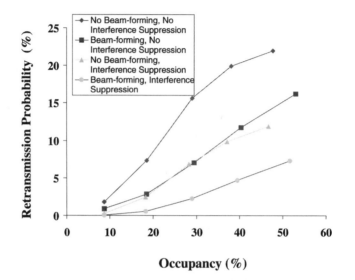

FIGURE 13.11
Average retransmission probability as a function of occupancy.

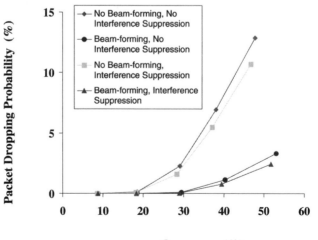

FIGURE 13.12
Average packet-dropping probability as a function of occupancy.

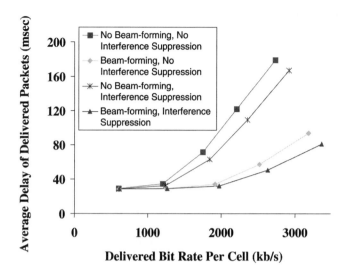

FIGURE 13.13
Average delay of the delivered packets as a function of the throughput per base station.

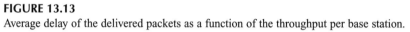

	EDGE with 1/3 reuse + Partial Loading	WCDMA + packet access	WOFDM + DPA
Maturity	++	+	-
Peak Rate	-	-	++
Spectral Efficiency	+	+	++
Inherent Diversity	-	++	++

FIGURE 13.14
Attributes of alternative air interface technologies.

13.5 CONCLUSIONS

The rapid growth of wireless voice subscribers, the growth of the Internet, and the increasing use of portable computing devices suggest that wireless Internet access will grow to be a major part of telecommunications. Today, the number of wireless data subscribers is small, with the most formidable obstacle to user acceptance being the performance limitations of existing services and products, including link performance (data rate, latency, and quality); network performance (access, coverage, spectrum efficiency, and quality of service); and price. Early wireless packet data access technologies providing about 10 Kbit/s transmission rates over wide areas are expected to evolve or be replaced with technologies providing 40–144 Kbit/s peak data rates with IS-136+, IS-95+, and General Packet Radio Service (GPRS) technologies. In the 2001 to 2003 time frame, the introduction of Enhanced Data for Global System for Mobile Communication (GSM) Evolution (EDGE) and Wideband Call Division Multiple Access (WCDMA) technologies will provide access rates up to about 384 Kbit/s. Wideband OFDM is expected to support peak bit rates of 2–5 Mbit/s in large cell environments and up to 10 Mbit/s in microcellular environments.

13.6 REFERENCES

[1] *IEEE Spectrum,* January 1997.
[2] http://www.ericsson.com/BR/market/
[3] Arthur D. Little, *Wireless Business and Finance,* July 17, 1996.
[4] I. Brodsky, "Countdown to Mobile Blast Off," *Network World,* February 19, 1996.
[5] J. F. DeRose, *The Wireless Data Handbook,* Quantum Publishing, 1994.
[6] M. Khan and J. Kilpatrick, "MOBITEX and Mobile Data Standards," *IEEE Commun. Mag.,* Vol. 33, No. 3, March 1995, pp. 96–101.
[7] A. K. Salkintzis and C. Chamzas, "Mobile Packet Data Technology: An Insight into MOBITEX Architecture," *IEEE Pers. Commun. Mag.,* Vol. 4, No. 1, February 1997, pp. 10–18.
[8] Cellular Telecommunications Industry Association, "Cellular Digital Packet Data System Specification," Release 1.0, July 1993.
[9] M. Streetharan and R. Kumar, *Cellular Digital Packet Data,* Artech House, 1996.
[10] M. Mouly and M.-B. Pautet, "Current Evolution of GSM Systems," *IEEE Pers. Commun. Mag.,* Vol. 2, No. 5, October 1995, pp. 9–19.
[11] N. R. Sollenberger, N. Seshadri, and R. Cox, "The Evolution of IS-136 TDMA for Third Generation Services," *IEEE Pers. Commun. Mag.,* (in press).
[12] "Data Service Options for Wideband Spread Spectrum Systems," *TIA/EIA/IS-707,* February 1998.
[13] http://www.metricom.com
[14] "Personal Handy Phone System," *Japanese Telecommunication System Standard,* RCR-STD, December 28, 1993.
[15] H. Ochsner, "DECT—Digital European Cordless Telecommunications," *Proc. of VTC'89,* pp. 718–721.
[16] A. R. Noerpel, Y.-B. Lin, and H. Sherry, "PACS: Personal Access Communications System—A Tutorial," *IEEE Pers. Commun. Mag.,* Vol. 3, No. 3, June 1996, pp. 32–43.
[17] B. Tuch, "An ISM Band Spread Spectrum Local Area Network: WaveLAN," Proc. *IEEE Workshop on Wireless LANs,* May 1991, pp. 103–111.
[18] http://www.proxim.com/proxim/products/
[19] IEEE P802.11, "Wireless LAN Medium Access Control (MAC) and Physical Layer (PHY) Specifications," *IEEE Standards Department,* 1997.
[20] ETSI TC-RES, "Radio Equipment and Systems (RES); High Performance Radio Local Area Network (HIPERLAN); Functional Specification," *ETSI 06921 Sophia Antipolis Cedex,* France, July 1995, draft prETS 300 652.

[21] "Special Issue on the HIgh PErformance Radio Local Area Network (HIPERLAN)," *Wireless Personal Commun.*, Kluwer, Vol. 3, No. 4, November 1996, and Vol. 4, No. 1, January 1997.

[22] IEEE P802.11a/D4.3, "High Speed Physical Layer in 5 GHz Band (supplement to IEEE Standard 802.11-1997)," *IEEE Standards Department, 1999.*

[23] "Broadband Radio Access Networks (BRAN): HIPERLAN type 2 Functional Specification," *ETSI Draft Standard, DTS/BRAN 030003-1, Sophia Antipolis, 1999.*

[24] T. Ojanpera and R. Prasad, "An Overview of Third-Generation Wireless Personal Communications: A European Perspective," *IEEE Pers. Commun. Mag.*, Vol. 5, No. 6, December 1998, pp. 59–65.

[25] "Requirements and Objectives for FPLMTS," *Radiocommunication Study Groups, ITU,* Rev. 3 to Doc. 8-1/TEMP/5-E, April 24, 1996.

[26] http://www.itu.int/imt/2-radio-dev/index.html

[27] "Special Issue on The European Path Toward UMTS," *IEEE Pers. Commun. Mag.*, Vol. 2, No. 1, February 1995.

[28] J. S. DaSilva, D. Ikonomou, and H. Erben, "European R&D Programs on Third-Generation Mobile Communication Systems," *IEEE Pers. Commun. Mag.*, Vol. 4, No. 1, February 1997, pp. 46–52.

[29] D. Grillo, N. Metzner, and E. D. Murray, "Testbeds for Assessing the Performance of a TDMA-Based Radio Access Design for UMTS," *IEEE Pers. Commun. Mag.*, Vol. 2, No. 2, April 1995, pp. 36–45.

[30] T. Stefansson and M. Allkins, "Real-Time Testbed for Assessing a CDMA-Based System," *IEEE Pers. Commun. Mag.*, Vol. 2, No. 5, October 1995, pp. 75–80.

[31] E. Nikula, A. Toskala, E. Dahlman, L. Girard, and A. Klein, "FRAMES Multiple Access for UMTS and IMT-2000," *IEEE Pers. Comm. Mag.*, Vol. 5, No. 2, April 1998, pp. 16–24.

[32] R. Prasad and T. Ojanpera, "An Overview of CDMA Evolution Toward Wideband CDMA," *IEEE Commun. Surveys,* Vol. 1, No. 1, Fourth Quarter 1998, pp. 2–29. (See http://www.comsoc.org/pubs/survey)

[33] T. Ojanpera and R. Prasad, *Wideband CDMA for Third Generation Mobile Communications,* Artech House, 1998.

[34] "Report on FPLMTS Radio Transmission Technology Special Group," (Round 2 Activity Report) Version E1.2, *Association of Radio Industries and Businesses (ARIB), FPLMTS Study Committee,* Japan, January 1997.

[35] S. Onoe, K. Ohno, K. Yamagata, and T. Nakamura, "Wideband-CDMA Radio Control Techniques for Third-Generation Mobile Communication Systems," *Proc. VTC'97,* May 1997, pp. 835–839.

[36] S. Glisic and B. Vucetic, *Spread Spectrum CDMA Systems for Wireless Communications*, Artech House, 1997.

[37] E. H. Dinan and B. Jabbari, "Spreading Codes for Direct Sequence CDMA and Wideband CDMA Cellular Networks," *IEEE Commun. Mag.*, Vol. 36, No. 9, Sept. 1998, pp. 48–54.

[38] J. Cai and D. J. Goodman, "General Packet Radio Service in GSM," *IEEE Commun. Mag.*, Vol. 35, No. 10, October 1997, pp. 122–131.

[39] "Working Document Towards Submission of RTT Candidate to ITU-R, IMT-2000 Process: The UWC-136 RTT Candidate Submission," *TR45.3 and U.S. TG 8/1,* June 1998. Document available at http://www.itu.org/imt/2-radio-dev/proposals/usa/tia/uwc-136.pdf.

[40] P. Schramm, H. Andreasson, C. Edholm, N. Edvardsson, M. Höök, S. Javerbring, F. Muller and J. Sköld, "Radio Interface Performance of EDGE, a Proposal for Enhanced Data Rates in Existing Digital Cellular Systems," *Proc., VTC'98,* pp. 1064–1068, May 1998.

[41] A. Furuskär, M. Frodigh, H. Olofsson, and J. Sköld, "System Performance of EDGE, a Proposal for Enhanced Data Rates in Existing Digital Cellular Systems," *Proc., VTC'98,* pp. 1284–1289, May 1998.

[42] "EDGE: Concept Proposal for Enhanced GPRS," *ETSI SMG2 Working Session on EDGE, TDoc SMG2 EDGE 006/99,* March 2–4, 1999, Toulouse, France.

[43] R. van Noblen, N, Seshadri, J. F. Whitehead, and S. Timiri, "An Adaptive Radio Link Protocol with Enhanced Data Rates for GSM Evolution," *IEEE Pers. Comm. Mag.*, Feb. 1999, pp. 54–64.

[44] "Incremental Redundancy Transmission for EDGE," *ETSI Tdoc SMG2 EDGE 053/98,* August 1998.

[45] J. C. Chuang, "Improvement of Data Throughput in Wireless Packet Systems with Link Adaptation and Efficient Frequency Reuse," *Proc., VTC'99,* May 1999, pp. 821–825.

[46] X. Qiu and J. C. Chuang, "Link Adaptation in Wireless Data Networks for Throughput Maximization under Retransmissions," *Proc., ICC'99,* June 1999, pp. 1664–1668.

[47] L. J. Cimini, Jr., J. C.-I. Chuang, and N. R. Sollenberger, "Advanced Cellular Internet Service (ACIS)," *IEEE Commun. Mag.,* Vol. 36, No. 10, Oct. 1998, pp. 150–159.

[48] J. G. Proakis, "Adaptive Equalization for TDMA Digital Mobile Radio," *IEEE Trans. Veh. Tech.,* Vol. 40, No. 2, May 1991, pp. 333–341.

[49] S. B. Weinstein and P. M. Ebert, "Data Transmission by Frequency-Division Multiplexing Using the Discrete Fourier Transform," *IEEE Trans. Commun. Tech.,* Vol. COM-19, No. 5, Oct. 1971, pp. 628–634.

[50] J. A. C. Bingham, "Multicarrier Modulation for Data Transmission: An Idea Whose Time Has Come," *IEEE Commun. Mag.,* Vol. 28, No. 5, May 1990, pp. 5–14.

[51] M. Alard and R. Lassale, "Principles of Modulation and Coding for Digital Broadcasting for Mobile Receivers," *EBU Tech. Rev.,* No. 224, Aug. 1987, pp. 168–190.

[52] P. S. Chow, J. C. Tu, and J. M. Cioffi, "A Multichannel Transceiver System for Asymmetric Digital Subscriber Line Service," *Proc. GLOBECOM'91,* pp. 1992–1996.

[53] G. J. Pottie, "System Design Issues in Personal Communications," *IEEE Pers. Commun.,* Vol. 2, No. 5, Oct. 1995, pp. 50–67.

[54] J. C.-I. Chuang and N. R. Sollenberger, "Spectrum Resource Allocation for Wireless Packet Access with Application to Advanced Cellular Internet Service," *IEEE J-SAC,* Vol. 16, No. 6, August 1998, pp. 820–829.

[55] Y. Li, J. C.-I. Chuang, and N. R. Sollenberger, "Transmitter Diversity for OFDM Systems and Its Impact on High-Rate Data Wireless Networks," *IEEE J-SAC: Wireless Commun.,* Vol. 17, No. 7, July 1999, pp. 1233–1243.

[56] J. Chuang, "An OFDM-Based System with Dynamic Packet Assignment and Interference Suppression for Advance Cellular Internet Service," *Kluwer J. Wireless Pers. Commun.,* 1999.

[57] Sony Corp., "BDMA, The Multiple Access Scheme Proposal for the UMTS Terrestrial Radio Air Interface (UTRA)," *ETSI SMG2,* June 23–27, 1997, London.

CHAPTER 14

Internet Protocols Over Wireless Networks

GEORGE C. POLYZOS
GEORGE XYLOMENOS

ABSTRACT

We discuss the problems that arise when standard Internet protocols are used over wireless links, such as degraded Transmission Control Protocol (TCP) performance when wireless errors are interpreted as congestion losses. We use case studies drawn from commercial Wireless Local Area Networks and cellular telephony systems to illustrate these problems. Then, we survey some proposed approaches to mitigating such problems and examine their applicability. Finally, we look at the future of wireless systems and the new challenges that they will create for Internet protocols and state some goals for further protocol enhancement and evolution.

14.1   INTRODUCTION

The Internet has expanded its reach over new telecommunications systems not long after each has become available, so it is not surprising that existing and emerging wireless systems are no exception. This ubiquity is partly due to the design of the Internet Protocol (IP), which seamlessly interconnects dissimilar networks into a global internetwork, offering a common interface to higher protocol layers. Despite the fact that satellite links have been long used on the Internet, the focus of Internet protocol development has been on wired media, with their decreasing error rates and increasing bandwidth. Supporting the simple services of IP over media with such nice characteristics has been rather straightforward.

Physical and economic factors cause wireless links to lag behind their wired counterparts, generally exhibiting higher error rates and lower data rates. Since these characteristics violate assumptions commonly made for wired media, more sophisticated and complex link layer protocols have been used over wireless links in an attempt to improve performance at the network and higher layers. A typical approach is to reduce error rate at the expense of data rate. As IP can accommodate any kind of link layer protocol, these techniques hide the peculiarities of underlying media without compromising compatibility with higher layers.

Emerging candidates for inclusion in the mainstream of the Internet are wireless communications systems such as Cellular and Cordless Telephony (CT) and Wireless Local Area Networks (WLANs). Our discussion also applies to satellite systems, but particularly in the case of geostationary satellite links, the much larger propagation delays encountered are a major consideration that we do not address here. Wireless systems present new challenges, such as a rapidly changing error behavior due to mobility and terrestrial obstructions and reflections. Cellular systems in addition suffer from communication pauses during *handoffs,* when mobile devices move between adjacent cells. Their performance shortcomings when employed on the Internet can be addressed by a synthesis of techniques for enhancing the performance of both wired and wireless links, modified to take into account their unique characteristics.

This article discusses the specific problems wireless links and mobility present to Internet protocol performance (Section 14.2), surveys approaches to enhancing their performance over such links (Section 14.3), and looks into the requirements future wireless systems and applications will probably introduce and how to address them (Section 14.4).

14.2   INTERNET PROTOCOLS AND WIRELESS LINKS

14.2.1   Internet Transport Layer Protocols

Transport layer protocols lie between user applications and the network. Although they offer user-oriented services, their design is based on assumptions about network characteristics. One choice offered by the Internet is the User Datagram Protocol (UDP), essentially a thin layer over IP. UDP offers a best-effort message delivery service, without any flow, congestion, or error control. Such facilities may be built on top of it, if needed, by higher-layer protocols or applications. Besides offering nearly direct access to IP, UDP is also useful for applications that communicate over Local Area Networks (LANs). Because wired LANs are typically extremely reliable and have plenty of bandwidth available, their lack of error and congestion control is unimportant.

Even though wired long-haul links have been exhibiting decreasing error rates (due to widespread use of optical fiber), the statistical multiplexing of increasing traffic loads over wide-area links has replaced errors with congestion as the dominant loss factor on the Internet. Congestion is caused by temporary overloading of links with traffic that causes transmission queues at network routers to build up, resulting in increased delays and eventually packet loss. When such losses occur, indicating high levels of congestion, the best remedy is to reduce the offered load to empty the queues and restore traffic to its long-term average rate [8].

The Transmission Control Protocol (TCP) is the other common transport layer protocol choice offered on the Internet, and the most popular one, since it supports many additional facilities compared to UDP. It offers a connection-oriented byte stream service that appears to applications similar to writing (reading) to (from) a sequential file. TCP supports reliable operation, flow and congestion control, and segmentation and reassembly of user data. TCP data segments are acknowledged by the receiving side in order. When arriving segments have a gap in their sequence, duplicate acknowledgments are generated for the last segment received in sequence. Losses are detected by the sender either by timing out while waiting for an acknowledgment, or by a series of duplicate acknowledgments implying that the next segment in the sequence was lost in transit. Since IP provides an end-to-end datagram delivery service, TCP resembles a Go-

Back-N link layer protocol transmitting datagrams instead of frames. On the other hand, IP can reorder datagrams, so TCP cannot assume that all gaps in the sequence numbers mean loss. This is why TCP waits for multiple duplicate acknowledgments before deciding to assume that a datagram was indeed lost.

During periods of low traffic or when acknowledgments are lost, TCP detects losses by the expiration of timers. Since Internet routing is dynamic, a timeout value for retransmissions is continuously estimated based on the averaged round-trip times of previous data/acknowledgment pairs. A good estimate is very important: Large timeout values delay recovery after losses, while small values may cause premature timeouts and thus retransmissions to occur when acknowledgments are delayed, even in the absence of loss. Recent versions of TCP make the key assumption that the vast majority of perceived losses are due to congestion [8], thus combining loss detection with congestion detection. As a result, losses cause, apart from retransmissions, the transmission rate of TCP to be reduced to a minimum and then gradually to increase so as to probe the network for the highest load that can be sustained without causing congestion. Since the link and network layers do not offer any indications as to the cause of a particular loss, this assumption is not always true, but it is sufficiently accurate for the low-error-rate wired links. A conservative reaction to congestion is critical in avoiding congestive collapse on the Internet with its ever-increasing traffic loads [8].

14.2.2 Protocol Performance Over a Single Wireless Link

The low-error-rate assumption, while reasonable for wired links, has disastrous results for wireless links. A representative WLAN, the WaveLAN, when transmitting UDP packets with 1400 bytes of user payload at a data rate of 1.5 Mbit/s over an 85-foot distance has an average Frame Error Rate (FER) of 1.55% [11]. The errors are usually clustered as in congestion losses. The main factors influencing FER are distance between hosts and frame size. Distance is generally imposed by the operating environment, but the frame size can be reduced to minimize the error rate. Reducing the frame size by 300 bytes for this distance (85 ft) cuts in half the measured FER [11], as shown in the first two columns of Table 14.1 (see [11] for a two-state error model accurately reflecting the behavior of this system). By encapsulating each UDP segment in one IP datagram and then in one WaveLAN frame, we have 48 bytes of header and trailer overhead per segment, so overhead as a percentage of total bandwidth used increases with shorter frames, as shown in the third column of Table 14.1. To find the frame size that maximizes data throughput, we have to combine FER and frame overhead for each frame size to get the percentage of total bandwidth used that consists of user data, shown in the fourth column of Table 14.1. In this case, when reducing frame size the increase in header overhead more than balances the scale in terms of effective data rate. Data throughput at the link layer is maximized with 1400-byte data segments, at roughly 95% of link capacity after discounting losses and overhead. As an example, if we transmit 1400-byte data segments, or 1448-byte Ethernet frames, at a peak bandwidth of 1.6 Mbit/s, FER is 0.0155 and the overhead factor is $48/1448 = 0.03315$. The data throughput then is $(1 - 0.0155) \times (1 - 0.03315) \times 1.6 = 1.523$ Mbit/s, about 95% of the 1.6 Mbit/s of bandwidth used.

TCP is expected to achieve lower data throughout than UDP, not only due to its extra 12 bytes of overhead per data segment (the difference between UDP and TCP header sizes), but also because reverse traffic (acknowledgments) must share the broadcast WLAN medium with forward (data) traffic. A secondary effect of sharing the link is that transmissions can be delayed due to collisions. Our measurements indicate that with the WaveLAN these collisions may sometimes go undetected, thus increasing the error rate visible to higher layers with bidirectional

Table 14.1 Wireless LAN Performance (UDP)

User Payload (bytes)	Datagram Error Rate (%)	Overhead (%)	Throughput (%)
200	0.097	19.355	80.567
500	0.194	8.759	91.064
800	0.388	5.660	93.974
1100	0.775	4.181	95.076
1400	1.550	3.315	95.186

(TCP) traffic [13] (Figure 14.1). Note also that host mobility increases error rates for this WLAN by about 30%, in the absence of handoffs [11].

Measurements of a TCP file transfer over a single WaveLAN link using 1400-byte data segments revealed that the throughput achieved is only 1.25 Mbit/s out of the 1.6 Mbit/s available on the link [11]; thus, a throughput reduction of 22% is caused by only a 1.55% frame error rate. This is due to TCP initiating congestion avoidance mechanisms that reduce its transmission rate, even though the loss was not due to congestion. Note that if errors were uniformly distributed rather than clustered, throughput would be 1.51 Mbit/s [11], only a 5.5% reduction from the nominal rate, since TCP performs worse with losses clustered within one transmission window.

Cellular links make even less of their nominal bandwidth available to the users, due to their 1–2% FER [9], quite large considering the short frames used. An IS-95 (CDMA cellular) link transmitting at full rate uses fixed-size 172-bit data frames, excluding link layer overhead. Since this frame size is insufficient even for TCP and IP headers, the link layer segments IP datagrams into multiple frames. For comparison with the WLAN case above, consider a UDP packet with 1400 bytes of user payload and 28 bytes of UDP/IP overhead. This datagram would be segmented into 68 frames, so at a 1% FER the probability that the datagram will make it across the link is 50.49%, while at a 2% FER this probability drops to 25.31%, assuming independent frame errors. Even if frame errors are heavily correlated (bursty), high average FER results in high datagram error rates. Frame errors are less bursty than bit errors in CT systems because

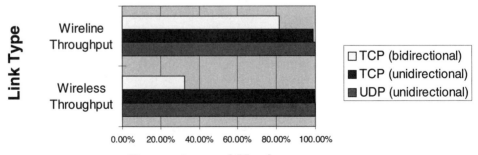

FIGURE 14.1
Wireless LAN performance: TCP traffic has to contend with ACKs on the single wireless channel. The inefficiency of Collision Avoidance versus Collision Detection is shown.

data bits from multiple frames are interleaved before transmission. After deinterleaving at the receiver, error bursts are spread uniformly over multiple frames so that usually the embedded error correction code of each frame can recover its contents. This technique reduces FER and randomizes frame errors so as to avoid audible speech degradation, but it adds considerable delay, as multiple frames need to be received before deinterleaving and decoding can take place. For example, the typical frame delivery delay on IS-95 is around 100 ms [9].

Reducing the datagram size reduces FER at the expense of increasing header overhead. However, TCP overhead can be reduced to 3–5 bytes per datagram by employing header compression, a technique customized for low-bandwidth serial links. This optimization is feasible *only* for the TCP/IP combination. Since CT systems use separate uplink and downlink channels, forward (data) and reverse (acknowledgment) traffic do not interfere as in the WLAN case. Rather surprisingly then, TCP offers potentially more bandwidth to the user than UDP in CT links, due to TCP header compression.

Assuming full frame utilization, 5-byte compressed TCP/IP headers, and a 2% frame error rate, Table 14.2 shows the percentage of the nominal link layer bandwidth that is available for user data. The datagram size varies from 4 to 64 link layer frames (i.e., 81–1371 data bytes), as shown in the first column, and independent errors are assumed for simplicity, with the datagram error rate shown in the second column. Independent errors are a pessimistic assumption, since clustered errors would affect fewer datagrams, but it is reasonable in view of the physical layer interleaving. While error rate increases with longer datagrams, the constant TCP/IP overhead becomes a smaller factor of the total bandwidth, as shown in the third column. Throughput is maximized here with 81-byte datagrams, at about 87% of link capacity, as error rate dominates overhead in the throughput calculation. As an example, when transmitting 81-byte data segments, that is, 86-byte datagrams, the error rate is $(1 - 0.98^4) = 0.07763$, since four link layer frames are used, and the overhead factor is $5/86 = 0.05814$. Since the total bandwidth of this system is 8.6 Kbit/s (each 171-bit frame takes 20 ms to transmit), the data throughput is $(1 - 0.07763) \times (1 - 0.05814) \times 8.6 = 7.47$ Kbit/s, which is about 87% of the bandwidth used. Note, however, that this throughput assumes perfect error recovery (i.e., TCP never waits for a timeout and never retransmits correctly received data), which is not the case in practice.

14.2.3   Protocol Performance Over Multiple Links

The preceding discussion focused on paths composed of a single wireless link. When multiple wireless links are traversed, errors accumulate accordingly. This is the case when users of separate cellular or WLAN systems communicate via the wired infrastructure. Making the assumption that the behavior of the two wireless links is uncorrelated, the cumulative datagram error rate for two WaveLAN links is 3.08%, while for two cellular links it is 14.92%, for the optimum

Table 14.2 Cellular Link Performance (TCP)

User Payload (bytes)	Datagram Error Rate (%)	Overhead (%)	Throughput (%)
81	7.763	5.814	86.874
167	14.924	2.907	82.603
339	27.620	1.453	71.328
683	47.612	0.727	52.008
1371	72.555	0.363	27.346

cases in the preceding tables (1400-byte UDP and 85-byte TCP payloads, respectively). Increased losses mean more frequent invocations of TCP congestion avoidance algorithms. Reducing transmission rates due to mistaking errors for congestion has the side effect of under-utilizing the wireless link. Since cellular links operating at bandwidths of around 10 Kbit/s are most likely the bottlenecks of an end-to-end path, underutilizing them means reduced end-to-end performance. In addition, cellular links usually serve one user with few transport connections, so their bandwidth will most likely be wasted, even though the user is typically billed based on (physical) connection time. WLANs, despite their higher bandwidths, are also likely to be the bottlenecks of end-to-end paths, but given enough users on the shared medium, the bandwidth released by one TCP connection may be absorbed by others.

Another problem with multihop paths is that TCP retransmissions traverse the path from the sender to the wireless link again, even if they already arrived successfully there previously. If a single wireless link is located on the receiver's end of the path, these retransmissions waste wired link bandwidth and delay recovery. If more than one link is wireless, datagrams are retransmitted over wireless links that were already crossed successfully, reducing their precious throughput (goodput). The combined effect of mistaking wireless loss for congestion and end-to-end recovery is more pronounced on longer paths that employ large TCP windows to keep data flowing through the network until acknowledgments return. This is because the TCP transmission window is reduced after losses are detected, leaving the end-to-end path underutilized until the window grows back to its appropriate size, which takes more time for long paths. Whenever multiple losses occur during a single window (more likely with larger windows), TCP performance is further reduced.

Figure 14.2 and Table 14.3 show the performance of TCP over a single hop (LAN) versus a multihop Wide Area Network (WAN) path (the results are taken from [2]), in absolute terms and as a percentage of the nominal bandwidth. One end of the path is on a wireless LAN transmitting 1400 byte frames with an FER of about 2.3%, a situation slightly worse than the one described in the preceding discussion. The nominal bandwidths hold in the absence of any congestion or wireless link losses, while the TCP throughput numbers are based on simulating

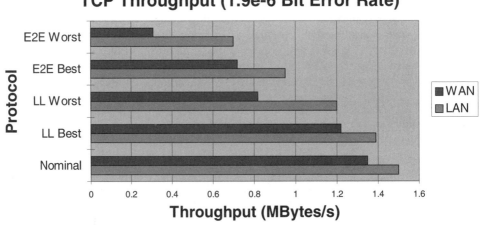

FIGURE 14.2
TCP Performance in multihop versus single-hop environments. (LL = link layer enhancements.)

Table 14.3 Throughput of LAN and WAN Connections (TCP)

	Nominal	Simple TCP		Improved TCP	
LAN	1.5 Mbit/s	0.70 Mbit/s	46.66%	0.89 Mbit/s	59.33%
WAN	1.35 Mbit/s	0.31 Mbit/s	22.96%	0.76 Mbit/s	56.29%

errors using a simple independent error model. The difference between the two TCP variants is that the improved protocol can recover from more than one error during a single-error recovery period. As a result, in the high-error environment of the wireless link, the improved variant achieves higher throughput and depicts smaller differences between the LAN and WAN cases. In all cases, the throughput degradation is at least 10 times the FER.

14.3 PERFORMANCE ENHANCEMENTS FOR INTERNET PROTOCOLS

14.3.1 Approaches at the Transport Layer

Most of the work on Internet protocol performance over wireless links has focused on TCP, since it is the most popular Internet transport protocol. The root cause of the reported problems is the TCP assumption that all losses are due to congestion, so that loss detection triggers congestion avoidance procedures. Thus, the frequent losses seen on wireless systems, whether due to communications errors or pauses during handoffs, cause TCP to underutilize the bandwidth-starved wireless links, dramatically reducing end-to-end performance. Longer paths further delay end-to-end recovery, aggravating these performance problems.

A direct approach to avoid TCP performance problems is to modify TCP itself, since it is TCP assumptions that cause the problems. In addition, TCP, being an end-to-end protocol, requires only the two communicating peers to upgrade their software in order to take advantage of improvements. Solutions depend on the cause of losses: handoffs or errors. During handoffs connectivity is temporarily lost, and many times a timer has to expire before recovery can be initiated. To avoid long communication pauses after handoffs, one approach is to invoke fast retransmission right after a handoff completes, instead of after a timeout. This requires signaling to notify the transport layer about handoff completion [6]. Invoking full congestion recovery procedures after every handoff still reduces throughput, so an alternative scheme attempts to detect whether loss is due to mobility or congestion by exploiting mobility hints from lower layers [10]. If the loss is classified as due to congestion, both slow start and congestion avoidance phases [8] are invoked. If the loss is classified as due to mobility, only slow start is invoked, reducing recovery time.

Unlike handoffs, where some congestion avoidance procedure is needed to probe the state of the new link, with losses due to errors we should skip congestion avoidance completely. Since these losses are local, end-to-end retransmissions unnecessarily delay recovery. One approach to improved error recovery is to split TCP connections at *pivot points,* that is, those routers on the path that are connected to both wireless and wired links (indirect TCP; see Figure 14.3). One instance of TCP executes over each wired part, while either another instance of TCP or a special-purpose transport layer protocol executes over each wireless part [1, 14]. These segments are bridged by a software agent at each pivot point that also translates between the different protocols, if required. As a result, losses at wireless segments do not trigger end-to-end recovery. When TCP is used over the wireless segments, it can recover fast due to the short paths

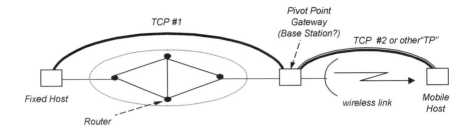

FIGURE 14.3
Illustration of a simple indirect TCP configuration.

involved. Alternatively, a transport protocol with more appropriate mechanisms (such as selective repeat) may improve performance [14]. The ideal scenario for this approach is a path with wireless endpoints, where the pivot points are the routers connecting the wireless network to the Internet. Only the pivot points and the wireless hosts need software upgrades. Handoffs in these schemes change the endpoint but not the pivot point, so that the agent at the pivot can establish new connections as needed and speed up recovery after a handoff. For UDP, datagrams lost during communications pauses can be retransmitted when connectivity is reestablished, as in M-UDP [4]. For TCP, the pivot agent can choke the remote sender by reducing the advertised window to zero, as in M-TCP [5]. This causes the sender to go into *persist* mode, during which it periodically probes the receiver's window while all pending timers are frozen. However, shrinking the advertised window violates TCP guidelines.

TCP modifications are not perfect solutions. Schemes that deal with handoff problems modify the transport layer only at the endpoints, but they still face performance degradation due to wireless errors. To detect handoffs, coupling is introduced between layers to solve an isolated and localized problem, with improvements that are applicable only to TCP. Split (or indirect) TCP approaches [1, 4] deal with both mobility and link errors, but they require modifications both at the endpoints and at the pivot points, the latter generally being beyond user control. New transport protocols compatible with TCP are needed to maximize performance over wireless segments, although the split connection idea is applicable to any protocol. The agents at the pivot points are complex: They must translate semantics and synchronize connections despite communications errors and pauses. Performance is questionable for wireless segments not at path ends or for multiple-link wireless segments. Finally, the end-to-end semantics of the transport layer may be violated, so applications that need end-to-end reliability must use additional protocols above TCP. Even worse, applications do not know that the end-to-end semantics are violated.

14.3.2 Approaches Below the Transport Layer

The main alternative to modifying TCP or other end-to-end protocols is to modify the link layer protocol that operates over the wireless link so as to hide losses using local recovery mechanisms. CT systems offer nontransparent Radio Link Protocols (RLPs) that enhance link reliability with this exact goal in mind. Another approach, applicable to both WLANs and CT systems, is adding some link layer functionality to IP to take care of local recovery [3]. IP datagrams carrying TCP data are buffered at the hosts that transmit them over wireless links, and are

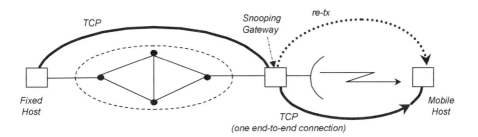

FIGURE 14.4
Illustration of snoop TCP.

retransmitted if they are not acknowledged within a short period of time or if multiple acknowl-edgments for previous data are received. The local error-recovery module snoops on all IP data-grams to gather TCP data and acknowledgment information. Buffered TCP data that need to be retransmitted are inserted into the data stream transparently to the receivers. By leveraging existing TCP messages, snoop TCP (see Figure 14.4) avoids additional control exchanges and simplifies integration with TCP. Conceptually, this is a link layer mechanism, as it involves single-link error recovery.

Interestingly, employing TCP acknowledgments for feedback, besides violating protocol lay-ering, causes this approach to work only in the direction toward the wireless host. In this direc-tion, TCP acknowledgments are received at the wired endpoint after a one-hop delay only, and retransmissions may be made before timers expire at the other end of the end-to-end path. In the reverse direction, TCP acknowledgments are returned after the round-trip delay for the whole path has nearly elapsed, so the wireless host cannot retransmit lost segments soon enough. To make retransmission effective in that direction, local control exchanges are needed (as required by protocol layering). This is also the case for wireless links that are not at the edges of end-to-end paths, where both sides need link layer control exchanges to initiate retransmissions on time. Overall, this local recovery scheme performs better than split transport layer schemes under wireless link errors [2]. Avoiding the violation of transport layer semantics comes at the cost of violating protocol layering, however. The advantages of TCP and link layer coupling is reduced link layer overhead and avoidance of conflicts between local and TCP retransmissions [7], but the scheme may require modifications whenever TCP is modified, and does not work for any other protocols.

CT system RLPs on the other hand, avoid layering violations, but run the risk of retransmit-ting data that the transport layer will retransmit anyway, hence the approach of the IS-95 RLP of limited recovery [9]. In addition, they may do more than what is required for applications that do not require complete reliability. Link layer schemes in general have the advantage over end-to-end schemes of working at the local level, with intimate knowledge of the underlying media and low round-trip delays that allow fast recovery, although they cannot deal with handoffs where multiple links are involved. The problem is how much to enhance the underlying link without getting in the way of higher layers. The goal is to offer adequate recovery to ease the task of reliable transports and to allow the realistic operation of unreliable transports, which assume only rare losses. Even if vendors supply fine-tuned link layer protocols with their devices, it is hard to design a single protocol that can cater to the needs of multiple existing and future transport layers and applications.

14.4   THE FUTURE: CHALLENGES AND OPPORTUNITIES

14.4.1   Wireless System Evolution

One of the most attractive characteristics of wireless systems is that they enable mobility. Cellular systems allow efficient sharing of the frequency spectrum via reuse and offer wide area mobility with reasonable power requirements. Since each cell is connected to other networks via its own base station, mobility between cells implies a need for handoffs of mobile devices between base stations. These handoffs cause pauses in communication while the mobile completes base station changes. When data are lost during handoffs, reliable Internet protocols such as TCP may be tricked into mistaking them for congestion [6]; this is the case even if they are not really lost but only delayed. Some future cellular systems are expected to employ smaller cells (*picocells*) to offer higher data rates and support more users at the same time. Picocells, due to their smaller area, will require a dense mesh of base stations, which will only be justifiable within buildings or other densely populated areas. Thus, current cellular systems with standard-sized cells will still provide coverage in areas with fewer users, while sparsely populated areas that do not warrant the cost of a terrestrial cellular infrastructure will be covered by satellite systems. The large area under a satellite beam will then form a *macrocell*. In low-orbit satellite systems handoffs will still occur, but mostly due to satellite rather than user movement. This will give rise to a hierarchical cell structure, as depicted in Figure 14.5. Higher-level cells are overlaid in areas with (presumably) more users by multiple lower-level cells. Users can use the highest bandwidth system available in each location and move from cell to cell within the same system (performing horizontal handoffs, as in existing cellular systems) or from one system to another (performing new-style vertical handoffs), depending on coverage.

Hierarchical systems will challenge us with additional handoff-induced problems. In the picocells, handoff frequency will increase, and since the per-user bandwidth will be higher, more data will potentially be lost during handoffs, urging for faster handoffs and fast, localized recovery. In parallel, since handoffs will be possible between different technologies, connections will face two levels of link performance variability. First, short-term performance variations will be caused by environmental changes such as fading, in the same manner as today, with details depending on the technology used at each link. Adapting to these link-specific variations is much easier locally, at the link layer, where the peculiarities of each medium are known. Second, medium-term performance variations will be caused by handoffs between different technologies (picocellular, cellular, and macrocellular). These handoffs will dramatically change the performance parameters of the wireless part of the end-to-end path, as each type of link will have

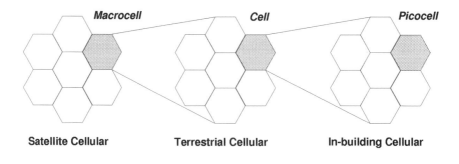

FIGURE 14.5
Overlay networks: hierarchical cellular system.

its own characteristics. Adapting to such variations is probably only feasible at an end-to-end layer. Since handoffs take place between two separate links, dealing with handoff outages is not a local task, although the link layer could help by providing information to higher layers, for example, notifications of handoffs and disconnections or current link properties after a vertical handoff. These systems illustrate how similar problems (horizontal and vertical handoffs) with different parameters may be best treated at distinct layers.

14.4.2 Goals for Protocol Evolution

If we examine the performance problems of existing protocols and the shortcomings of proposed enhancements for wireless links, a few goals for further research emerge.

- Masking wireless errors with nontransparent RLPs is not adequate for all types of transport protocols and applications. For example, UDP-based real-time applications typically prefer sending new packets to retransmitting old ones. Similarly, transport layer modifications, despite using the most appropriate mechanism for any given transport protocol, are protocol specific and not easily extensible to hierarchical cellular systems. We thus need a multiprotocol approach to wireless link enhancements.

- While mobility problems cannot be dealt with at the link layer (whose scope is isolated links), wireless link losses can be efficiently recovered from using local mechanisms [2]. Besides handling handoff pauses, higher layers will have to adapt to the medium-term link variabilities caused by (vertical) handoffs to different networks in hierarchical cellular systems. Thus we need to deal with each wireless link problem at the appropriate layer.

- Since applications have varying requirements, multiple transport layers should be supported on the Internet without having to worry about the details of each specific link. Isolating transport layer protocols from link peculiarities can be achieved by link layer protocols that support flexible and general services rather than specific transport protocols. Thus protocol layering and isolation should be adhered to during design.

- The proliferation of wireless links will eventually force many existing protocols to review their design assumptions and modify their mechanisms. New protocols may also emerge that are more flexible in dealing with wireless links and mobility, by exploiting adaptive mechanisms. Such protocols would be able to exploit advanced link layer services in order to offer enhanced functionality to their own users. Thus lower-layer protocols should consider future higher-layer needs.

In our view, neither link layer nor transport layer approaches are sufficient by themselves to solve the problems presented here. While transport protocols are more aware of end-to-end application requirements, link protocols are better positioned to handle local issues in a fast and transparent manner. A link layer handling link-specific problems, such as wireless errors, isolates higher layers from physical layer peculiarities and evolution, adapting its mechanisms to prevailing conditions. Intimate knowledge of link-specific attributes, including actual hardware status, also allows the link layer to inform higher layers of interesting events (e.g., handoffs) and performance metrics of a link, so that they can in turn adapt their mechanisms accordingly. For example, in the Linux OS we could use pseudodevice files to export performance metrics and software signals for event notifications. Upward propagation of information through the protocol stack could thus ease the introduction of more adaptive and versatile protocols at each layer. On the other hand, each higher-layer protocol would like to see different enhancements to the raw link offered by the link layer. Instead of offering a different protocol for each type of service, a single link layer could support multiple services that cater to various requirements and needs, all sharing the link simultaneously [12]. Choosing between multiple services requires propagation of additional information downward through the protocol stack. We thus advocate a more synergistic approach between layers, where generic end-to-end requirements are supported by customized local mechanisms. We believe that making the interfaces between protocol layers richer in information content could pave the way for smarter and better-performing future Internet protocols.

14.5   SUMMARY

We presented the performance problems faced by Internet protocols when deployed over wireless networks. We also discussed existing performance enhancements at various protocol layers. A scenario for the future evolution of wireless communications was then presented, and we concluded by presenting some design goals for further protocol enhancement to address the shortcomings of existing approaches and the new requirements imposed by emerging systems. Overall, we believe that better cooperation and communication between protocol layers are key requirements for improving Internet protocol performance in the future.

14.6   REFERENCES

[1]   A. V. Bakre and B. R. Badrinath. Implementation and performance evaluation of Indirect TCP. *IEEE Transactions on Computers,* 46(3):260–278, March 1997.

[2]   H. Balakrishnan, V. N. Padmanabhan, S. Seshan, and R. H. Katz. A comparison of mechanisms for improving TCP performance over wireless links. *IEEE/ACM Transactions on Networking,* 5(6): 756–769, December 1997.

[3]   H. Balakrishnan, S. Seshan, and R. H. Katz. Improving reliable transport and handoff performance in cellular wireless networks. *Wireless Networks,* 1(4):469–481, 1995.

[4]   K. Brown and S. Singh. M-UDP:UDP for mobile celullar networks. *Computer Communications Review,* 26(5):60–78, October 1996.

[5]   K. Brown and S. Singh. M-TCP:TCP for mobile celullar networks. *Computer Communications Review,* 27(5):19–43, October 1997.

[6]   R. Caceres and L. Iftode. Improving the performance of reliable transport protocols in mobile computing environments. *IEEE Journal on Selected Areas in Communications,* 13(5):850–857, June 1995.

[7]   A. DeSimone, M. C. Chuah, and O. C. Yue. Throughput performance of transport-layer protocols over wireless LANs. In *Proceedings of the IEEE GLOBECOM '93,* pages 542–549, December 1993.

[8]   V. Jacobson. Congestion avoidance and control. *Computer Communication Review,* 18(4):314–329, August 1988. (SIGCOMM '88 Symposium: Communications Architectures and Protocols, Stanford, CA, August 1988.)

[9]   P. Karn. The Qualcomm CDMA digital cellular system. In *Proceedings of the USENIX Mobile and Location-Independent Computing Symposium,* pages 35–39, August 1993.

[10] P. Manzoni, D. Ghosal, and G. Serazzi. Impact of mobility on TCP/IP: An integrated performance study. *IEEE Journal on Selected Areas in Communications* 13(5):858–867, June 1995.

[11] B. D. Noble, M. Satyanarayanan, G. T. Nguyen, and R. H. Katz. Trace-based mobile network emulation. *Computer Communication Review,* 27(4):51–61, October 1997. (Proceedings ACM SIGCOMM '97 Conference, Cannes, France, September 1997.)

[12] G. C. Polyzos and G. Xylomenos. Enhancing Wireless Internet Links for Multimedia Traffic. In *Proceedings International Workshop on Mobile Multimedia Communications (MoMuC'98)*, Berlin, Germany, October 1998.

[13] G. Xylomenos and G. C. Polyzos. TCP and UDP performance over a wireless LAN. In *Proceedings IEEE Conference on Computer Communications (INFOCOM '99)*, New York, March 1999, pages 439–446.

[14] R. Yavatkar and N. Bhagawat. Improving end-to-end performance of TCP over mobile internetworks. In *Proceedings of the IEEE Workshop on Mobile Computing Systems and Applications*, pages 146–152, December 1994.

CHAPTER 15

Transcoding of the Internet's Multimedia Content for Universal Access

RICHARD HAN
JOHN R. SMITH

15.1 INTRODUCTION

Transcoding is a technique that transforms multimedia, for example, text, images, audio, and video, from the original format in which the multimedia was encoded into a second alternative format. Within the emerging world of wireless connectivity to the Web, transcoding of Internet multimedia enables Internet content providers, for example, web sites, and Internet Service Providers (ISPs) to transform Web content so that a wider collection of Internet-enabled client devices, such as cellular phones and personal digital assistants (PDAs), can have access to that content. As a second equally important benefit, transcoding of Internet multimedia enables Web sites and ISPs to increase the perceived speed of access to that content by additional compression of text, image, audio, and video over slow Internet access links.

As shown in Figure 15.1, the Internet is experiencing a rapid diversification in the types of network links used to connect to the Internet, for example, 56K modems, cable modems, digital subscriber loops (DSLs), cellular data links such as the European-Asian Global System for Mobile Communications (GSM) standard, wireless data services such as Mobitex, and wireless Local Area Networks (LANs). Transcoding enables recompression of multimedia content, that is, reduction in the content's byte size, so that the delay for downloading multimedia content can be reduced. Compression-based transcoding over links with the most limited bandwidth can dramatically improve the interactivity of Web access. An additional advantage of reducing the byte size of multimedia content is reducing the cost over tariffed access links that charge per kilobyte of data transferred.

At the same time, the Internet is experiencing a rapid proliferation in the kinds of client devices used to access the Internet, for example, desktop workstations, laptops, Web-enabled screen phones, Internet-enabled cellular phones, integrated television/Web set-top boxes, and handheld Palm PDAs with Web access. Transcoding enables the transformation of multimedia

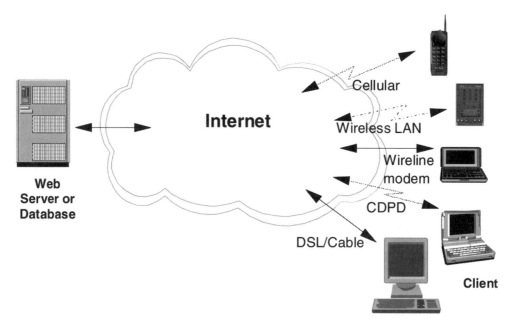

FIGURE 15.1

The emerging world of the wireless Web consists of a wide variety of fixed and mobile clients connected to the Internet over a wide range of link speeds, all trying to access multimedia content stored on a remote Web server or database. (CDPD = Cellular Digital Packet Data; DSL = Digital Subscriber Line.)

content so that each of these increasingly diverse client devices can have access to the Internet's rich multimedia content, albeit in transcoded form.

15.1.1 Adapting to Bandwidth Heterogeneity

A common motivation for applying transcoding is to reduce the delay experienced when downloading multimedia-rich Web pages over Internet access links of limited bandwidth, for example, modem links and wireless access links [Liljeberg95, Fox96a, Bickmore97, Fleming97, Han98, Smith98a, Bharadvaj98]. The transcoding function is typically placed within an HTTP proxy that resides between the content provider's Web server and the client Web browser. The transcoding proxy reduces the size in bytes of the multimedia content via lossy compression techniques [e.g., images are more aggressively compressed, HyperText Markup Language (HTML) text is summarized]. The proxy then sends the recompressed multimedia content over the modem/wireless access link down to the client. The reduction in byte size over the access link, typically a bottleneck, enables an often significant reduction in the perceived response time. Transcoding has also been applied to continuous media to reduce a video's bit rate and thereby enable clients connected over bandwidth-constrained links to receive the modified video stream [Amir95, Yeadon96, Tudor97]. The terms *distillation, summarization, dynamic rate shaping*, and *filtering* are frequently used as synonyms for transcoding when reducing latency (via compression over low-bandwidth links) is the primary motivation for transcoding.

While transcoding reduces response time over very slow links, compression-based transcoding can actually increase the response time over high bandwidth links [Han98]. The operation of transcoding adds latency, especially for images, because decompression and compression are collectively compute-intensive. Over high-bandwidth access links, the decrease in download

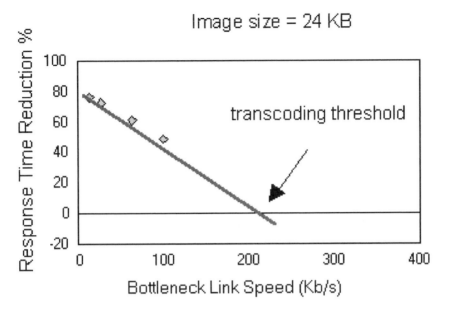

FIGURE 15.2
For slow access link speeds, compression-based transcoding can dramatically reduce response times. For fast-link bandwidths, the slight reduction in delay due to compression may be less than the delay added by compute-intensive transcoding. (From [Han98].)

time achieved by reducing the byte size may be less than the increase in delay caused by transcoding (decompression and recompression), thereby resulting in inflated response times. Figure 15.2 summarizes this behavior and suggests that transcoding should only be performed when the network poses a sufficient bottleneck. The precise conditions are described in later sections.

In addition to compression, a transcoding proxy can also reduce response time by choosing an output transcoded format that minimizes client-side decoding delay. For mobile clients that are limited by Central Processing Unit (CPU) speed, memory, power consumption, and Operating System (OS) capabilities, client-side decoding delay can comprise a significant portion of the overall response time. Consequently, a transcoding proxy that intelligently transcodes into an output format that reduces client-side decoding delay can minimize overall delay. However, there are tradeoffs similar to those encountered in the use of compression, namely, that minimizing client-side decoding delay does not necessarily minimize overall delay. This is because sending a low-complexity format to the end device can actually result in sending more data (e.g., bitmaps instead of compressed images) over the access link. Comparisons of the tradeoffs in response time obtained by moving decoding complexity away from the PDA's browser toward a transcoding proxy have been analyzed [Fox98b, Han99]. An early study found that partitioning most of a text-only Web browser's complexity away from a PDA into a proxy enabled response times on the order of several seconds (in conjunction with caching and prefetching) [Bartlett94]. General architectures have been proposed that partition application functionality [Watson94] or distribute code objects [Joseph95] between the network (a server or transcoding proxy) and a mobile client, especially as a means of dynamically adapting to changing network and client conditions [Nakajima97, Noble97]. The commercial Palm VII PDA conducts wireless access to the Internet via a partitioned application structure; streamlined "query" applications on the PDA request information from selected Web servers, whose content is then transcoded by an intermediate "Web clipping" proxy to a PDA-specific format [PalmVII99].

The degree to which compression can and should be applied by a transcoder is bounded from above and from below. Intuitively, transcoding should only be applied when the response time can be lowered. This condition puts a lower bound on the compression ratio, that is, some minimum degree of compression must be applied to ensure that response time is reduced. There is also an upper bound on how much compression should be applied, imposed by the requirement that the quality of the transcoded content be subjectively acceptable to the end user. Our experience with lossy image compression techniques suggests that compression ratios of 6 to 10 times can be achieved for typical Web images without losing their intelligibility. Beyond this range, more ambitious compression ratios can lead to an unacceptable degradation of image quality.

Compression-based transcoding includes both compression within a media type as well as summarization across media types. Lossless compression of text has been applied in a transcoding proxy [Floyd98, Bharadvaj98]. Lossy compression of an image can lead not only to an image of reduced byte size, but also to a brief text description of that image, for example, the description of an image contained in the ALT attribute of an IMG tag within an HTML page. Similarly, lossy compression of a movie can lead to several possibilities: a more aggressively compressed video sequence, an audio-only stream, a small set of image stills corresponding to key video frames, a text summary, or just the title of a movie.

A convenient means of summarizing the multiple ways to transcode, that is, conversion and recompression of content both between media types as well as within a media type, is depicted in the "InfoPyramid" data model of Figure 15.3 [Mohan99]. Media objects are classified according to their fidelity (vertical axis) and modality (horizontal axis). Increased compression corresponds to moving upward toward the pyramid's apex (in this case, upward movement corresponds to lower not higher fidelity). Conversion of a media object across media types corresponds to left/right translation across modalities. For example, compression of a video stream to another video stream of lower fidelity involves sliding up several levels of the

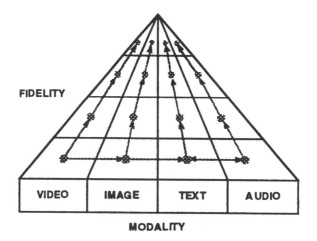

FIGURE 15.3

The "InfoPyramid" model conceptually organizes transcoding of media objects into the operations of translation (conversion across modalities) and summarization (compression that reduces fidelity). (From [Mohan99].)

pyramid within the same modality. Compression of a video stream into a text summary involves both a rightward translation into the text modality as well as a translation up the pyramid as appropriate. The InfoPyramid model is flexible enough to support multiple modalities within a media type, for example, GIF [Nelson96] and Joint Photographic Experts Group (JPEG) [Pennebaker93] within the image category and multiple spoken languages and codec formats within the audio category. As explained in a later section, the InfoPyramid serves as a conceptual data structure that is used by a transcoding proxy for representing each media object within a multimedia document, so that the manipulation routines of the transcoder can easily translate or summarize these media objects.

While the type of transcoding described in this section uses recompression to adapt to links with different bandwidth characteristics, it has also been proposed that transcoding be applied to adapt to links with different loss/error characteristics [Swann98, de los Reyes98]. In this form of transcoding, the encoding format of the multimedia is adapted by a transcoder into a format that is more error-resilient to packet losses or bit corruption within a packet's payload.

15.1.2 Adapting to Client Heterogeneity

A second common motivation for applying transcoding is to resolve mismatches between a client's decoding capabilities and a content provider's encoding formats, so that multimedia can be conveyed to a variety of clients. Codec mismatches are often caused by hardware limitations on mobile clients, software restrictions at the client or content provider, and constraints imposed by the end user or environment in which the mobile client is operated. To compensate for codec incompatibilities, transcoding is applied to convert the original format of multimedia into a form that can be viewed or played back by a client. Consequently, transcoding expands the number and types of fixed and mobile clients with access to existing content. The terms translation, content repurposing, content adaptation, reformatting, data transformation, media conversion, clipping, universal access, format conversion, and filtering are often used as synonyms for transcoding when it is applied to adapt to client heterogeneity.

Transcoding is currently being applied in a variety of contexts to adapt to different client/user requirements. Speech recognition and speech synthesis represent two traditional examples of transcoding technologies in which a conversion process is required in order to resolve codec mismatches and convey useful information to the client/user. They can be classified as speech-to-text transcoding and text-to-speech transcoding, respectively. In addition, transcoding is being applied to transform Web pages (images and HTML text) so that handheld PDAs can browse the Web [Gessler95, Fox98a, Han98, Smith98a]. Since PDA clients are constrained by their small screen size (e.g., 160×160 screen), often limited display depth (e.g., 2 bits of grayscale per pixel), limited memory, relatively slow processor, and limited support of media types (e.g., audio or video may not be supported), then transcoding provides a way to overcome these constraints so that images and text can be properly displayed on these handheld devices. Color Graphics Interface Format (GIF) and JPEG images have been transcoded and prescaled to 2-bit grayscale bitmaps by a proxy [Fox98b, Han99].

In the future, we envision an expansion of the role of transcoding to cope with the rising number of codec mismatches caused by an increasingly heterogeneous world of Internet access devices. First, increasingly diverse client-side hardware for Internet access devices will likely drive the need for transcoding. For example, if a Palm handheld with Internet access lacks a speaker, then a transcoder capable of recognizing that situation could convert speech to text to enable a reader to follow an Internet multicast radio program despite the lack of audio output. Conversely, if a cellular phone with Internet access physically lacks a screen, then a transcoder capable of synthesizing audio from an HTML page would enable speech-based browsing of the

Web. As new devices emerge, new applications of transcoding will likely be spurred, for example, a futuristic audio/video wristwatch limited to hardware decompression of only one video format would require a transcoder to convert other forms of video to the decoding format supported by the watch.

A second likely role for transcoding will be adaptation of multimedia to suit human user preferences or needs as well as to suit environmental restrictions. For example, if a user has an impaired sense, for example, severely impaired vision or hearing, then transcoding can be applied to transform multimedia content into a media type satisfactory to the user, for example, speech-to-text transcoding for the hearing impaired and text-to-speech transcoding for the visually impaired. In addition, the end user's environment can also impose constraints. Consider an automobile driver who would like to browse the Web. The act of driving makes visual interaction with the browser quite dangerous, so that speech browsing will likely be the most effective mode of interaction while driving. However, when the car is stopped, then the interaction mode should revert back to supporting visual interaction, for example, for maps. A text-to-speech transcoder would enable a user to surf the Web while his or her car is in motion.

A third category in which we anticipate the development of more transcoding solutions is based on resolving software codec mismatches between the client's decoder and the content provider's encoder. Web servers and Web browsers will likely need to transition to incorporate new software standards, for example, eXtensible Markup Language (XML), MP3, Portable Network Graphics (PNG), JPEG2000, and MPEG4. Invariably, some client devices will be behind web sites in standards implementation, that is, their browsers will support only legacy standards, while other client devices will be ahead of web sites in standards adoption, that is, some Web servers will generate multimedia encoded with legacy standards. Downloadable plug-ins for the browser offer one means for correcting mismatches in formats. Backward compatibility of browsers and encoders, in conjunction with a protocol similar to what modems execute for negotiating down to a mutually compatible format, offers a second solution for resolving codec mismatches. However, we believe that there will be many cases in which the client device's browser will be unable to be upgraded via plug-ins (or at least the upgrading process will be sufficiently difficult to dissuade a large number of users), or will not support a large number of legacy formats, thereby compromising backward compatibility. For example, a software "microbrowser" [Gardner99] embedded within a Web-enabled Palm handheld or cellular phone will likely only support a small fixed set of standards, perhaps only one, due to limitations in CPU, memory, power, and cost. In this case, we believe that transcoding provides a convenient alternative that is capable of resolving mismatched formats, so that the client can view or play back the content provider's multimedia.

15.2   END-TO-END VS. PROXY-BASED TRANSCODING DESIGNS

Multimedia content that needs to be adapted for presentation to diverse clients over diverse access links can be either transcoded on-the-fly by a proxy or pretranscoded offline by a server. In on-the-fly or real-time transcoding, a request for content triggers the content provider to serve the original document, which is then transformed on-the-fly by a transcoding entity downstream to create a newly adapted version of the document. As illustrated in Figure 15.4, the real-time transcoding entity can be placed without loss of generality in a proxy that is located at some intermediate point between a server and a requesting client. The proxy is typically situated in the network on the opposite side of the access link as the client. In this way, compression-based transcoding can reduce download times over the often bottlenecked access link. For ISPs that wish to offer a Web transcoding service, the transcoding proxy can be placed within the ISP

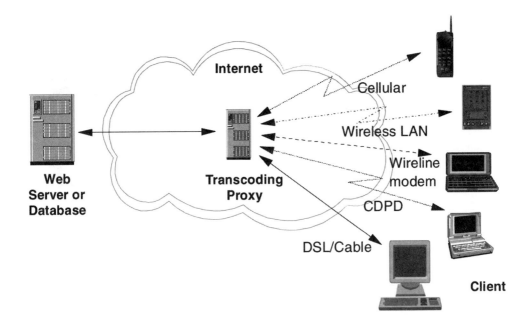

FIGURE 15.4
Proxy-based transcoding of multimedia content involves on-the-fly transformations of a server's content by an intermediate proxy for presentation to diverse clients connected to the Internet over diverse access links.

access cloud. For web sites that desire to provide universal access to their content, the transcoding proxy can be placed adjacent to the Web server on the content provider's premises. Other terms for a transcoding proxy include mixer [Schulzrinne96] and gateway [Amir95, WAPForum99].

In contrast to the proxy-based approach, the end-to-end design philosophy for transcoding prepares multiple pretranscoded versions of a multimedia document offline, that is, a server authors multiple versions of a multimedia document prior to receiving any requests for that document. Each version is specially tailored for a given client and access link speed. Consequently, no intermediate proxy is needed to transform the data. As illustrated in Figure 15.5, when a request for a multimedia document is received from a particular client over a given link speed, the content provider's server simply selects the appropriate pretranscoded version to serve, thereby avoiding on-the-fly transformations. A simple example of server-side authoring is the support by certain web sites for text-only or low-graphics versions of Web pages. Accessing these text-only versions currently requires manual direction from the end user. An alternative would be to employ an automated mechanism that transparently conveys client characteristics from a browser to the server, so that the server can automatically select the appropriate version of a Web page to return to the client. The User-Agent header field of an HTTP request automatically enables browsers to identify themselves to a server. Also, platform capabilities can be automatically identified to the server via content negotiation (e.g., Accept header) within HTTP 1.1 [Fielding97].

Table 15.1 summarizes the tradeoffs associated with implementing either end-to-end server-side preauthoring or proxy-based real-time transcoding. One of the primary motivations for

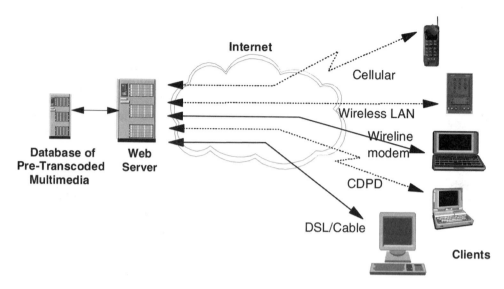

FIGURE 15.5

End-to-end transcoding generates multiple versions of a multimedia document prior to a client request. Each version is adapted for a specific client and access link. In response to a client request, the server returns the appropriate pretranscoded version, thereby avoiding on-the-fly transcoding by a proxy.

proxy-based implementations of transcoding is that the existing infrastructure of servers and clients doesn't have to be modified to implement a transcoding service. The client browser merely needs to be redirected to access the Web through an HTTP transcoding proxy. Also, existing Web servers need not be changed. Ideally, any news or entertainment site on the Web could be accessed via transcoding technology without burdening content providers with the need to customize each site for display on multiple client devices. In contrast, end-to-end server-side authoring requires modifications to each Web server to generate multiple client-specific versions of a given Web page a priori.

While proxy-based implementations appear to offer an easy means to deploy transcoding into the existing Web server and Web client infrastructure, proxy-based solutions introduce their own problems. First, proxies often perform compute-intensive decompression and recompression of images or video. The latency introduced by such transcoding is often nonnegligible, for example, on the order of a hundred milliseconds on an unloaded image transcoding proxy [Han98]. If the purpose of applying transcoding is to improve the response time over low-bandwidth links, then transcoding can under certain conditions worsen the overall response time. As we scale to many users, proxies may become quickly overloaded such that compute-intensive transcoding becomes completely impractical, that is, the cost of maintaining a farm of proxies to meet transcoding obligations may become too high. However, several techniques have been offered to alleviate the burden of computation on transcoding proxies. Fast transcoding of video and images occurs in the compressed or frequency domain [Acharya98, Amir95, Assuncao98, Chong96]. Keeping transcoding operations in the frequency domain for DCT-based standards like JPEG and Motion Picture Experts Group (MPEG) simplifies computation. DCT coefficients are requantized at a coarser level, possibly to zero, thereby achieving compression-based transcoding without having to perform computationally intensive inverse transforms to and from the pixel domain. Another approach is to reuse motion vectors obtained from the decoding stage of a video transcoder in the subsequent encoding stage, so that the encoder need not recalculate

Table 15.1 End-to-End Serving of Pretranscoded Web Pages vs. Proxy-Based Real-Time Transcoding of Web Pages

Properties	End-to-End Serving of Pretranscoded Content	Proxy-Based Real-Time Transcoding
Modifications to existing infrastructure	Upgrade servers to generate multiple pretranscoded versions of a Web page, server informed of client characteristics	No modification at server, client redirected to point to proxy, client needs to advertise its characteristics to proxy
Latency	Little additional delay	Compression reduces delay, but compute-intensive transcoding adds delay
Scalability in terms of processing	Offline transcoding can be performed when convenient	Highly compute-intensive transcoding, often under real-time deadline pressure
Security	Supports end-to-end encryption	Proxy must be trusted to decrypt, decompress, recompress, and re-encrypt
Semantic understanding	Server knows semantic importance of each media object in a page or document	Proxy has incomplete knowledge of page composer's intents; proxy can be aided by hints from server
Degradation	No additional degradation	Noise/analog degradation accumulates with each decompression-lossy compression cycle
Efficiency	Send appropriately sized images and text over backbone	Excessive image/text/audio/video sent over backbone between server and proxy
Scalability in terms of upgrading to a new format	Each server needs to be upgraded to support $N + 1$ encoders from N encoders	Each proxy needs to be upgraded to add new encoder and decoder (i.e., $2(N + 1)$ codecs)
Legacy support	Backward compatibility may be limited	Transitional role supporting legacy client hardware and compression standards, phased introduction of new standards
TCP semantics	Unaffected	Often broken by a proxy

the motion vectors via a computationally intensive search [Bjork98]. A performance analysis of some of these reduced-complexity transcoding techniques has been conducted for MPEG video [Yeadon96]. Despite these methods to reduce complexity, on-the-fly transcoding still exposes the user to the latency cost of the transcoding operation. In contrast, server-side authoring effectively pretranscodes all multimedia content offline, so that the end user never directly experiences the latency of transforming content. Multiple versions of Web pages are prepared a priori without interfering with the servicing of requests. The implication is that server-side authoring is far more scalable in terms of accomplishing the same goal of adaptation for far lower hardware costs than a proxy-based approach.

Another drawback of proxy-based transcoding is security related. If multimedia content is encrypted, then a proxy must be a trusted entity allowed to decrypt the content so that decompression and recompression can be applied to the protected content [Haskall98]. If the transcoding proxy resides within the content provider's site, then the web site can be designed to invoke a transcoding proxy before that content has been encrypted, thereby sidestepping the security concerns. However, if transcoding proxies are applied over the wide area by untrusted third-party ISPs, then trusting the third-party transcoding proxy becomes an important security issue. In contrast, server-side authoring supports end-to-end encryption; there is no need to alter content in the middle of the network, so that decryption is only performed at the client endpoint. A second problem with proxies occurs when an end user is required to navigate through a company firewall to reach the broader Internet. In this case, the user can be forced to redirect the browser to point to the firewall proxy. Users connected to the company intranet via slow access links may benefit from a transcoding service. If such a transcoding service were offered within the firewall, then the user could redirect the browser to point at the transcoding proxy, which in turn could access the firewall. However, if such a transcoding service were to be offered in a second third-party proxy outside of the firewall, most browsers would be incapable of specifying that two proxies be chained, that is, that after the HTTP GET request is sent to the firewall, it should then be sent to an external transcoding proxy.

Another disadvantage of the proxy architecture is that the proxy has imperfect knowledge concerning the relative importance of various media objects within a multimedia Web page, which can lead to an imperfectly transcoded Web page. For example, when a proxy needs to condense a page for display on a PDA, the proxy must decide what to summarize, what to filter out, and how much summarization should be applied. If the proxy is given no direct hints about priority from the page's designer, then the proxy must fall back on suboptimal inferences and ad hoc heuristics to transcode the page. The end result may be a transcoded page that fails to convey the important material in the document. Even if an HTML/XML document conveys hints about the relative importance of text and images in a page, the desired message may not be fully expressed in the transcoded output. In contrast, server-side authoring permits the web-site designer to author the page for a PDA precisely as the designer had intended.

The process of decompression and lossy recompression associated with transcoding proxies inherently accumulates quantization noise each time the cycle is invoked. This effect is most noticeable when there is a concatenation of transcoding proxies. Analog noise accumulates with each transcoding operation, so that the final image or audio suffers cumulative degradation, even though individually each proxy can claim that the degradation it introduced was relatively minor [Chang94, Wilkinson98]. For example, cellular phone-to-cellular phone calls suffer a transcoding of speech from cellular vocoder to Plain Old Telephone Service (POTS) 8-bit Pulse Code Modulation (PCM) and then from POTS back to the cellular vocoder standard. The resulting accumulated degradation of two cascaded transcoding operations severely reduces the quality of the speech. One solution that is shown to reduce the impairments introduced by transcoding is for the encoder to reuse the decoder's information, for example, the quantization tables, motion vectors, and I/P/B frame classification obtained from decoding an MPEG-2 frame are reused by the transcoder's MPEG-2 encoder [Tudor97]. Another study has found that minimizing the distortion of frequency-domain MPEG transcoding can be approximated by a variant of dropping the highest DCT coefficients [Eleftheriadis95].

Transcoding proxies also lead to excessive information being sent over the backbone network from server to proxy. For example, the entire Web page is typically transmitted from server to proxy, only to have the proxy significantly filter the objects on that page via compression. A more efficient design would be to send only the images and text that are necessary for final reconstruction on the client, so as not to waste backbone bandwidth. This efficient approach is taken by server-side authoring, which sends images and text that have already been compressed to account for the destination client.

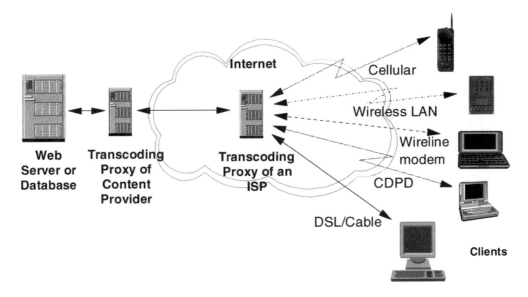

FIGURE 15.6
A chain of transcoding proxies can lead to an accumulation of noise, additive delay, and heightened security risks.

As shown in Figure 15.6, a chain of proxies will exacerbate many of the problems mentioned previously. Analog degradation will accumulate. Latency will also accumulate across proxies. And security problems will only grow as more intermediaries have to be trusted. Figure 15.6 illustrates one scenario among many that could lead to a concatenation of proxies. A content provider may place a proxy-based transcoder within its premises, while an ISP may decide to place a transcoder within its network. Neither proxy may understand what kind of transcoding is being employed by the other proxy, resulting in duplicate independent transcoding through cascaded proxies.

Despite these numerous disadvantages, proxies have the advantage (in addition to ease of deployment) of offering transitional technology that allows new standards to be incrementally introduced and permits legacy standards to be indefinitely supported. When a server adds a new standard for images, audio, video, or text, clients need not immediately adopt the new standard in order to access the newly encoded content, provided that there is a proxy that can translate the new format into a legacy format. It is unlikely that all web sites that adopt a new standard, especially noncommercial ones, will retain full backward compatibility, that is, retain multiple versions of an image (in JPEG2000 and GIF) or an audio stream (MP3 and WAV format). In the absence of a proxy, it would therefore be difficult to incrementally deploy new standards without isolating certain users with legacy decoding software.

Some additional tradeoffs in the design of general-purpose proxy-based systems have been enumerated [Zenel97, Border99], including how proxies can break the semantics of TCP.

15.3 ARCHITECTURE OF A TRANSCODING PROXY

Since transcoding is commonly implemented in a proxy configuration, it is helpful to examine the internal architecture of a transcoding proxy. As shown in Figure 15.7, a transcoding Web

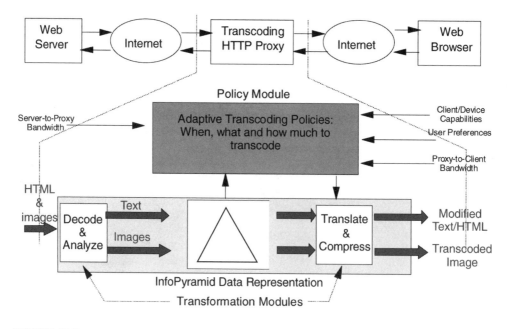

FIGURE 15.7
Internal architecture of a Web transcoding proxy.

proxy is constructed by integrating a transcoding subsystem into a HyperText Transfer Protocol (HTTP) proxy. The transcoding subsystem can be separated into two primary components: the transformation modules and the policy module [Han98]. The transformation modules modify the downstream data (i.e., HTML pages and GIF and JPEG images) that are being returned as responses to the client Web browser. The policy module makes the decision concerning what policy (i.e., the transcoding algorithm along with its parameters) to apply to the Web data. This policy module uses as input the following criteria: (1) the characteristics of the data (e.g., byte size of the images, current encoding efficiency, structural role in the HTML page) as determined by an InfoPyramid analysis of the arriving content; (2) the current estimate of the bandwidths on the server-to-proxy and proxy-to-client links; (3) the characteristics of the client that have been communicated back to the proxy, particularly the client's input/output (I/O) capabilities and hardware constraints (effective processing speed, memory, power, etc.); and (4) the user's preferences that have been communicated back to the proxy, particularly the desired delay or received quality.

The policy module generates a set of transcoding parameters, or transcoding vector, that controls the extent and types of compression performed by the transformation modules. For example, transcoding an image requires multiple parameters to be specified: how much to scale/downsample; whether to quantize to 24-bit color, 8-bit colormapped, 8-bit grayscale, or monochrome; how to build the quantization table and in which domain—pixel or frequency; whether to quantize via a JPEG image's quality factor Q [Chandra99], etc. Given N parametrizable compression options, then the N-tuple space of possible combinations becomes quite large, thereby increasing the difficulty of choosing the optimal parameter settings. However, if a cost is assigned to each N-tuple, then the resulting well-behaved cost function can be analytically optimized to find the best transcoding policy, an approach that is described later in Section 15.5.

User preferences can be expressed to the proxy's policy module in a variety of ways. Figure 15.8 demonstrates one example in which the level of a slide bar is used to express the tradeoff between quality and delay desired by the user. The "user preference" slide bar is a means for the user to interact with the transcoding proxy to dynamically control the tradeoff between image quality and download time. The user sets a level or index value on the slide bar, which is then fed back to the transcoding proxy's policy module. Ultimately, this index value maps onto a set of transcoding parameters that are passed to the data transformation modules. If the policy module interprets the index value to mean that the user is fixing the quality of the image at the specified level, then an image will be transcoded according to a fixed set of transcoding parameters associated with this level. If instead the policy module interprets the level of the slide bar to set a fixed delay, that is, specifying an upper bound on the tolerable delay, then the policy module has some freedom to dynamically adjust the set of transcoding parameters used to manipulate

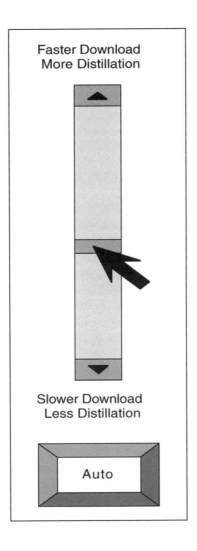

FIGURE 15.8
User preferences in terms of quality and response time are specified using a slide bar and are then communicated to the proxy.

an image, within the constraints imposed by the upper bound on delay. This latter interpretation has been named *automated transcoding* [Han98].

The architecture of the transformation subsystem is best revealed by inspecting an image conversion library. In such a library, the input compressed image is first decompressed, then functions such as rotation, scaling, and colorspace conversion and cropping are applied to the decompressed image, followed finally by compression into the new output format. If there are N image processing standards to be supported, and each format needs to be converted into every other format, then in the worst case a transcoder may need to support N^2 separate transformational modules. However, a typical strategy that reduces the required number of conversion modules is to devise a common intermediate format. Each decoder will decompress an image into the common intermediate format. Each encoder will compress an image from the common intermediate. The result is that only $2N$ modules are required, N for encoding and N for decoding. In practice, there can be a small number of intermediate formats. For example, an image conversion package may support four intermediate formats: 24-bit Red Green Blue (RGB), 8-bit colormap, 8-bit grayscale, and 1-bit monochrome.

15.4   TO TRANSCODE OR NOT TO TRANSCODE

The next two sections develop an analytic framework for evaluating transcoding policies. The goal is to determine under what combination of networking and proxy loading conditions it is beneficial to transcode. Both store-and-forward and streamed proxy-based architectures are analyzed [Han98, Han99].

15.4.1   A Store-and-Forward Image Transcoding Proxy

For the specific case of image transcoding within in a store-and-forward proxy, the goal is to determine under what conditions transcoding reduces response time.

A store-and-forward image transcoder is defined as an image transcoder that must wait to accumulate an entire input image before transcoding can begin on this image and then must wait to generate a transcoded image in its entirety before it is made available to be output, that is, the input image cannot be read partially nor can the output image be written partially by the transcoder. Store-and-forward proxies are a logical first step in analysis, because they are conceptually simple to model. Moreover, many standalone image-processing packages are essentially store-and-forward transcoders.

Consider the model outlined in Figure 15.9 of standard Web image retrieval in the absence of a proxy. Let T_o denote the download time of sending a Web object of size S (bits) from the Web server to the Web client in the absence of transcoding. We assume in our comparative analysis that the upstream delay to send an HTTP GET request is the same both with and without a proxy, and therefore focus only on the difference in downstream image download delays between having and not having a proxy.

The network can be conceptually divided into two segments based on the proposed insertion point of the proxy. Define the bandwidth or overall bit transmission rate from the Web server to the proposed proxy insertion point as B_{sp}, and similarly define the bandwidth from the proposed insertion point to the Web client as B_{pc}. For the purposes of the following discussion, caching is assumed to not be supported at the proxy.

The download time T_o from Web server to Web client in the absence of transcoding consists of the sum of three terms. First, D_{prop} is the propagation latency from the server to the client, that is, the time required for the first bit of the image to propagate from the server to the

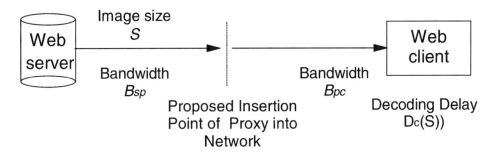

FIGURE 15.9
Model of standard Web image retrieval in the absence of a transcoding proxy.

client. Second, a Web image incurs a transmission delay equal to the spread in time between arrival of its first and last bits. Let $\min(B_{pc}, B_{sp})$ denote the bottleneck bandwidth between the client and the server. In the absence of a proxy, the first and last bits of an image will be spread in time by $S/\min(B_{pc}, B_{sp})$. This spread corresponds to the effective transmission time of the image over the concatenated server-to-proxy-to-client connection.

A third component of the download time is the delay $D_c(S)$ incurred by decoding and displaying an image on the mobile client. This factor models the Web client, for example, a Web-enabled PDA, as a black box, and measures only the cumulative delay, or equivalently the "effective" processing speed of the mobile client. Both the raw CPU speed as well as measured software inefficiencies in the Web browsing application and operation system are accounted for by $D_c(S)$. At times, this factor can introduce substantial latency and can therefore affect the image transcoding decision. Consequently, the overall image download time in the absence of a transcoding proxy can be expressed as:

$$T_0 = D_{prop} + \frac{S}{\min(B_{pc}, B_{sp})} + D_c(S) \tag{1}$$

Next, a store-and-forward transcoding proxy is inserted between the Web server and Web client, as shown in Figure 15.10. Let T_p denote the download time of sending the same Web image from the server through a store-and-forward transcoding proxy and then onward to the Web client. T_p consists of the sum of five terms. First, the server-to-client propagation latency D_{prop} experienced by the image is the same as defined in the preceding discussion, given the

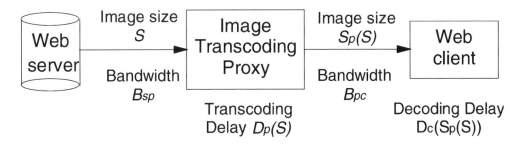

FIGURE 15.10
Model of store-and-forward image transcoding proxy.

same network transmission path. Second, the image download time from the server to the proxy is given by S/B_{sp}. Third, the proxy introduces a delay of $D_p(S)$, which is the time required to transcode the image. Fourth, the image download time from proxy to client is given by $S_p(S)/B_{pc}$. Finally, the decoding and display time of the transcoded image on the mobile client is given by $D_c(S_p(S))$. This factor is similar to $D_c(S)$, except that $D_c(S_p(S))$ measures the latency incurred by decoding and displaying the transcoded image rather than the original image. Consequently, the overall image download time through a store-and-forward transcoding proxy can be expressed as:

$$T_p = D_{prop} + D_p(S) + \frac{S}{B_{sp}} + \frac{S_p(S)}{B_{pc}} + D_c(S_p(S)) \qquad (2)$$

Transcoding will reduce response time only if $T_p < T_o$. That is,

$$D_p(S) + \frac{S}{B_{sp}} + \frac{S_p(S)}{B_{pc}} + D_c(S_p(S)) < \frac{S}{\min(B_{pc}, B_{sp})} + D_c(S), \text{ or}$$

$$D_p(S) + \frac{S}{B_{sp}} + \frac{S_p(S)}{B_{pc}} + [D_c(S_p(S)) - D_c(S)] < \frac{S}{\min(B_{pc}, B_{sp})} \qquad (3)$$

The above inequality precisely characterizes the conditions under which transcoding will reduce response time, and therefore is the key expression used by the transcoding proxy to determine whether each incoming image should be transcoded, how much compression is needed, and, indirectly, to what compression format an image should be transcoded. Except for S, the size of the original image that can be determined from the content-length header of the HTTP response message, the rest of the parameters in the above inequality need to be predicted when the image arrives at the proxy and before initiating transcoding.

Prior work has shown that predicting $D_p(S)$ and $S_p(S)$ is a complex task that depends upon the statistical distribution of images to be transcoded (e.g., sample of images from popular web sites), the efficiency of the software implementation of image compression and decompression algorithms, the set of transcoding parameters used to compress images, as well as the hardware/OS platform on which the transcoding is executed [Han98]. It was shown, for example, that JPEG-to-JPEG transcoding times and output sizes are a roughly linear function of the area of the image. For this specific case, given the dimensions of a JPEG image, it should then be possible to predict both $D_p(S)$ and $S_p(S)$, and thereby accurately evaluate Inequality 3.

Several special cases arise from Inequality 3. First, suppose $B_{pc} > B_{sp}$, that is, the Internet backbone's server-proxy connection, is the bottleneck. In this case, Inequality 3 reduces to

$$D_p(S) + \frac{S_p(S)}{B_{pc}} + D_c(S_p(S)) < D_c(S) \qquad (4)$$

If the effective processing speed of the mobile client is slow enough, that is, the right-hand side $D_c(S)$ of Inequality 4 is large enough, then there is sufficient margin for transcoding to be of some use; there is sufficient freedom to adjust the left-hand-side parameters $D_p(S)$, $S_p(S)$, and $D_c(S_p(S))$. However, as mobile clients become faster due to inevitable software and hardware improvements, $D_c(S_p(S))$ converges toward $D_c(S)$ and it becomes difficult to adjust the left-hand-side parameters without degrading the image beyond the bounds of intelligibility.

Second, consider the more typical case when the proxy-client access link is the bottleneck, that is, $B_{pc} < B_{sp}$. In this case, store-and-forward proxy-based transcoding is useful only if

$$D_p(S) + \frac{S_p(S)}{B_{pc}} + D_c(S_p(S)) < D_c(S) + \frac{S}{B_{pc}} - \frac{S}{B_{sp}} \tag{5}$$

Finally, let us consider the third case in which a transcoding is placed directly adjacent to a Web server, rather than at some intermediate node in the network. This situation is neatly summarized by setting $B_{sp} = \infty$. Consequently, the image transcoding decision's inequality reduces to the following:

$$D_p(S) + \frac{S_p(S)}{B_{pc}} + D_c(S_p(S)) < \frac{S}{B_{pc}} + D_c(S) \tag{6}$$

15.4.2 A Streamed Image Transcoding Proxy

For the specific case of image transcoding within in a streamed proxy, the goal is to determine under what conditions transcoding both avoids buffer overflow as well as reduces response time. The streamed proxy must manage any impedance mismatches between flow rates into and out of each buffer to avoid overflow. The following analysis focuses on streamed transcoding of still images, though the analysis is generalizable to streamed audio and video. A streamed transcoding proxy can output a partial image even before the entire image has been read into the transcoder. This gives streamed transcoding an advantage in terms of response time over store-and-forward transcoding, since the transcoding delay $D_p(S)$ is virtually nonexistent, or at least very small. The analysis of delay is deferred to first focus on the problem of avoiding buffer overflow with streamed proxies.

If the input is modeled as a fixed-rate stream of bits, then the transcoder takes a small group of G input bits (from a total image of size S) and transcodes them to a small group G_p of output bits (eventually producing a total transcoded image of size $S_p(S)$), where $G < S$, and $G_p < S_p(S)$. We explore variable rate analysis later on. As shown in Figure 15.11, the streamed proxy system is modeled as having three buffers of relevance: a proxy input buffer, a proxy output buffer, and a mobile client input buffer.

To avoid overflowing the input buffer between the arriving bits and the proxy, each input group must be processed before the next input group arrives, that is, groups must be transcoded/emptied out of the input buffer at a rate faster than they enter. (This strict definition of

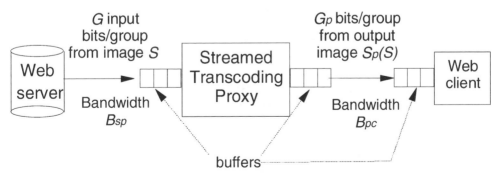

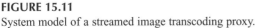

FIGURE 15.11
System model of a streamed image transcoding proxy.

overflow prohibits overflow at any time, but less stringent conditions can apply.) Equivalently, the group transcoding delay must be less than the interarrival times between groups, namely, $D_p(S)/S/G < G/B_{sp}$, or

$$D_p(S) < \frac{S}{B_{sp}} \qquad (7)$$

To avoid overflowing the buffer between the proxy and the proxy-client transmission link, the transcoded output image group size G_p must be transmitted over the proxy-client link at a rate faster than output groups are produced by the transcoder. Equivalently, the time to transmit each output group must be less than the interarrival time between output groups. Assuming that Inequality 7 is satisfied, then the interarrival time between output groups is the same as the interarrival time between input groups. Therefore, avoiding buffer overflow requires $G_p/B_{pc} < G/B_{sp}$, or

$$\gamma > \frac{B_{sp}}{B_{pc}} \qquad (8)$$

where γ = group image compression ratio G/G_p, which is assumed to be on average equivalent to the overall image compression ratio.

At the client, avoiding buffer overflow requires that the mobile client decode and display transcoded image groups faster than transcoded groups arrive at the client. Equivalently, the time to decode and display a proxy group G_p should be less than the time to generate a proxy group G_p. Assuming Inequality 8 holds, the spacing between groups is preserved through the proxy, that is, the interarrival times of proxy groups are the same as the interarrival times of groups at the input to the proxy. Therefore, $D_c(S_p(S))/S_p(S)/Gp < G/B_{sp}$, or rearranging terms,

$$D_c(S_p(S)) < \gamma \cdot \frac{S_p(S)}{B_{sp}}, \text{ or } D_c(S_p(S)) < \frac{S}{B_{sp}} \qquad (9)$$

If the mobile client is especially slow, then Inequality 9 tells us that even if the streamed transcoding proxy satisfies its buffer overflow requirements, the downstream mobile client will not be able to process the input groups fast enough to avoid buffer overflow. In this case, no transcoding should be performed.

In summary, the streamed image transcoder should only perform transcoding from a buffer overflow standpoint when Inequalities 7, 8, and 9 are satisfied.

If the server-proxy link is the bottleneck, that is, $B_{sp} < B_{pc}$, then Inequality 8 reduces to $\gamma > N$, where N is a number less than 1. Normally, the compression ratio is always greater than 1, so Inequality 7 will always be satisfied. Hence, only Inequalities 7 and 9 must be satisfied for transcoding to not be disadvantageous. In fact, when the server-proxy link is the bottleneck, Inequality 8 could be interpreted as providing an upper bound on the ratio of expansion allowed for a transcoded image, namely, $1/\gamma < B_{pc}/B_{sp}$. Expansion of an image may occasionally be necessary when format conversion is mandatory, for example, the mobile client only supports one image decoding format. The above inequality allows us to determine when such format conversion will increase the chances of buffer overflow, and when format conversion will not cause buffer overflow. For example, if $B_{sp} = 1$ bps, $B_{pc} = 2$ bit/s, and $G = 1$ bit, then Inequality 8 says that the output group G_p can expand to a maximum of 2 bits.

If the proxy-client link is the bottleneck, that is $B_{sp} > B_{pc}$, then Inequality 8 says that the image compression ratio γ must be greater than the ratio of server-proxy to proxy-client bandwidths in order for transcoding to be worthwhile. In addition, Inequalities 7 and 9 must still be satisfied.

Note that Inequalities 7, 8, and 9 are tight bounds that assume that the buffer must never be allowed to overflow. Looser constraints may be derived given that images are of finite length, rather than the continuous stream assumed in the analysis. More relaxed constraints would permit more time for transcoding or allow less aggressive compression.

Returning to the topic of delay in streamed transcoding, intuitively each streamed transcoder introduces only a small delay offset, corresponding to the initial group of bits that are processed. All subsequent bits in the stream are translated in time only by the initial small delay offset. Since this offset is often negligible compared to the length of the stream, then essentially streamed transcoding introduces negligible delay. For example, suppose a video sequence lasts 10 minutes in length, and suppose that a streamed video transcoder takes three frame times to transcode the initial frame. Since the streamed transcoder must process the video frames at the input frame rate (e.g., 30 frames/sec) to strictly avoid overflowing its buffers, then the transcoded frames will arrive at the receiver at the same rate as the input frame rate. The last frame will be delivered to the end user only three frame times later than if there were no transcoder in the path. The overall time to send the video sequence with a streamed transcoder would be 10 minutes plus a negligible offset lag of three frame times (e.g., ~100 ms).

A more analytical inspection of latency in streamed proxies is derived next. Assuming that Inequalities 7, 8, and 9 are enforced, then a key consequence is that the input arrival rate of image groups is preserved at each buffer throughout the path. As a result, the spread between the first and last image bits is always S/B_{sp}. The next step is to examine the delay introduced for the first image bit. The first bit accumulates the same propagation delay D_{prop} as in the previous section. In addition, there is a small component of transcoding delay D_G introduced by the streamed proxy due to processing of the first group of G bits in the stream. Finally, the mobile client also introduces a small delay D_{Gp} while processing the first group of G_p bits that it receives in the stream. The overall download time T'_p for a user in the streaming case will be given by

$$T'_p = D_{prop} + \frac{S}{B_{sp}} + D_G + D_{Gp} \tag{10}$$

Similarly, the revised download time T'_0 without a proxy, given a streamed client that introduces a small delay D'_G while processing the first group of G bits at the client, would be $T'_0 = D_{prop} + S/\min(B_{pc}, B_{sp}) + D'_G$. Streamed transcoding will reduce response time when $T'_p < T'_0$, namely,

$$\frac{S}{B_{sp}} + D_G + D_{Gp} < \frac{S}{\min(B_{pc}, B_{sp})} + D_G \tag{11}$$

Since the quantities D_G, D'_G, and D_{Gp} are typically very small, Inequality 11 can be approximated with the following inequality:

$$\frac{S}{B_{sp}} < \frac{S}{\min(B_{pc}, B_{sp})} \tag{12}$$

From Inequality 12, it is tempting to conclude that streamed transcoding either reduces response time or never adds to the response time. However, in reality, if $B_{sp} < B_{pc}$, then streamed transcoding always increases response time, because the last frame of video will arrive at a small offset delay $(D_G + D_{Gp} - D'_G)$, components ignored by Inequality 12, later than if it would have arrived without streamed transcoding. The derivation of Inequality 12 was meant to emphasize that streamed proxies ideally introduce negligible delay when compared to the length of the audio/video sequence being transcoded. If $B_{sp} > B_{pc}$, then an untranscoded stream will become backed up at the proxy-client bottleneck, necessitating a large buffer in the network and incurring a large delay. In this case, streamed transcoding is in a position to reduce the delay because it can compress the bit stream sufficiently so that the stream doesn't back up and delay doesn't become dominated by the proxy-client access link.

The earlier derivations of Inequalities 7, 8, and 9 assumed a constant bit-rate source. Often, video sources such as MPEG encoders generate a variable bit-rate stream. These inequalities can be modified to accommodate variable bit-rate sources. Let the stream into the proxy consist of a sequence of variably sized groups of bits $G(t)$, the stream out of the proxy consist of a sequence of variably sized groups of bits $G_p(t)$ ($G_p(t)$ results from $G(t)$), and the current bandwidth vary with time $B_{sp}(t)$ and $B_{pc}(t)$. Let $G(t+)$ be the next input group arriving at the proxy while the current group $G(t)$ is being transcoded. To avoid buffer overflow for each variably sized group, the delay to transcode the current group $D_p(G(t))$ must be less than the interarrival time $G(t+)/B_{sp}(t)$ between $G(t+)$ and $G(t)$. Following this reasoning, each of the Inequalities 7, 8, and 9 are modified to become the inequalities

$$D_p(G(t)) < \frac{G(t+)}{B_{sp}(t)}, \quad \gamma(t) > \frac{B_{sp}(t)}{B_{pc}(t)}, \text{ and } \quad D_c(G_p(t)) < \frac{G(t+)}{B_{sp}(t)} \qquad (13)$$

where $\gamma(t) = G(t+)/G_p(t)$. These conditions strictly prohibit instantaneous overflow, that is, no buffer at any time is allowed to violate Inequality 13, because otherwise the buffers at the proxy and receiver would eventually overflow given an infinite stream. In practice, the conditions can be relaxed to permit occasional overflow, since it is quite possible that a streamed transcoder that cannot finish transcoding group A before the next group of bits B arrives may be able to "catch up" by finishing group A, then quickly finishing the transcoding of group B before group C arrives. Consequently, by the time group C arrives, the streamed transcoder's buffer will be emptied of overflow bits. Prior work has shown that buffering can be reduced for variable-rate video transcoding by leveraging off of some of the low-complexity transcoding techniques discussed in Section 15.2 [Kan98]. Another interpretation of buffering in video transcoders places a buffer within the transcoder, between the decoder and encoder, rather than at the input and outputs of the transcoder, as shown in Figure 15.11, for the purposes of reducing buffering delay [Morrison97].

15.5 TRANSCODING POLICIES FOR SELECTING CONTENT

The analysis of the previous section was confined to the limiting assumptions of proxy-based architectures, a limited set of input parameters (network bandwidths, predicted delays, and predicted output sizes), and a single media type (images). The general decision of transcoding seeks to answer the broader question: given many multimedia objects in a Web page, which objects should be removed, which objects should be compressed, and to what extent should these objects be summarized? The general set of transcoding policies used for selecting and transcoding content therefore must consider the following: both server-side authoring issues as well as proxy-based architectures; a larger set of input factors that influence the transcoding decision (e.g., detecting the purpose and importance of a media object from analysis of its content, position within a page, relation/references to other objects, etc.); and multiple media types (text, audio, and video) in addition to images. The next two sections explore the complexities of developing general-content selection policies for offline transcoding as well as real-time proxy-based transcoding.

15.5.1 Optimal Policies for Offline Pretranscoding

In offline pretranscoding systems, the content server uses the InfoPyramid to manage and select the different versions of the media objects. When a client device requests a multimedia document, the server selects and delivers the most appropriate version of each of the media objects. The selection can be made on the basis of the capabilities of the client devices, such as display size, display color depth, network bandwidth, and client storage.

The selection of the different versions of media objects within a multimedia document can be optimized by maximizing the total content value given the constraints of the client devices. Define a multimedia presentation $M = [D, L]$ as a tuple consisting of a multimedia document D and a document layout L. Define the multimedia document D as the set of media objects O_{ij}, as follows:

$$D = \{(O_{ijn})\} \qquad \text{Multimedia document} \qquad (14)$$

where (O_{ijn}) gives the nth media object, which has modality i and fidelity j. The document layout L gives the relative spatial and temporal location and size of each media object. Define an InfoPyramid IP of a media object as a collection of the different versions of the media object Oij, as follows:

$$IP = \{O_{ij}\} \qquad \text{InfoPyramid} \qquad (15)$$

Define an InfoPyramid document IPD as a set of InfoPyramid objects $\{O_{ij}\}$, as follows:

$$IPD = \{IP_n\} = \{O_{ij}\}_n \qquad \text{InfoPyramid document} \qquad (16)$$

To optimize the selection, the InfoPyramid uses content value scores $V(O_{ij})$ for each of the media objects O_{ij}, as shown in Figure 15.12. The content value scores can be based on automatic measures, such as entropy, or loss in fidelity that results from translating or summarizing the content. For example, the content value scores can be linked to the distortion introduced from

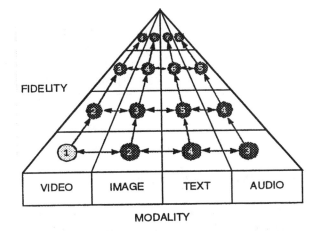

FIGURE 15.12
Example of the reciprocal content value scores assigned for different media object versions of a video in the InfoPyramid.

compressing the images or audio. Otherwise, the content value scores can be tied directly to the methods that manipulate the content, or be assigned manually.

Figure 15.12 illustrates examples of the relative reciprocal content value scores of different versions of a video object. In this example, the original video (lower left) has the highest content value. The manipulation of the video along the dimension of fidelity or modality reduces the content value. For example, converting the video to a sequence of images results in a small reduction in content value. Converting the video to a highly compressed audio track produces a higher reduction in the content value.

Given a multimedia document with media N objects, let $\{O_{ij}\}_n$ give the InfoPyramid of the nth media object. Let $V((O_{ij})_n)$ give the relative content value score of the version of the nth media object with modality i and fidelity j, and let $D((O_{ij})_n)$ give its data size. Let D_T give the total maximum data size allocated for the multimedia document by the client device. The total maximum data size may, in practice, be derived from the user's specified maximum tolerable delay and the network conditions, or from the device constraints in storage or processing.

The content selection process selects media object versions O_{ij}^* from each InfoPyramid to maximize the total content value for a target data size D_T, as follows:

$$\sum_n V((O_{ij}^*)_n) = \max(\sum_n V((O_{ij})_n))$$

$$\sum_n D((O_{ij}^*)_n) \leq D_T \tag{17}$$

where $(O_{ij})_n$ gives for each n, the optimal version of the media object, which has fidelity i and modality j.

Alternatively, given a minimum acceptable total content value V_T, the content select process selects media object versions O_{ij}^* from each InfoPyramid to minimize the total data size, as follows:

$$\sum_n D((O_{ij}^*)_n) = \min(\sum_n D((O_{ij})_n))$$

$$\sum_n V((O_{ij}^*)_n) \geq V_T \tag{18}$$

where, as above, $(O_{ij}^*)_n$ gives for each n, the optimal version of the media object, which has fidelity i and modality j.

By extending the selection process, other constraints of the client devices can be considered. For example, the content selection system can incorporate device screen size S_T, as follows: let $S((O_{ij})_n)$ give the spatial size of the version of the nth media object with modality i and fidelity j. Then, add the constraint

$$\sum_n S((O_{ij}^*)_n) \leq S_T \tag{19}$$

to the optimization process. In the same way, additional device constraints such as color depth, streaming bandwidth, and processing power can be included.

Using the InfoPyramid for managing the versions of the media objects, the number of different versions of each multimedia document is combinatorial in the number of media objects (N) and number of versions of each media object (M), and is given by M^N. To solve the optimization problems of Equations 17 and 18, the constrained optimization problems can be converted into the equivalent Lagrangian unconstrained problems, as described in [Mohan99].

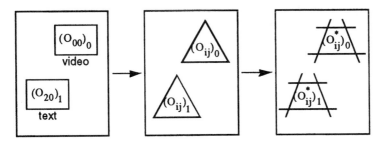

FIGURE 15.13

Example of content selection for a multimedia document D consisting of two media objects: (a) $(O_{00})_0$ (video) and $(O_{20})_1$ (text), (b) InfoPyramid document with $(O_{ij})_0$ and $(O_{ij})_1$, and (c) selected versions of the media objects $(O_{ij}^*)_0$ and $(O_{ij}^*)_1$.

The optimization solution is based on the resource allocation technique proposed in [Shoham88] for arbitrary discrete functions. This is illustrated by converting the problem in Equation 18 into the following unconstrained problem:

$$\min\{\Sigma_n D((O_{ij})_n) - \lambda(V_T - V((O_{ij})_n))\} \tag{20}$$

The optimal solution gives that for all n, the selected versions of the media objects $(O_{ij}^*)_n$ operate at the same constant tradeoff λ in content value $V((O_{ij})_n)$ versus data size $D((O_{ij})_n)$. To solve the optimization problem, a search is conducted over values of λ.

The content selection for an example multimedia document is shown in Figure 15.13. The multimedia document has two media objects: a video object $= (O_{00})_0$ and a text object $= (O_{20})_1$.

For each media object $(O_{ij})_n$ where $n \in \{0,1\}$, an InfoPyramid $\{O_{ij}\}_n$ is constructed, which gives the different versions of the media object. The selection process selects the versions $(O_{ij}^*)_0$ and $(O_{ij}^*)_1$, respectively, to maximize the total content value.

Consider four versions of each media object with content values and data sizes given in Table 15.2. By iterating over values for the tradeoff λ in content value and data size, the content selection table of Table 15.3 is obtained, which shows the media object versions that maximize the total content value $\max(\Sigma_n V((O_{ij})_n))$ for different total maximum data sizes D_T, as given in Equation 18.

Table 15.2   Summary of Different Versions of Two Media Objects, $(O_{ij})_0$ and $(O_{ij})_1$

$(O_{ij})_n$	$V((O_{ij})_n$	$D((O_{ij})_n)$	Modality
$(O_{00})_0$	0.75	0.25	Video
$(O_{10})_0$	0.5	0.1	Image
$(O_{11})_0$	0.25	0.05	Image
$(O_{20})_1$	1	0.5	Text
$(O_{21})_1$	0.5	0.1	Text
$(O_{32})_1$	0.75	0.25	Audio
$(O_{33})_1$	0.25	0.05	Audio

Table 15.3 Summary of the Selected Versions of the Two Media Objects $(O_{ij})_0$ and $(O_{ij})_1$ Under Different Total Maximum Data Size Constraints D_T

D_T	$(O_{ij}^*)_0$	$(O_{ij}^*)_1$	$\Sigma_n V((O_{ij}^*)_n)$	$\Sigma_n D((O_{ij}^*)_n)$
1.5	$(O_{00})_0$	$(O_{20})_1$	2	1.5
1.25	$(O_{00})_0$	$(O_{32})_1$	1.75	1.25
1	$(O_{01})_0$	$(O_{20})_1$	1.75	0.75
0.6	$(O_{10})_0$	$(O_{20})_1$	1.5	0.6
0.35	$(O_{01})_0$	$(O_{21})_1$	1.25	0.35
0.35	$(O_{10})_0$	$(O_{32})_1$	1.25	0.35
0.2	$(O_{10})_0$	$(O_{21})_1$	1	0.2
0.1	$(O_{11})_0$	$(O_{32})_1$	0.5	0.1

15.5.2   Policies for Real-Time Transcoding

The second form of scalable multimedia delivery to pervasive devices is real-time transcoding. Adaptive selection discussed in the previous section is most appropriate when the InfoPyramid is used to store and manage different versions of the media objects. However, in many cases, the multimedia content is already stored in legacy formats and is served by traditional, nonadaptive content servers. For example, this is usually the case for multimedia documents on the Web. One way to solve this problem is by using active proxies for transcoding the content on-the-fly to adapt it to client devices. For real-time transcoding, the InfoPyramid is used as a transient structure in transcoding the media objects to the most appropriate modalities and fidelities.

There are many methods for image transcoding, including methods for image size reduction, image color reduction, image compression, and format conversion. Image transcoding functions are needed that perform both image manipulation and image analysis. The objective of image analysis is to obtain information about the images that improves the transcoding.

Image content analysis is motivated by the need to differentiate the selection of the transcoding methods based on the input image. Content analysis can be important for developing high quality efficient transcoders. Image content analysis permits transcoding of images depending on their individual content characteristics.

For example, it is desirable to transcode graphics and photographs differently with regard to size reduction, color reduction, and quality reduction. To classify the images, the image analysis procedures classify the images into image type and purpose classes, as described in [Smith98a]. Consider the following image type classes:

$$T = BWG, BWP, GRG, GRP, SCG, CCG, CP, \tag{21}$$

where BWG = black-and-white (b/w) graphics, BWP = b/w photographs, GRG = gray graphics, GRP = gray photographs, SCG = simple color graphics, CCG = complex color graphics, and CP = color photographs. Consider the following image purpose classes:

$$P = ADV, DEC, BUL, RUL, MAP, INF, NAV, CON, \tag{22}$$

where ADV = advertisements, DEC = decorations, BUL = bullets, RUL = rules, MAP = maps, INF = informational images, NAV = navigation images, and CON = content-related images [Paek98].

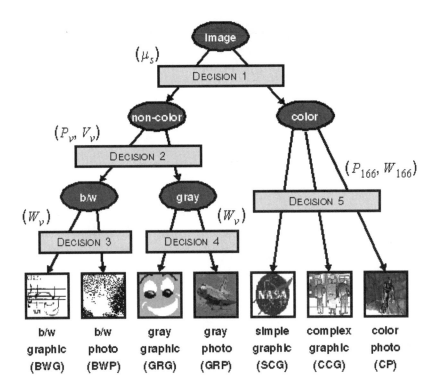

FIGURE 15.14
Image type decision tree consisting of five decision points for classifying the images into image type classes.

The image type classification system extracts color features of the images and utilizes a decision tree classifier, as described in [Smith98a]. The decision tree classifies the images along the dimensions of color content (color, gray, b/w), and source (photographs, graphics). Distinguishing between b/w, gray, and color requires image analysis because of artifacts introduced in the image production, and compression often obfuscates the image type.

The image type decision tree consists of five decision points, each of which utilizes a set of features extracted from the images. Keeping in mind the need for real-time transcoding, the features are extracted only as needed for the tests to minimize processing. The image features are derived from several color and texture measures computed from the images. The classification parameters for these measures were obtained from a training set of 1282 images retrieved from the WebEach image $X[m, n]$, which has three color components corresponding to the Red Green Blue (RGB) color channels, as follows: $X_{rgb} = (x_r, x_g, x_b)$ where $x_r, x_g, x_b \in \{0, 255\}$. The decision tree performs the following tests for each image X.

Table 15.4. The Color vs. Noncolor Test Uses Mean Saturation per Pixel μ_s

	Test 1	#	$E(\mu_s)$	(μ_s)
	Noncolor	464	2	5.6
	Color	818	63	46.2

Table 15.5. The b/w vs. Gray Test Uses Intensity Entropy P_v and Variance V_v

Test 2	#	$E(P_v)$	$\sigma(P_v)$	$E(V_v)$	$\sigma(V_v)$
B/W	300	1.4	1.1	11,644	4993
Gray	164	4.8	2.1	4,196	2256

The first test distinguishes between color and noncolor images using the measure of the mean saturation per pixel μ_s. The saturation channel y_s of the image is computed from X from $y_s = \max(x_r, x_g, x_b) - \min(x_r, x_g, x_b)$. Then, $\mu_s = 1/MN \sum_{m,n} y_s[m, n]$ gives the mean saturation, where M, N are the image width and height, respectively. Table 15.4 shows the mean $E(\mu_s)$ and standard deviation $\sigma(\mu_s)$ of the saturation measure for the set of 1282 images. The mean saturation μ_s discriminates well between color and noncolor images, since the presence of color requires $\mu_s > 0$, while strictly noncolor images have $\mu_s = 0$. However, due to noise, a small number of saturated colors often appear in noncolor images. For example, for the 464 noncolor images, $E(\mu_s) = 2.0$.

The second test distinguishes between b/w and gray images using the entropy P_v and variance V_v of the intensity channel y_v. The intensity channel of the image is computed as from $y_v = 0.3x_r + 0.6x_g + 0.1x_b$. Then, the intensity entropy is given by $P_v = -\sum_{k=0}^{255} p[k]\log_2 p[k]$, where $P(k) = \dfrac{1}{MN} \sum_{m,n} \begin{cases} 1 & k = y_v[m, n] \\ 0 & \text{otherwise} \end{cases}$.

The intensity variance is given by $V_v = \dfrac{1}{MN}\sum_{m,n}(y_v[m, n] - \mu_v)^2$, where $\mu_v = \dfrac{1}{MN}\sum_{m,n} y_v[m, n]$.

Table 15.5 shows the statistics of P_v and V_v for 464 noncolor images. For b/w images the expected entropy P_v is low and the expected variance V_v is high. The reverse is true for gray images.

The third test distinguishes between b/w graphics and b/w photographs using the minimum of the mean number of intensity switches in horizontal and vertical scans of the image. The mean number of intensity switches in the horizontal direction μ_{sw}^h is defined by

$$\mu_{sw}^h = \frac{1}{MN} \sum_{m,n} \begin{cases} 1 & y_v[m - 1, n] \neq y_v[m, n] \\ 0 & \text{otherwise} \end{cases} \tag{23}$$

Table 15.6 The BWG vs. BWP Test Uses Intensity Switches W_v

Test 3	#	$E(W_v)$	$\sigma(W_v)$
BWG	90	0.09	0.07
BWP	210	0.47	0.14

Table 15.7 The GRG vs. GRP Uses (W_v) and Intensity Entropy P_v

Test 4	#	$E(W_v)$	$\sigma(W_v)$	$E(P_v)$	$\sigma(P_v)$
GRP	80	0.4	0.26	3.3	1.8
GRP	84	0.81	0.16	0.16	1.4

Table 15.8   The SCG vs. CCG vs. CP Test Uses Mean Saturation μ_s, HSV Entropy P_{166}, and HSV Switches W_{166}

Test 5	#	$E(W_v)$	$\sigma(W_v)$	$E(P_{166})$	$\sigma(P_{166})$	$E(W_{166})$	$\sigma(W_{166})$
SCG	492	69.7	50.8	2.1	0.8	0.24	0.16
CCG	116	71.2	46.2	3.1	1	0.36	0.16
CP	210	42.5	23.5	3.3	0.7	0.38	0.15

The vertical switches μ_{sw}^v are defined similarly from the transposed image y_v'. Then, the intensity switch measure is given by $W_v = \min(\mu_{sw}^h, \mu_{sw}^v)$.

The fourth test distinguishes between gray graphics and gray photographs using the intensity switch measure W_v and the intensity entropy P_v. Table 15.6 shows the mean $E(W_v)$ and standard deviation $\sigma(W_v)$ of the intensity switch measure for 300 b/w and 164 gray images. The switch measure distinguishes well between b/w graphics and photographs, since it typically has a much lower value for b/w graphics. The gray graphics are found to have a lower switch measure and lower entropy than the gray photographs.

The fifth test distinguishes between simple color graphics, complex color graphics, and color photographs. The images are transformed to HSV and vector quantized. The process generates a 166-HSV color representation of the image λ_{166}, where each pixel refers to an index in the HSV color lookup table.

The test uses the 166-HSV color entropy P_{166} and mean color switch per pixel W_{166} measures. In the computation of the 166-HSV color entropy, $p[k]$ gives the frequency of pixels with color index value k. The color switch measure is defined as in the test three measure, except that it is extracted from the 166-HSV color image y_{166}. The test also uses the measure of mean saturation per pixel μ_s. Table 15.8 shows the statistics for μ_s, P_{166}, and W_{166} for 818 color images. Color graphics have a higher expected saturation $E(\mu_s)$ than color photographs. But color photographs and complex color graphics have higher expected entropies $E(P_{166})$ and switch measures $E(W_{166})$ in the quantized HSV color space.

Web documents often contain information related to each image that can be used to infer information about them [Rowe97]. An image-purpose classification system can use this information in concert with the image type information to classify the images into image-purpose classes. The system can make use of five contexts for the images in the Web documents: C = BAK, INL, ISM, REF, and LIN, defined in terms of HTML code as follows:

- BAK: background, that is, <body backgr = ... >
- INL: inline, that is, <img src = ... >
- ISM: ismap, that is, <img src = ... ismap>
- REF: referenced, that is, <a href = ... >
- LIN: linked, i.e., <a href = ...><img src = ... ></a>

The system can also use a dictionary of terms extracted from the text related to the images. The terms are extracted from the "alt" tag text, the image Universal Resource Locator (URL) address strings, and the text nearby the images in the Web documents. The system can make use of terms such as D = "ad", "texture", "bullet", "map", "logo", and "icon". The system can also extract a number of image attributes, such as image width (w), height (h), and aspect ratio ($r = w/h$).

The system can classify the images into the purpose classes using a rule-based decision tree framework described in [Paek98]. The rules map the values for image type $t \in T$, context, $c \in C$,

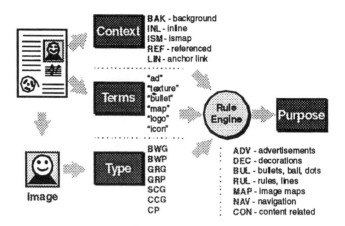

FIGURE 15.15
Purpose detection uses image-type information, multimedia document context, and related text to classify images into image-purpose classes.

terms d ∈ D, and image attributes a ∈ {w, h, r} into the purpose classes. The following examples illustrate some examples of the image-purpose rules:

- p = ADV←t = SCG, c = REF, d = "ad"
- p = DEC←c = BAK, d = "texture"
- p = MAP←t = SCG, c = ISM, w > 256, h > 256
- p = BUL←t = SCG, r > 0.9, r < 1.1, w < 12
- p = RUL←t = SCG, r > 20, h < 12
- p = INF←t = SCG, c = INL, h < 96, w < 96

To provide feedback about the embedded images for text browsers, the system can generate image summary information. The summary information contains the assigned image type and purpose, the Web document context, and related text. The system can use an image subject classification system that maps images into subject categories (s) using key terms (d), that is, d→s, which is described in [Smith97]. The summary information can be made available to the transcoding engine to allow the substitution of the image with text.

The system can transcode the images using a set of transcoding policies. The policies apply the appropriate transcoding functions for the constraints in delivery and display, processing, and storage of the client devices.

Referring to Figure 15.9, the transcoding system can provide a set of transcoding functions that manipulate the images along the dimensions of image size, fidelity, and color, and that substitute the images with text or HTML code. For example, the transcoding system can reduce the amount of data needed to represent the images and speed up download times. The transcoding system can also reduce the size of the images to fit the images onto the client display screens. The transcoder can also change the storage format of the image to gain compatibility with the image-handling methods of the client device. Some example transcoding functions include the following.

- *Size:* Minify, crop, and subsample. For example the full-resolution 256 × 256 image can be spatially reduced to generate a smaller 192 × 192 image.

- *Fidelity:* JPEG compress, GIF compress, quantize, reduce resolution, enhance edges, contrast stretch, histogram equalize, gamma correct, smooth, sharpen, and de-noise. For example the full-resolution image can be compressed, in addition to being spatially reduced, to further reduce the amount of data to 23KB.
- *Color content:* Reduce color, map to color table, convert to gray, convert to b/w, threshold, and dither. For example, the 24-bit RGB color image can undergo color reduction to generate an 8-bit RGB color image with only 256 colors. The image can undergo further color reduction to generate a 4-bit gray image with only 16 levels of gray. The image can undergo even further color reduction to generate a 1-bit b/w image. The color reduction can involve dithering to make the photograph look good in b/w.
- *Substitution:* Substitute attributes (*a*), text (*d*), type (*t*), purpose (*p*), and subject (*s*), and remove image. For example, the image can be replaced with the term *bridge*.

15.6   A SAMPLE SET OF TRANSCODING POLICIES

Previous sections developed an abstract analytical framework for determining the conditions under which it is beneficial to transcode for store-and-forward as well as streamed proxies. This section describes a set of practical transcoding policies that have been implemented in a store-and-forward image transcoding HTTP proxy. These policies are derived from much experimentation, but are also in part based on the abstract modeling from previous sections [Han98]. The implemented policies adapt the transcoding to the client's device type, user-quality preferences, and image content, but do not adapt to changing network bandwidth, nor do they perform prediction of the image transcoding delay and output byte size.

The sample set of transcoding policies is summarized in pseudocode in Figure 15.16. The transcoding proxy first makes a distinction in line 1 between transcoding to a laptop and transcoding to a handheld PDA. Unless otherwise notified, the proxy assumes that it is transcoding to a personal computer (PC) or laptop, namely, a client device that fully supports GIF and JPEG decoding.

Given a PC or laptop, the proxy next checks in line 2 to see if the image is sufficiently large for transcoding. Images smaller than a threshold of 1000 bytes are deemed to be not worth the savings in download time brought about by compression and hence are not transcoded. This threshold is obtained by applying Inequality 3's test condition. The following assumptions were made: B_{pc} and B_{sp} are fixed; the client decoding delays $D_c(S_p(S))$ and $D_c(S)$ are small; and the access link is the bottleneck, $B_{pc} < B_{sp}$. Rearranging terms in Inequality 3 results in the input byte size S having to satisfy the following inequality for transcoding to be beneficial:

$$S > \frac{D_p(S) + \dfrac{S_p(S)}{B_{pc}}}{\left(\dfrac{1}{B_{pc}} - \dfrac{1}{B_{sp}}\right)} \tag{24}$$

Let $B_{pc} = 50$ Kbit/s, $B_{sp} = 1$ Mbit/s. For small images of input length about 500 bytes, average transcoding times of approximately 40 ms have been measured. Assume that it is typically hard to squeeze significant compression out of small images without losing intelligibility, say no more than a factor of 2 to 1, so that input and output sizes are closely related, and that the transcoding time does not vary much within this small range. Consequently, producing $S_p(S) = 500$ output bytes starting from a small image will take about $D_p(S_p(S)) = 40$ ms of processing delay. When we substitute values, Inequality 24 shows that the input byte size S must exceed about 800 bytes for transcoding to reduce response time, hence the threshold of 1000 bytes.

If (not to Palm) /* e.g. to laptop/PC */   *line 1*
    If (input byte size > 1000 bytes) /* static evaluation of Ineq. 3 */   *2*
      If (input is GIF)   *3*
        if (well-compressed GIF)   *4*
          GIF->GIF as f(user preferences)   *5*
        else   *6*
          GIF->JPEG as f(user preferences)   *7*
      else /* input is JPEG */   *8*
        JPEG->JPEG as f(user preferences)   *9*
    If (output byte size > input byte size)   *10*
      send original image   *11*
    else   *12*
      send transcoded image   *13*
else /* to Palm */   *14*
  GIF/JPEG -> Palm 2-bit grayscale as f(user preferences)   *15*

FIGURE 15.16

A sample set of transcoding policies implemented in an image transcoding proxy.

Next, in line 3, the transcoding policy makes a distinction based on the input image format, such that input GIF images are transcoded differently than input JPEG images. If the input image is in GIF format, then a further check is made in line 4 to see if the image is already well compressed in GIF format. The GIF format is well suited to compressing graphics-type artificially rendered images that have very few colors, such as a map or a cartoon. GIF is poorly suited to compressing natural images with many shades of color, such as a photograph of a person. Conversely, JPEG is well suited to compressing natural images with many colors, and is poorly suited to compressing artificially rendered graphics that have embedded text and sharp edges. If the input GIF is already well compressed, then the image should be transcoded from GIF to GIF (line 5), because transcoding the image from GIF to JPEG will likely result in an expansion of the image byte size. Otherwise, if the input is not already well compressed in GIF format, then GIF should be transcoded to JPEG (line 7), assuming that the image has enough natural variation to make it suitable for JPEG encoding.

The test for determining whether the GIF is already well compressed is based on calculating the bits per pixel value bpp of the input image. Extracting the image dimensions height h and width w from the image header, and given the image file length S, then bpp = $S/h \cdot w$. Experimentation with a variety of input GIF images found that $bpp < 0.1$ was a good indicator of well-compressed GIF maps and logos from GIF images. GIF images with more color invariably exceeded this threshold value.

If the input image format is in JPEG format (line 9), then the transcoder performs JPEG-to-JPEG transcoding. The proxy never converts a JPEG to a GIF format, because a JPEG image typically has many colors, and will compress poorly in GIF format. In fact, GIF encoding of a decoded JPEG image will likely result in expansion rather than compression of the image. While it is possible to apply color quantization to reduce the number of colors to an acceptable level for GIF encoding, such quantization algorithms as median-cut are time-consuming or introduce too many quantization effects to be considered here.

In lines 5, 7, and 9, once the output format has been determined, the image is transcoded according to the user preferences (e.g., slide-bar value) received from the user. The same slider index value can be mapped onto different transcoding vectors depending on the output format. Transcoding to GIF format is limited to compression via scaling and colormap reduction. Transcoding to JPEG format expands the options for compression by allowing quantization of frequency coefficients, in addition to scaling. Occasionally, the choice of transcoding vector may actually result in expansion of the image. In this case, the original image is returned to the requesting client (line 11), rather than the transcoded image.

When transcoding to a Palm PDA, GIF images and JPEG images are transcoded to a custom output format, namely a 2-bit grayscale bitmap [Han98]. The degree of scaling that is applied to the bitmap is a function of the user's input preference (line 15). The HTTP Palm client informs the HTTP transcoding proxy of its characteristics by modifying the Accept field in the HTTP GET request to say "image/palm".

The set of transcoding policies described by Figure 15.16 represents but one sample among a constellation of possible policy rules, and are clearly subject to modification if the assumptions used in their development are changed. For example, it was assumed in the development of the policies above that the client's decoding delay could be ignored. Some mobile clients, such as handheld PDAs, can have limited processing speeds an order of magnitude or more slower than laptops, due to power, cost, and other considerations. In addition, software inefficiencies in the browsing application written for the PDA, such as inefficient image processing operations or excessive memory copying, can introduce substantial delay that is on the order of minutes [Han99]. A proxy that is aware of the "effective" processing speed of the PDA, including software inefficiencies and hardware speed, can implement a transcoding policy that minimizes the response time by minimizing the amount of processing and decoding performed on the mobile client. Prescaling of GIFs to fit a PDA's small screen size can dramatically reduce response times by a factor of 10 or more in comparison with performing all image processing on the handheld PDA. Most of the savings in delay here are attributed to reducing the number and complexity of the tasks that need to be executed on Palm handheld computers, rather than reducing the download time by sending fewer bits.

15.7   RELATED ISSUES

Caching and prefetching are commonly used techniques that, like compression-based transcoding, are employed to improve the response time for downloading a Web page. Caching and prefetching technology have been integrated with a transcoding proxy in a variety of projects [Loon97, Fox98a, Gessler95, Nakajima97, Fleming97, Liljeberg95]. In prefetching, a proxy parses each HTML page returned from the server before passing it back to the client. The proxy prefetches any inline images, or other important links that the proxy believes that the client will request next, and caches the prefetched data inside the proxy cache. Subsequent HTTP GET requests from the client are then quickly answered from the proxy's cache. In some instances, the transcoding proxy caches the transcoded version of a multimedia object, rather than the original [Fox98a]. An image-caching method for Internet proxies has been developed that reduces the resolution of infrequently accessed images to conserve storage space and bandwidth [Ortega97].

Several commercial systems apply transcoding technology to improve Internet access for pervasive devices. Intel's Quick Web [Quickweb], Spyglass's Prism [Spyglass], and ProxiNet [Proxinet] each employ a transcoding proxy to compress the Internet's multimedia content and speed up download time. In addition, the Palm VII PDA provides wireless access to the Web.

The Palm VII currently achieves wireless connectivity via the Mobitex wide-area wireless data networking standard. A "Web clipping" proxy provides a transcoding service that enables the PDA's user to view a selected subset of Web pages that have been transcoded by the intermediate proxy [PalmVII99]. On the horizon are Web-enabled cellular phones with microbrowsers capable of surfing the Web [Gardner99]. Such microbrowsers will likely speak the Wireless Application Protocol (WAP) Forum's version of hypertext called the Wireless Markup Language (WML) [WAPForum99]. A WAP gateway will reside in the network between the WML microbrowser and the HTML/XML back-end Web servers, transcoding non-WML Web pages into the corresponding WML approximation. The underlying motivation for WML was to develop a more efficient alternative to standard Web access protocols like HTTP, TCP, and HTML when wireless access from highly constrained mobile clients is considered. The open market will determine whether such a wireless-specific alternative will be successful.

In the area of content authoring, special markup languages, such as WML ([HDML97]), Synchronized Multimedia Integrated Language (SMIL) ([Bouthillier98]), and multimedia formats such as MPEG-4 ([Avaro97]), are being developed with pervasive devices or content adaptation in mind. At present, WML is suited for developing text-only applications for devices such as pagers and cellular phones. SMIL provides a language for creating synchronized multimedia presentations. SMIL supports client adaptation by providing switch statements, such as for high and low bandwidth. By representing video streams by collections of individual and independent media objects, MPEG-4 allows for better adaptation of the delivery of video to different bandwidth conditions.

In many multimedia presentations, maintaining synchronization among the individual media objects is important. Synchronization presents a problem when the media objects are distributed or need to be transcoded. Candan et al. [Candan96], and Vogel et al. [Vogel94] investigated synchronized presentation issues for distributed multimedia documents. Candan et al. included modeling parameters for maintaining synchronization under transcoding, such as image conversion between JPEG and GIF formats. Methods have been investigated for adapting multimedia presentations to client devices by optimizing media object selection for the device constraints [Mohan98]. Weitzman et al. investigated grammars for dynamic presentation and layout of multimedia documents under different presentation conditions [Weitzman94].

Universal access is an important functionality of digital libraries [Smith99a, Kling94]. Since digital libraries are designed to serve large numbers of diverse users, they need to accommodate the special needs of the users, constraints of the client devices, and bandwidth conditions. In some cases, the users, such as those that are hearing or sight impaired, need speech-only or text-only access to the digital library. To provide universal access, the digital libraries need to either store and deliver different versions of the content or perform media conversion before content delivery. Recently, proposals have been made to MPEG-7 to establish standard meta-data to assist in media conversion for universal access [Smith99b, Smith99c, Christopoulos99].

There is continuing standards activity in the area of communicating hints from the user to the transcoding proxy as well as from the client to the transcoding proxy. A proposal to the W3C called Composite Capability/Preferences Profiles (CC/PP) describes a framework for clients to specify their device capabilities to the proxy [Reynolds98]. A second proposal to the W3C suggests means of annotating Web pages to embed in the hypertext hints to the proxy concerning the importance or priority of various multimedia objects in the page [Hori99].

Several recent projects have focused on improving Internet access for pervasive devices (e.g., see [Bickmore97], [Vanderheiden97], [Bharadvaj98], [Fox98a]. Fox et al. developed a system for compressing Internet content at a proxy to deal with client variability and improve end-to-end performance [Fox96b]. Other methods have been investigated for real-time transcoding images [Smith98b] and general Internet content [Li98]. Scalability of multiple transcoding proxies has been studied [Fox96b].

15.8 ACKNOWLEDGMENTS

We wish to thank David Messerschmitt as well as the MobiCom 98 panel for their insights regarding proxy versus end-to-end issues.

15.9 REFERENCES

[Acharya98] S. Acharya and B. Smith, "Compressed Domain Transcoding of MPEG," *IEEE International Conference on Multimedia Computing and Systems,* 1998, pp. 295–304.

[Amir95] E. Amir, S. McCanne, and H. Zhang, "An Application-Level Video Gateway," *ACM Multimedia,* 1995, pp. 255–.

[Assuncao98] P. Assuncao and M. Ghanbari, "A Frequency Domain Video Transcoder for Dynamic Bit Rate Reduction of MPEG 2 Bit Streams," *IEEE Transactions on Circuits Systems Video Technology,* Vol. 8, no. 8, December 1998, pp. 953–967.

[Avaro97] O. Avaro, P. Chou, A. Eleftheriadis, C. Herpel, and C. Reader, "The MPEG-4 Systems and Description Languages: A Way Ahead in Audio Visual Information Representation," *Signal Processing: Image Communication, Special Issue on MPEG-4,* Vol. 4, no. 9, May 1997, 385–431.

[Bartlett94] J. Bartlett, "W4—the Wireless World Wide Web," *IEEE Workshop on Mobile Computing Systems and Applications,* 1994, pp. 176–178.

[Bharadvaj98] H. Bharadvaj, A. Joshi, and S. Auephanwiriyakul, "An Active Transcoding Proxy to Support Mobile Web Access," *IEEE Symposium on Reliable Distributed Systems,* 1998, pp. 118–123.

[Bickmore97] T. Bickmore and B. Schilit, "Digestor: Device-Independent Access to the World Wide Web," Sixth International World Wide Web Conference, *Computer Networks and ISDN Systems,* Vol. 29, no. 8–13, September 1997, pp. 1075–1082.

[Bjork98] N. Bjork and C. Christopoulos, "Transcoder Architectures for Video Coding," *IEEE Transactions on Consumer Electronics,* Vol. 44, no. 1, February 1998, pp. 88–98.

[Border99] J. Border, M. Kojo, J. Griner, and G. Montenegro, "Performance Enhancing Proxies," *Internet Engineering Task Force Draft,* June 25, 1999 (see http://www.ietf.org/ under Internet Drafts).

[Bouthillier98] L. Bouthillier, "Synchronized Multimedia on the Web," *Web Techniques Magazine,* Vol. 3, no. 9, September 1998.

[Candan96] K. S. Candan, B. Prabhakaran, and V. S. Subrahmanian, "CHIMP: A Framework for Supporting Distributed Multimedia Document Authoring and Presentation," *Proc. ACM Intern. Conf. Multimedia (ACMMM),* Boston, MA, November 1996, pp. 329–339.

[Chandra99] S. Chandra, C. Ellis, and A. Vahdat, "Multimedia Web Services for Mobile Clients Using Quality Aware Transcoding," *ACM International Workshop on Wireless Mobile Multimedia (WOW-MOM),* 1999, pp. 99–108.

[Chang94] S. Chang and A. Eleftheriadis, "Error Accumulation of Repetitive Image Coding," *IEEE International Symposium on Circuits and Systems,* Vol. 3, 1994, pp. 201–204.

[Chong96] U. Chong and S. Kim, "Wavelet Transcoding of Block DCT Based Images Through Block Transform Domain Processing," *Proceedings of SPIE, Wavelet Applications in Signal and Image Processing IV,* Vol. 2825, pt. 2, 1996, pp. 901–908.

[Christopoulos99] C. Christopoulos, T. Ebrahimi, V. V. Vinod, J. R. Smith, R. Mohan, and C.-S. Li, "Universal Access and Media Conversion," *MPEG-7 Applications Proposal, number MPEG99/M4433 in ISO/IEC JTC1/SC29/WG11,* Seoul, Korea, March 1999.

[de los Reyes98] G. de los Reyes, A. Reibman, J. Chuang, and S. Chang, "Video Transcoding for Resilience in Wireless Channels," *International Conference on Image Processing (ICIP),* Vol. 1, 1998, pp. 338–342.

[Eleftheriadis95] A. Eleftheriadis and D. Anastassiou, "Constrained and General Dynamic Rate Shaping of Compressed Digital Video," *International Conference on Image Processing (ICIP),* Vol. 3, 1995, pp. 396–399.

[Fielding97] R. Fielding, J. Gettys, J. Mogul, H. Frystyk, and T. Berners-Lee, *HTTP/1.1, RFC 2068,* January 1997.

[Fleming97] T. Fleming, S. Midkiff, and N. Davis, "Improving the Performance of the World Wide Web Over Wireless Networks," *GLOBECOM,* Vol. 3, 1997, pp. 1937–1942.

[Floyd98] R. Floyd, B. Housel, and C. Tait, "Mobile Web Access Using eNetwork Web Express," *IEEE Personal Communications,* Vol. 5, no. 5, October 1998, pp. 47–52.

[Fox96a] A. Fox and E. Brewer, "Reducing WWW Latency and Bandwidth Requirements by Real Time Distillation," *Fifth International World Wide Web Conference, Computer Networks and ISDN Systems,* Vol. 28, no. 7–11 May 1996, pp. 1445–1456.

[Fox96b] A. Fox, S. D. Gribble, E. A. Brewer, and E. Amir, "Adapting to Network and Client Variability Via On-Demand Dynamic Distillation," *ASPLOS-VII,* Cambridge, MA, October 1996.

[Fox98a] A. Fox, S. Gribble, Y. Chawathe, and E. Brewer, "Adapting to Network and Client Variation Using Active Proxies: Lessons and Perspectives," *IEEE Personal Communications,* Vol. 5, no. 4, August 1998, pp. 10–19.

[Fox98b] A. Fox, I. Goldberg, S. Gribble, D. Lee, A. Polito, and E. Brewer, "Experience with Top Gun Wingman: A Proxy Based Graphical Web Browser for the 3Com PalmPilot," *Middleware '98, IFIP International Conference on Distributed Systems Platforms and Open Distributed Processing,* September 1998, pp. 407–424.

[Gardner99] W. Gardner, "Web Microbrowser Market Is Still Up for Grabs," *Portable Design,* October 1999, pp. 18–.

[Gessler95] S. Gessler and A. Kotulla, "PDA's as Mobile WWW Browsers," *Computer Networks and ISDN Systems,* Vol. 28, nos. 1 and 2, December 1995, pp. 53–59.

[Han98] R. Han, P. Bhagwat, R. LaMaire, T. Mummert, V. Perret, and J. Rubas, "Dynamic Adaptation in an Image Transcoding Proxy for Mobile Web Browsing," *IEEE Personal Communications,* Vol. 5, no. 6, December 1998, pp. 8–17.

[Han99] R. Han, "Factoring a Mobile Client's Effective Processing Speed Into the Image Transcoding Decision," *ACM International Workshop on Wireless Mobile Multimedia (WOWMOM),* 1999, pp. 91–98.

[Haskell98] P. Haskall, D. Messerschmitt, and L. Yun, "Architectural Principles for Multimedia Networks," book chapter no. 15 in *Wireless Communications: Signal Processing Perspectives,* ed. H. Poor and G. Wernell, Prentice Hall, 1998, pp. 229–281.

[HDML97] Unwired Planet, "Handheld Device Markup Language Specification," *Technical Report Version 2.0,* Unwired Planet, Inc., April 1997.

[Hori99] M. Hori, R. Mohan, H. Maruyama, and S. Singhal, "Annotation of Web Content for Transcoding," *Technical Report NOTE-annot-19990524,* W3C, July 1999.

[Joseph95] A. Joseph, et al., "Rover: A Toolkit for Mobile Information Access," *ACM Symposium on Operating Systems Principles,* 1995.

[Kan98] K. Kan and K. Fan, "Video Transcoding Architecture with Minimum Buffer Requirement for Compressed MPEG 2 Bitstream," *Asia Pacific Broadcasting Union (ABU) Technical Review,* no. 177, July/August 1998, pp. 3–9.

[Kling94] R. Kling and M. Elliott, "Digital Library Design for Usability," *Proc. Conf. Theory and Practice of Digital Libraries,* College Station, TX, June 1994.

[Li98] C.-S. Li, R. Mohan, and J. R. Smith, "Multimedia Content Description in the InfoPyramid," *IEEE Proc. Int. Conf. Acoust., Speech, Signal Processing (ICASSP),* Seattle, WA, May 1998. Special session on Signal Processing in Modern Multimedia Standards.

[Liljeberg95] M. Liljeberg, T. Alanko, M. Kojo, H. Laamanen, and K. Raatikainen, "Optimizing World-Wide Web for Weakly Connected Mobile Workstations: An Indirect Approach," *Second International Workshop on Services in Distributed and Networked Environments,* 1995, pp. 132–139.

[Loon97] T. Loon and V. Bharghavan, "Alleviating the Latency and Bandwidth Problems in WWW Browsing," *USENIX Symposium on Internet Technologies and Systems,* 1997, pp. 219–230.

[Mohan99] R. Mohan, J. Smith, and C. Li, "Adapting Multimedia Internet Content for Universal Access," *IEEE Transactions on Multimedia,* Vol. 1, no. 1, March 1999, pp. 104–114.

[Morrison97] G. Morrison, "Video Transcoders with Low Delay," *IEICE Transactions on Communications,* Vol. E80-B, no. 6, June 1997, pp. 963–969.

[Nakajima97] T. Nakajima and A. Hokimoto, "Adaptive Continuous Media Applications in Mobile Computing Environments," *IEEE International Conference on Multimedia Computing and Systems,* 1997, pp. 152–160.

[Nelson96] M. Nelson and J. Gailly, *The Data Compression Book,* Second Edition, M&T Books, 1996.

[Noble97] B. Noble et al., "Agile Application-Aware Adaptation for Mobility," *Proc. 16th ACM Symposium on Operating System Principles,* Saint-Malo, France, October 5–8, 1997.

[Ortega97] A. Ortega, F. Carignano, S. Ayer, and M. Vetterli, "Soft Caching: Web Cache Management Techniques for Images," *Workshop on Multimedia Signal Processing (IEEE),* Princeton, NJ, June 1997, pp. 475–480.

[Paek98] S. Paek and J. R. Smith, "Detecting Image Purpose in World-Wide Web Documents," *IS\&T/SPIE Symposium on Electronic Imaging: Science and Technology—Document Recognition,* San Jose, CA, January 1998.

[PalmVII99] Palm Computing, http://www.palm.com, White Paper on Palm VII.

[Pennebaker93] W. Pennebaker, and J. Mitchell, *JPEG Still Image Data Compression Standard,* Chapman & Hall, 1993.

[Proxinet] ProxiNet, http://www.proxinet.com

[Quickweb] Intel Quick Web, http://www.intel.com/quickweb

[Reynolds98] F. Reynolds, J. Hjelm, S. Dawkins, and S. Singhal, "A User Side Framework for Content Negotiation: Composite Capability/Preference Profiles (CC/PP)," *Technical Report,* November 1998, *http://www.w3.org/TR/NOTE-CCPP, W3C.*

[Rowe97] N. C. Rowe and B. Frew, "Finding Photograph Captions Multimodally on the World Wide Web," *Technical Report Code CS/Rp,* Dept. of Computer Science, Naval Postgraduate School, 1997.

[Schulzrinne96] H. Schulzrinne, S. Casner, R. Frederick and V. Jacobson, "RTP: A Transport Protocol for Real-Time Applications," *Internet Engineering Task Force Request for Comment (RFC),* no. 1889, January 1996.

[Shoham88] Y. Shoham and A. Gersho, "Efficient Bit Allocation for an Arbitrary Set of Quantizers," *IEEE Trans. Acoust., Speech, Signal Processing,* Vol. 36, no. 9, September 1988.

[Smith97] J. R. Smith and S.-F. Chang, "Visually Searching the Web for Content," *IEEE Multimedia Mag.,* Vol. 4, no. 3, pp. 12–20, July–September 1997.

[Smith98a] J. Smith, R. Mohan, and C. Li, "Content-based Transcoding of Images in the Internet," *Proceedings of the International Conference on Image Processing (ICIP),* Vol. 3, 1998, pp. 7–11.

[Smith98b] J. R. Smith, R. Mohan, and C.-S. Li, "Transcoding Internet Content for Heterogeneous Client Devices," *Proc. IEEE Int. Symp. on Circuits and Syst. (ISCAS),* June 1998, Special session on Next Generation Internet.

[Smith99a] J. R. Smith, "Digital Video Libraries and the Internet," *IEEE Communications Mag.,* Vol. 37, no. 1, January 1999, pp. 92–99, Special issue on the Next Generation Internet.

[Smith99b] J. R. Smith, C.-S. Li, and R. Mohan, "InfoPyramid Description Scheme for Multimedia Content," *MPEG-7 Proposal, number MPEG99/P473 in ISO/IEC JTC1/SC29/WG11,* Lancaster, UK, February 1999.

[Smith99c] J. R. Smith, C.-S. Li, R. Mohan, A. Puri, C. Christopoulos, A. B. Benitez, P. Bocheck, S.-F. Chang, T. Ebrahimi, and V. V. Vinod, "MPEG-7 Content Description Scheme for Universal Multimedia Access," *MPEG-7 Proposal, number MPEG99/M4949 in ISO/IEC JTC1/SC29/WG11,* Vancouver, BC, July 1999.

[Spyglass] Spyglass-Prism, http://www.spyglass.com/products/prism

[Swann98] R. Swann and N. Kingsbury, "Bandwidth Efficient Transmission of MPEG II Video Over Noisy Mobile Links," *Signal Processing, Image Communications,* Vol. 12, no. 2, April 1998, pp. 105–115.

[Tudor97] P. Tudor and O. Werner, "Real Time Transcoding of MPEG 2 Video Bit Streams," *Proceedings of International Broadcasting Conference,* September 1997, pp. 296–301.

[Vanderheiden97] G. C. Vanderheiden, "Anywhere, Anytime (+Anyone) Access to the Next-Generation WWW," *Proc. 6th Int. WWW Conf,* 1997.

[Vogel94] A. Vogel, B. Kerherve, G. von Bockmann, and J. Gecsei, "Distributed Multimedia and QOS: A Survey," *IEEE Multimedia Mag.,* Vol. 2, no. 2, pp. 10–18, Summer 1994.

[WAPForum99] Wireless Application Protocol (WAP) Forum, http://www.wapforum.org, White Paper.

[Watson94] T. Watson, "Application Design For Wireless Computing," *IEEE Workshop on Mobile Computing Systems and Applications,* 1994, pp. 91–94.

[Weitzman94] L. Weitzman and K. Wittenburg, "Automatic Presentation of Multimedia Documents Using Relational Grammars," *Proc. ACM Intern. Conf. Multimedia (ACMMM),* San Francisco, CA, November 1994 pp. 443–451.

[Wilkinson98] J. Willkinson, "Understanding MPEG Concatenation and Transcoding," *Asia Pacific Broadcasting Union (ABU) Technical Review,* no. 177, July/August 1998, pp. 3–9.

[Yeadon96] N. Yeadon, F. Garcia, F. Hutchison, and D. Shepherd, "Continuous Media Filters for Heterogeneous Internetworking," *SPIE Multimedia Computing and Networking (MMCN),* Vol. 2667, 1996, pp. 246–257.

[Zenel97] B. Zenel and D. Duchamp, "General Purpose Proxies: Solved and Unsolved Problems," *Sixth Workshop on Hot Topics in Operating Systems,* May 1997, pp. 87–92.

CHAPTER 16

Multicasting: Issues and Networking Support

UPKAR VARSHNEY

16.1 INTRODUCTION

Many applications such as multimedia communications, video conferencing, distance learning, computer-supported collaborative work (CSCW), distributed games, and video-on-demand require multipoint group communications to enable the participation of multiple sources and receivers. This group communication can be achieved by sending a copy of the message to one receiver at a time, sending everyone the same message including to those who are not members of a group, or by selectively and simultaneously sending only to the group members. These approaches have been termed replicated unicasting, broadcasting or flooding, and multicasting, respectively. These can be illustrated by the example shown in Figure 16.1. If only A and C need to communicate, unicasting is just fine. If everyone in the network needs to communicate, broadcasting can be deployed. However, if only some of them, say A, B, and D, need to be part of a group, then the network should limit the traffic among these without broadcasting to everyone by using multicast communications.

Multicasting is more efficient than replicated unicasting or broadcasting for multipoint group communications. However, incorporating the support for multicasting is difficult in current and emerging networks, as many of them are not designed to support multicast communications. There has been a significant amount of work toward supporting multicast communications in existing and emerging networks for the last several years. The support for multicasting in the Internet Protocol (IP) was first proposed in [Deering, 1989] and now has been included in many routers, hosts, and clients [Jamison, 1998]. Also, several solutions have been proposed for supporting multicast communications in emerging Asynchronous Transfer Mode (ATM) networks. Emerging wide-area networks such as the very-high-speed Backbone Network Service (vBNS) have built-in support for multicast communications. Attempts are also being made to extend the support for multicast communications in emerging wireless and mobile networks.

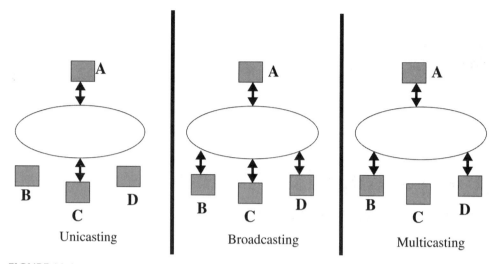

FIGURE 16.1
Unicasting, multicasting, and broadcasting.

Besides being an interesting challenge to the research community, multicasting has also become a topic of great interest to Internet Service Providers (ISPs), competitive local exchange carriers (CLECs), content providers, and businesses with multiple sites requiring simultaneous updates. ISPs could use multicasting to support content distribution services and thus be able to charge more to their customers for this premium service. Businesses could use multicasting to distribute software and data updates to branch offices and stores worldwide [Lawton, 1998]. There are many important issues, technical and nontechnical, in multicasting that need to be addressed before multicasting can be widely deployed as shown in Table 16.1.

16.2   MULTICASTING SUPPORT

Many networks that are inherently broadcast in nature such as Ethernet or other shared-media Local Area Networks (LANs) and satellites can use broadcasting techniques to support multicast communications. The point-to-point networks may support it by using several point-to-point (unicast) communications. However, multicasting can be supported more efficiently as a "logical" multicast tree made up of switches/routers and network links. The multicast tree may be a rooted tree, where a source is at the root and all possible receivers connect to the root via branches or shortest possible paths between the source and each of the receivers. This works well for cases

Table 16.1 Interested Parties and Major Issues in Multicasting

Interested Parties	Major Issues in Multicast Support
ISPs/CLECs	Maturity of multicast software
	Pricing/division of revenue with other providers
Corporate networks	Maturity of multicast software
	Possibility of congestion affecting other applications
University networks	Cost
	Amount of traffic

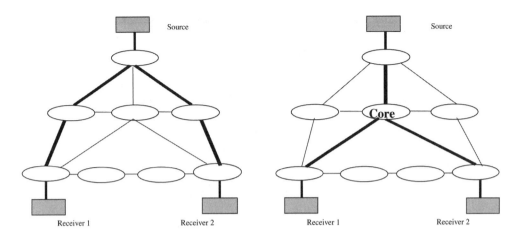

FIGURE 16.2
Rooted and core-based tree for multicasting.

involving few sources. For a large number of sources and receivers, the rooted tree may not perform well. Or the multicast tree may be a shared tree, where a core or center is selected and the
receivers send "join" messages to the center. The senders send packets toward the center and these
can be routed to receivers. To provide fault tolerance, multiple cores or centers can be employed.
The rooted and core-based multicast trees are shown in Figure 16.2.

Irrespective of the choice of the multicast tree, the support for multicast involves creating,
maintaining, and updating efficient multicast trees. The trees can be created dynamically as network users join and leave the multicast group. The tree can be maintained by some of the network nodes, mainly those sitting in the paths between sources and receivers. The tree needs to
be updated after changes in connectivity or the load of the underlying network topology. How
frequently the tree should be updated may be implementation dependent, as the overhead of tree
updating should be compared with the inefficiency of the multicast tree caused by changes in
connectivity or load. In addition to this, there are other issues in multicast support, including
group addressing format and allocation of group addresses, managing group membership,
packet delivery, traffic management and signaling protocols, and the support for quality of service (QoS). The multicasting can be implemented at one of the several different layers of protocols, including application such as in video conferencing, transport as in ATM, network as in IP,
and at the link layer as in shared-medium LANs [Fahmy et al. 1997]. Typically, it is more efficient to perform multicasting at lower layers than upper layers.

16.3   MULTICASTING IN IP-BASED NETWORKS

The Internet Protocol (IP) is the network layer protocol used in the Internet or Transmission
Control Protocol/Internet Protocol (TCP/IP)-based networks and it provides best effort and connectionless service. The reliability, if required, is provided by upper-layer protocols. Unlike
unicasting addresses, a single address is used for the entire multicast group. This is done by
using class D of IP addresses that range from 224.0.0.0 to 239.255.255.255. This type of address
really represents a list of receivers and not the individual receivers. Some of these addresses are
well-known multicast addresses and the rest are transient addresses. The well-known addresses
are published over the Internet, while transient addresses are allocated only for the duration of

a certain multicast session. Before starting a multicast session, a class-D address is chosen by the group initiator [Diot et al., 1997]. In IP multicasting, the members can be located anywhere on the network and the sender does not need to know about the other members. The sender does not have to be a member of the multicast group it is transmitting to. To join a multicast group, a host informs a local multicast router, which in turn contacts other multicast routers, and so on. After the local multicast router joins the multicast session, the router periodically checks if any of the hosts are still members of that multicast group. As long as at least one host on its network remains a member, the multicast router continues to be a part of that group. If it hears no responses, it assumes that no one is interested in the multicast group and it stops being part of that multicast group. The multicast router to host communication uses Internet Group Multicast Protocol (IGMP), which is also used for managing group memberships with other multicast routers [Deering, 1989]. The Internet multicast also makes use of time-to-live (TTL) field to limit how far (how many hops) a packet can traverse to a receiver [Diot et al., 1997]. The IGMPv2 adds a low-latency leave to IGMP to allow a more prompt pruning of a group after all members in the subnetwork leave.

IP multicasting in the Internet has been implemented using MBone, a virtual overlay network that has been operational since 1992 [Eriksson, 1994]. Since most IP routers currently do not support multicast routing, the forwarding of multicast datagrams between "islands" of multicast-capable subnetworks is handled by "multicast routers" through tunnels, as shown in Figure 16.3. The tunnels are implemented by encapsulating IP packets destined for a multicast address within an IP packet, with the unicast address of the next multicast-capable router along the path. IP multicasting does not use TCP, but uses User Datagram Protocol (UDP) to avoid the acknowledgment implosion problem caused by the increased number of acknowledgments from multiple receivers for every message. That is why IP multicasting is considered a best-effort service; however, upper-layer protocols (application or reliable transport protocols) can be employed to provide reliable multicasting service. The other interesting issue is how to deal with traffic congestion issues in IP multicast, as no TCP is being used. Therefore, applications should attempt to deal with congestion and the amount of multicast data that can be put over an IP-based network.

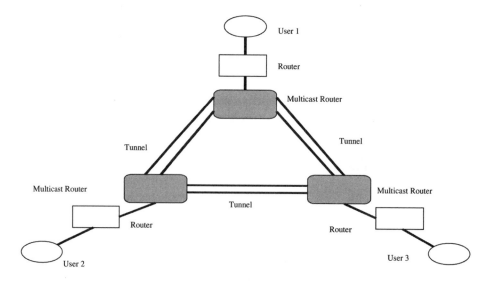

FIGURE 16.3
IP multicasting in Mbone.

Table 16.2 Routing Protocols for IP Multicasting

Routing Protocol	Tree Type	Information Collection	Comment
DVMRP	Source-based	Using exchange between routers	Not scalable
MOSPF	Source-based	Link state database	Not scalable
PIM	RP-rooted/source-based	From routing table	Supports both sparse and dense mode

16.3.1   Routing Protocols for IP Multicast

There are several routing protocols that can be used to support IP multicast, such as Distance Vector Multicast Routing Protocol (DVMRP), Multicast Open Shortest Path First (MOSPF), and Protocol Independent Multicast (PIM), as shown in Table 16.2.

Both the DVMRP and MOSPF build a source-based multicast tree, but the way they collect information is different. In DVMRP [Waitzman et al., 1998] multicast routers exchange information on reverse path distances, while MOSPF uses routing information from the link state database, allowing multicast routers to build efficient source-based tree without flooding as used in DVMRP. Therefore, MOSPF is more efficient than DVMRP, but more computation, although on demand, is required.

Both the DVMRP and MOSPF do not scale well, so another routing protocol has been proposed. It is termed Protocol Independent Multicast (PIM) and has two modes of operation, sparse and dense, based on how the multicast users are distributed. In sparse-mode PIM, some points on the network are designated as rendezvous points (RPs) and an RP-rooted tree is constructed as follows. The highest IP-addressed router is chosen as the designated router (DR) on a network. The receiver's DR sends explicit join messages to the RP. The sender's DR sends register messages to the RP, which sends a join to sources. The packets will follow the RP-rooted shared tree, but the receiver (or router) may switch to the source's shortest path tree. The densemode PIM is essentially the same as DVMRP, except that unicast routers are imported from existing routing tables rather than incorporating a specific unicast routing algorithm. That is why it is termed Protocol Independent Multicasting. Many vendors are supporting multicast communications in their routers and other equipment, such as the support for DVMRP and MOSPF in Ascend's IP navigator software and the support for PIM in Cisco 7507 routers as used in vBNS [Jamison et al., 1998].

MBone uses UDP for end-to-end transmission control, IGMP for group management, and DVMRP for multicast routing. Another example of IP multicasting implementation is vBNS that uses native IP multicasting without using any overlay networks [Jamison et al., 1998].

16.3.2   Multimedia Support and IP Multicasting

Multimedia applications may require low latency and a certain minimum amount of bandwidth. Supporting such QoS for unicast has been an interesting research problem for many years and a significant amount of work has been done to provide solutions for QoS problems. However, these solutions may not be applied to multicast communications, as different receivers may have different QoS requirements, capabilities, and constraints. Resource reservation is difficult to set up in a multipoint environment; however, a guarantee of minimum bandwidth on a multicast route is not a problem [Zhang et al., 1993]. The Resource Reservation Protocol (RSVP) can be

Table 16.3 Some Multimedia Applications on Mbone

Application	Purpose	Comment
vic	Video conferencing	Uses RTP version 2
vat	Audio conferencing	
nv	Video conferencing	Allows slow frame rates
ivs	Audio/video	Simple
Freephone	Audio conferencing	Special coding
wb	Shared workspace	
sdr	Advertisement and joining of multicast conferences on MBone	Similar to popular application called sd

applied for multicasting and is capable of supporting different QoS for different receivers. It defines three categories of services: guaranteed service (bounds on throughput and delay), controlled load service (by approximating the performance of an unloaded datagram network), and best-effort service. It maintains a soft state and sends periodic "refresh" messages to maintain the state along the reserved paths, allowing for dynamic adaptation to group membership and changing routes [Zhang et al., 1993].

16.3.3   Multimedia Multicasting Applications on The MBone

There are many multicasting applications that have been designed and are currently being used over the MBone, as shown in Table 16.3. Audio and video applications do not receive any reliability or transport layer ordering, and these applications implement an application-level congestion control scheme. Audio packets are reordered in application playout buffer.

Shared Workspace wb is the most well-known shared workspace application on the MBone. The communication system provides reliable but not ordered multicast. An application-level recovery is performed if an out-of-order packet is received. Session Directory (sd) is not a group application but provides the possibility to perform multicast address allocation (randomly chosen). The sd has been obsoleted by sdr, which is designed to allow the advertisement and joining of multicast conferences on the MBone. More information on MBone, the applications currently supported, and how to join a multicast session over the MBone can be found from the references provided in the "For Further Reading" section.

16.4   MULTICASTING IN ATM NETWORKS

Unlike IP multicasting, ATM multicasting is in its early phases due to the fact that, unlike IP, ATM is connection-oriented and the support for multicasting requires setting up multipoint connections. In addition to this, ATM requires that the QoS negotiation be performed at the time of connection setup by the source and switches. ATM does not support different QoS for different receivers or the support of QoS renegotiation.

Possible connections in ATM are shown in Figure 16.4. As far as group communications is concerned, ATM only supports point-to-multipoint connections currently [ATM User Network Interface (UNI) 3.1, 1995]. However, multipoint-to-multipoint connections can be set up as a group of point-to-multipoint connections, but then it's not scalable, as the number of these connections increases with the number of sources. ATM has to deal with many issues important for supporting multicast communications. These issues are signaling, routing, connection admission

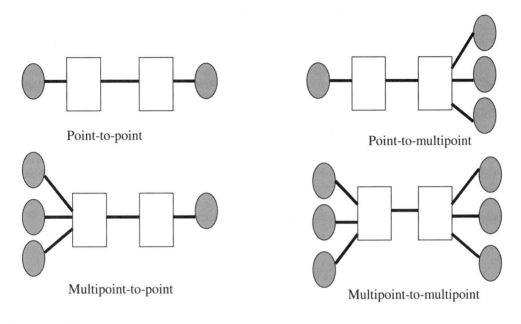

FIGURE 16.4
Possible connections in ATM.

control, and traffic management for multicast communications. ATM UNI 3.1 signaling standard [UNI 3.1, 1995] supports the source-based tree approach of multicast data distribution. The root-initiated joins, where only the root can set up the point-to-multipoint connections, add leaves, and send ATM cells, are supported. Since that is not scalable, support for a receiver-initiated or leaf-initiated join will be supported in UNI 4.0. However, that standard will not support the use of pure multipoint-to-point or multipoint-to-multipoint connections [Fahmy et al., 1997].

There are several traffic management issues in multicasting over ATM networks. These include the amount of feedback from receivers (causing feedback implosion). The feedback implosion problem is caused by feedback from several receivers to a sender. To avoid this problem, the feedback from the different leaves should be consolidated so that the amount of feedback information does not increase proportional to the number of leaves.

One problem in ATM multicasting is the inability of a receiver to distinguish cells from different concurrent senders, and this is one reason why multipoint-to-point or multipoint-to-multipoint connections have not been supported in existing ATM networks. This is called the cell interleaving problem [5] and is caused because ATM Adaptation Layer 5 (AAL5), the most widely used AAL, does not have multiplexing identifiers or sequence numbers. Therefore, AAL5 cannot indicate to upper layers that these cells are coming from different senders. Possible solutions to this problem are use of AAL3/4 (involving more overhead), use of a multicast server, and packet-level buffering requiring more memory and increased delay.

16.4.1 Multicasting Schemes for ATM Networks

There are three distinct ways of doing multicasting in ATM. These are the use of a rooted tree at every source [implemented as Virtual Path/Virtual channel (VP/VC) mesh], the use of a multicast server, and the use of a shared tree. The first approach involves the use of a significant number of Virtual Paths/Virtual Channels (VPs/VCs) and is not a scalable solution. The multicast

Table 16.4 Some Multicast Schemes for ATM Networks

Scheme	Basic Technique	Type of Connection	Application
MARS	Multicast server	Point-to-multipoint	IP over ATM
SEAM	Core-based routing	Multipoint-to-multipoint	ATM
SMART	Token-based transmission	Multipoint-to-multipoint	ATM

server and shared tree approaches have attracted a lot of attention (Table 16.4) and we will discuss them in more detail.

16.4.1.1 Multicast Address Resolution Server (MARS)

ATM-based IP hosts and routers can use a Multicast Address Resolution Server (MARS) [Talpade and Ammar, 1997] to support IP multicast over ATM point-to-multipoint connection service. The MARS maps the layer 3 multicast addresses into the ATM addresses of the group members. The endpoints establish and manage point-to-multipoint VCs to the group, as shown in Figure 16.5. They also establish a point-to-point bidirectional VC to the MARS for control traffic that includes IGMP join/leave and address resolution requests. The MARS asynchronously notifies hosts of group membership changes using a point-to-multipoint VC. A multicast server can receive data from senders and multicast it to the receivers on a point-to-multipoint VC.

16.4.1.2 Scalable and Efficient ATM Multicast (SEAM)

SEAM [Grossglauser and Ramakrishnan, 1997] is a scalable multipoint-to-multipoint solution based on a core-based shared tree, as shown in Figure 16.6. It allows member-initiated joins in addition to core-initiated joins. A unique identifier is used (called group handle) that identifies all cells associated with a given multicast group. A "shortcutting" approach is also employed to allow cells to take the shortest path along the shared tree. To handle the cell interleaving problem, VC merge is used.

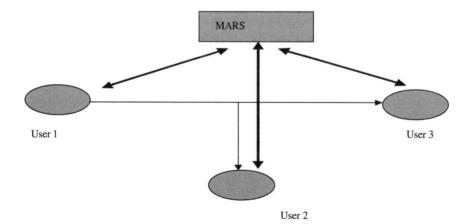

FIGURE 16.5
Multicasting using MARS.

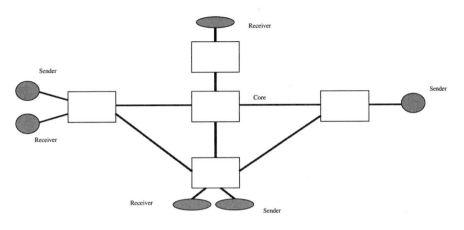

FIGURE 16.6
ATM multicasting using SEAM.

16.4.1.3   *Shared Many-to-Many ATM Reservations (SMART)*

Shared Many-to-Many ATM Reservations (SMART) [Gauthier et al., 1997] is a scalable solution as it uses a single shared tree. It serializes the communications of various senders to a set of receivers using a token scheme, thus eliminating the cell interleaving problem. The token scheme is based on allowing a sender to send to the group only after it has acquired a token.

16.5   IP MULTICASTING OVER ATM

With increasing deployment of ATM in wide-area networks such as vBNS and other backbone networks, and due to the widespread and continued support of TCP/IP by vendors, users, and developers, it is of significant importance that support for IP multicast over ATM be considered [Armitage, 1997]. Due to the inherent differences in IP and ATM technologies (including connection-oriented versus connectionless), when IP multicast is supported over ATM many interesting issues arise. Since IP uses variable-size packets (576–65,536 bytes) when mapping IP packets to ATM cells, issues are which AAL to use and the impact of a single-cell loss on the entire IP packet.

16.5.1   Problems in RSVP Over ATM

ATM does not support variegated VC, where different receivers can be provided with different quality of service. These VCs could allow different, dynamic, and varying needs of receivers in IP multicast. The RSVP allows heterogeneous receivers and reservation parameter negotiation, while ATM does not. The issues in RSVP over ATM include how to map IP integrated services to ATM services and how to manage ATM VCs. The first issue can be addressed by mapping IP services as follows: map guaranteed service to Constant Bit Rate (CBR) or Variable Bit Rate real time (VBR-rt), controlled load to VBR-nrt or Available Bit Rate (ABR) with a minimum cell rate, and the best-effort service to Unspecified Bit Rate UBR/ABR.

16.5.2   IP Multicast Over ATM in VBNS

The vBNS is implemented as an IP-over-ATM network and unlike MBone, it supports native IP multicast service [Jamison, et al., 1998]. It uses Protocol Independent Multicast (PIM) dense-mode configuration. The DVMRP unicast routing is used, allowing vBNS to support delivery of MBone traffic. The vBNS backbone will move to a sparse-mode PIM that is more suited for multicast groups with sparsely populated membership.

16.6   RELIABLE MULTICAST TRANSPORT PROTOCOLS

Since multicasting is typically done at lower layers (network or data-link layers) reliability can not be guaranteed, as protocols employed at these layers are not designed for end-to-end reliability. So for applications requiring reliable multicast, protocol support is provided. The multicast communications may involve multiple receivers and the transport protocols that use positive receiver acknowledgment (such as TCP) may lead to an acknowledgment implosion problem. One way to alleviate this problem is to use transport protocols where the receiver sends an acknowledgment only if the packet is not received successfully. The issues in transport protocols for multicasting include the acknowledgment implosion problem, how to compute timeouts, how to manage group and addressing information, how to provide routing support, and how to implement fairness. There are several transport or upper-layer protocols that have been proposed to provide some sort of reliable multicast. Real Time Protocol (RTP)[Schulzrinne et al., 1996] is not a transport protocol, as it simply combines the common functions needed by real-time applications for both unicast and multicast.

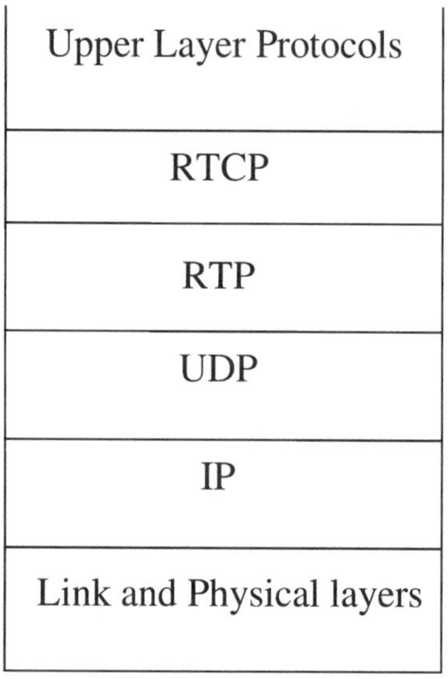

| Upper Layer Protocols |
| RTCP |
| RTP |
| UDP |
| IP |
| Link and Physical layers |

FIGURE 16.7
Protocol stacks for RTP/RTCP.

RTP runs on top of UDP and provides payload-type identification, sequence numbering, and time stamping. The Real Time Control Protocol (RTCP) [Schulzrinne et al., 1996] works in conjunction with RTP, identifies RTP sources, monitors delivery, and provides information to the application. The protocol stacks for RTCP/RTP running over UDP/IP for reliable multicasting is shown in Figure 16.7. Another protocol, Reliable Multicast Transport Protocol (RMTP) [Paul et al., 1997], provides sequenced lossless delivery using the selective repeat of packets. It allows receivers to join at any point during the connection lifetime, and the congestion avoidance is implemented using slow start as used in TCP. The RMTP uses a hierarchical tree of designated receivers that cache data, handle retransmission, and interact with the ultimate receivers. These designated receivers propagate and consolidate ACKs up the tree and thus avoid the ACK implosion problem. The Reliable Adaptable Multicast Protocol (RAMP) uses immediate (not delayed) receiver-initiated, negative acknowledgment-based unicast error notification combined with originator-based unicast retransmission. The RAMP is useful for ATM, as packet loss in ATM is due more to receiver errors and buffer overflows than retransmission errors.

16.7   MULTICASTING IN WIRELESS NETWORKS

In the last few years, wireless and mobile networks have attracted significant attention due to their potential to provide communications and computing support for an increasingly mobile workforce. Mobile devices (computing-oriented handheld devices such as PalmPilot and communications-oriented devices such as smart phones) have become smaller and more powerful and are beginning to support many new features such as Internet access, brokerage services, and location tracking, as shown in Figure 16.8.

Some of these features, and also emerging wireless applications, will benefit if the underlying wireless and mobile networks support multicasting. Many of the emerging wireless and mobile networks deploy widely different technology, protocols, and wireless links, and therefore the support for multicasting is an even bigger challenge.

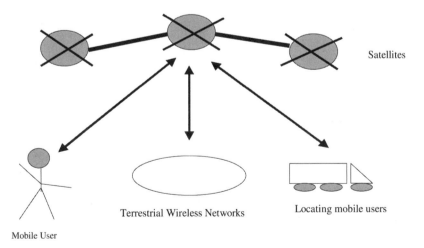

FIGURE 16.8
A possible scenario for mobile and wireless networking.

16.7.1   Issues in IP Multicasting Over Wireless

Multicast support becomes even more difficult in this case due to the mobility of users and to the varying performance of wireless links. The issues in IP multicasting over wireless networks include building a dynamic multicast tree involving wireless links, managing group membership and location tracking of mobile users, routing packets, and the support for QoS and error control. One major issue here is the increased amount of overhead created due to the join operation, multicast tree computation, and polling of mobile users by routers. The IGMP packets are also subject to loss and may be delayed. The amount of overhead that is required to support quality of service using RSVP would also increase significantly due to host mobility. One interesting approach for multicasting in wireless networks is the use of mobile IP. With this, a mobile host can participate in multicast using home agent or foreign agent. A comparison of these approaches can be found in [Xylomenos and Polyzos, 1997].

16.7.2   Multicast Support in Wireless ATM

One interesting and promising approach to support multimedia on wireless networks is to extend ATM technology to wireless networks. This has been termed Wireless ATM and has attracted significant attention in the research and development community. Since many of the multimedia applications require multicasting, some efforts should be made to incorporate support for multicasting in wireless ATM networks. Various solutions that have been proposed for multicasting in ATM (such as VC mesh and multicast server) assume stationary ATM users. Several issues are important for multicasting in wireless ATM networks. These include the cost of updating and maintaining the multicast tree, the impact of admission control, and scalability. In general, multicasting schemes proposed for ATM networks can be extended to support mobile users. One point that must be made here is that due to mobility, the updating of multicast trees would be a major issue, as it has the potential of wasting a significant amount of network bandwidth if proper updating was not performed after handoffs of mobile users. But updating of the tree, especially in case of the fast mobility or many users, may require a significant amount of overhead. If a shared tree is used for multicasting, then after a handoff, the length of the multicast tree (and hence the amount of network resources in use) may increase if the tree is not properly updated. A multicast tree may also be subject to admission control after a mobile user moves to a new location where the base station cannot provide the requested amount of bandwidth or resources. The type of connection rerouting scheme used in wireless ATM networks will also impact the scalability of a multicast tree in terms of number of users, network size, and frequency of change.

16.8   SUMMARY AND THE FUTURE OF MULTICASTING

Multicasting is required by several multimedia applications and therefore it is of significant importance. There has been some progress in support for multicasting in IP, ATM, and wireless networks. The factors affecting the widespread deployment of multicast communications include the maturity of multicast software; congestion control and reliable multicast support; support for multicasting by vendors; interest in multicasting by CLEC, ISP, and content providers, development of new econometric models for division of revenue among multiple networks or ISPs; ways to charge and measure the use of multicasting; and the amount of traffic actually caused by multicast communications on corporate networks or the Internet.

16.9 DEFINITIONS OF KEY TERMS

Asynchronous Transfer Mode (ATM): An extremely fast packet-switching technology that has been designed to carry all types of information using 53-byte fixed packets.

ATM Adaptation Layer (AAL): Provides for the conversion of non-ATM packets to ATM cells and vice versa.

Broadcasting: A traditional way to achieve group communications. It involves transmitting the information to everyone on the network.

CLEC (Competitive Local Exchange Carrier): Phone companies that are allowed to compete with baby bells by the 1996 telecom reform act.

DVMRP (Distance Vector Multicasting Routing Protocol): A routing protocol that is used for IP multicasting.

IGMP (Internet Group Multicast Protocol): A protocol used for communications between a host and a router and among multicast routers.

IP (Internet Protocol): The most popular network layer protocol used in the TCP/IP-based networks such as the Internet.

MARS (Multicast Address Resolution Server): A proposed standard for providing IP multicasting over ATM networks.

MBone (Multicast Backbone): A virtual overlay network that connects islands of multicast routers using logical channels called tunnels.

Mobile IP: Extends the Internet Protocol to support mobility of users.

MOSPF (Mobile Open Shortest Path First): A routing protocol used for IP multicasting.

Multicasting: An efficient way to achieve group communications without using broadcasting.

PIM (Protocol Independent Multicast): A routing protocol for IP multicasting.

QoS (Quality of Service): Attributes that a source or receiver can specify for its transmission or reception. It typically includes packet delay and loss.

Resource Reservation Protocol (RSVP): A protocol designed to support different qualities of service for different receivers in a TCP/IP-based network.

RTP/RTCP: Protocols used to provide reliable multicasting.

SEAM: A proposed scheme to provide multipoint-to-multipoint ATM multicasting using a core-based tree.

SMART: A proposed token-based scheme to support multipoint-to-multipoint ATM multicasting.

Transmission Control Protocol (TCP): The most popular transport layer protocol for providing reliability using acknowledgment and flow control.

TTL: A part of the Internet Protocol packet overhead. It is used to limit how many hops a packet can go.

Unicasting: The normal one-to-one communications.

User Datagram Protocol (UDP): A simple transport protocol that does not provide reliability unlike TCP.

VBNS: A backbone network that connects major research universities and runs at 2.4 Gbit/s. It provides native IP multicasting.

16.10 REFERENCES

Armitage, G., 1997, IP Multicasting Over ATM Networks, *IEEE Journal on Selected Areas in Communications*, Vol. 15, no. 3, pp. 445–457, April 1997.

ATM User Network Interface (UNI) Specification Version 3.1, Prentice Hall, Englewood Cliffs, NJ, 1995.

Deering, S., 1989, Host Extension for IP Multicasting, *Internet RFC* 1112, August 1989.

Diot, C., W. Dabbous, and J. Crowcroft, 1997, Multipoint Communication: A Survey of Protocols, Functions, and Mechanisms, *IEEE Journal on Selected Areas in Communications,* Vol. 15, no. 3, pp. 277–290, April 1997.

Eriksson, H., 1994, MBONE: The Multicast Backbone, *Communications of the ACM,* Vol. 37, no. 8, pp. 54–60, August 1994.

Fahmy, S., R. Jain, S. Kalyanaraman, R. Goyal, B. Vandalore, X. Cai, S. Kota, and P. Samudra, 1997, A Survey of Protocols and Open Issues in ATM Multipoint Communications, Technical Report, Ohio State University, August 1997 (available at www.cis.ohio-state.edu/~jain/papers/mcast.htm).

Gauthier, E., J. Boudec, and P. Oechslin, 1997. SMART: A Many to Many Multicast Protocol for ATM, *IEEE Journal on Selected Areas in Communications,* Vol. 15, no. 3, April 1997.

Grossglauser, M., and K. K. Ramakrishnan, 1997, SEAM: Scalable and Efficient ATM Multicast, *Proc. IEEE INFOCOM '97.*

Jamison, J., R. Nicklas, G. Miller, K. Thompson, R. Wilder, L. Kunningham, and C. Song, 1998, vBNS: Not Your Father's Internet, *IEEE Spectrum,* Vol. 35, no. 7, pp. 38–46, July 1998.

Lawton, G., 1998, Multicasting: Will It Transform the Internet, *IEEE Computer,* July 1998.

Paul, S., K. Sabnani, J. Lin, and S. Bhattacharyya, 1997, Reliable Multicast Transport Protocol, *IEEE Journal on Selected Areas in Communications,* Vol. 15, no. 3, pp. 407–421, April 1997.

Schulzrinne, H., S. Casner, R. Frederick, and V. Jacobson, 1996, RTP: A Transport Protocol for Real-time Applications, *Internet RFC* 1889, January 1996.

Talpade, R., and M. Ammar, 1997, Multicast Server Architectures for MARS-Based ATM Multicasting, *Internet RFC* 2149, May 1997.

Waitzman, D., C. Partridge, and S. Deering, 1988, Distance Vector Multicast Routing Protocol, *Internet RFC* 1075, November 1988.

Xylomenos, G., and G. Polyzos, 1997, IP Multicast for Mobile Hosts, *IEEE Communications Magazine,* Vol. 35, no. 1, pp. 54–58, January 1997.

Zhang, L., S. Deering, D. Estrin, S. Shenker, and D. Zappala, 1993, RSVP: A New Resource Reservation Protocol, *IEEE Networks,* Vol. 7, pp. 8–18, September 1993.

16.11   FOR FURTHER READING

Almeroth, K., and M. Ammar, 1996. The Use of Multicast Delivery to Provide Scalable and Interactive Video-on-Demand Service, *IEEE Journal on Selected Areas in Communications,* Vol. 14, no. 6, pp. 1110–1122, August 1996.

IP Multicast Software Using Anonymous ftp from ftp://parcftp.xerox.com/pub/net-research/ipmulti

Laubach, M., 1994, Classical IP and ARP Over ATM, *Internet RFC* 1577, January 1994.

MBone FAQ at www.cs.columbia.edu/~hgs/internet/mbone-faq.html

MBone software list at www.merit.edu/~mbone/index/titles.html

McCanne S., and V. Jacobson, 1995, Vic: A Flexible Framework for Packet Video, *Proc. ACM Multimedia,* pp. 5111–522, November 1995.

Perkins, C. E., 1997, Mobile IP, *IEEE Communications Magazine,* Vol. 35, no. 5, pp. 84–999, May 1997.

Towsley, D., J. Kurose, and S. Pingali, 1997. A Comparison of Sender-Initiated and Receiver-Initiated Reliable Multicast Protocols, *IEEE Journal on Selected Areas in Communications,* Vol. 15, no. 3, pp. 398–406, April 1997.

Valko, A., 1999, Cellular IP: A New Approach to Internet Host Mobility, *Computer Communications Review,* Vol. 29, no. 1, pp. 50–65, January 1999.

Varshney, U., 1997, Supporting Mobility Using Wireless ATM, *IEEE Computer,* Vol. 30, no. 1, pp. 132–134, January 1997.

Varshney, U., and S. Chatterjee, 1999, Architectural Issues in IP multicast Over Wireless Networks, in *Proc. IEEE Wireless Communications and Networking Conference,* New Orleans, LA, Vol. 1, pp. 41–45, September 1999.

Varshney, U., 1999. Connection Rerouting for Wireless ATM Networks, In *Proc. Hawaii International Conference on Systems Sciences,* Maui, Hawaii, p. 306 Abstract, full paper in CD-ROM, January 1999. IEEE Computer Society Press.

Vat for MBone at http://www-nrg.ee.lbl.gov/vat

Wb for MBone at ftp://ftp.ee.lbl.gov/conferencing/wb

INDEX

Figures and tables are indicated by an italic *f* or *t*.

311